AF698160

WITHDRAWN

ECOLOGY OF PLANT PATHOGENS

ECOLOGY OF PLANT PATHOGENS

Edited for the British Society for Plant Pathology by

J.P. Blakeman

Plant Pathology Research Division
Department of Agriculture for Northern Ireland
and
Queen's University of Belfast
Newforge Lane
Belfast BT9 5PX, UK

and

B. Williamson

Scottish Crop Research Institute
Invergowrie
Dundee DD2 5DA, UK

CAB INTERNATIONAL

SB
731
.E36
1994

CAB INTERNATIONAL
Wallingford
Oxon OX10 8DE
UK

Tel: Wallingford (0491) 832111
Telex: 847964 (COMAGG G)
Telecom Gold/Dialcom: 84: CAU001
Fax: (0491) 833508

© CAB INTERNATIONAL 1994. All rights reserved. No part of this publication may be reproduced in any form or by any means, electronically, mechanically, by photocopying, recording or otherwise, without the prior permission of the copyright owners

A catalogue entry for this book is available from the British Library.

ISBN 0 85198 876 8

Typeset by Solidus (Bristol) Limited
Printed and bound in the UK by Biddles Ltd, Guildford

Contents

Contributors ix

Preface xiii

I: ECOLOGICAL DETERMINANTS OF PLANT PATHOGEN POPULATIONS

1 All Creatures Unitary and Modular
J.H. Andrews 3

2 The Biological Environment and Pathogen Population Dynamics: Uncertainty, Coexistence and Competition
M.W. Shaw and J.C. Peters 17

3 The Microchemistry of Plant/Microorganism Interactions
T. Shepherd 39

4 Evaluation of Plant Pathogens in Complex Ecosystems
J.M. Lenné, D.M. Teverson and M.J. Jeger 63

5 Humidity and Fungal Diseases of Plants – Problems
J.G. Harrison, R. Lowe and N.A. Williams 79

II: METHODOLOGY

6 Interaction of *Bacillus* Species with Phytopathogenic Fungi – Methods of Analysis and Manipulation for Biocontrol Purposes
S.G. Edwards, T. McKay and B. Seddon 101

7 Methods for Detecting *Armillaria mellea*
R.T.V. Fox, H.M. Manley, A. Culham, K. Hahne and A.I. Tiffin 119

8 Identification of Colonizing Forms of *Trichoderma* Species Isolated from Mushroom Compost
P.R. Mills and S. Muthumeenakshi 135

9 Use of the Polymerase Chain Reaction for Identifying Plant Pathogens
E. Ward 143

10 Quantifying *Polymyxa betae* and BNYVV in Soil
G. Tuitert 161

III: ASPECTS OF AUTECOLOGY

11 Spore Dispersal: Environmental and Biological Factors
H.A. McCartney 171

12 Latency and Quiescence in Survival and Success of Fungal Plant Pathogens
B. Williamson 187

13 Aspects of the Autecology of the Take-all Fungus
D. Hornby 209

14 Autecology of Foliar Pseudomonads
S.S. Hirano and C.D. Upper 227

15 Autecology and Evolution of the Witches' Broom Pathogen (*Crinipellis perniciosa*) of Cocoa
G.W. Griffith, E. Bravo-Velasquez, F.J. Wilson, D.M. Lewis and J.N. Hedger 245

16 The Ecology of African Cassava Mosaic Geminivirus
D. Fargette and J.M. Thresh 269

IV: MICROBIAL INTERACTIONS

17 Straw Disposal and Cereal Diseases
J.F. Jenkyn, R.J. Gutteridge and M. Jalaluddin 285

18 The Cereal *Fusarium* Complex
D.W. Parry, T.R. Pettitt, P. Jenkinson and A.K. Lees 301

19 Fungal Interactions on Living and Necrotic Leaves
J. Köhl and N.J. Fokkema 321

20 Biological Control of *Erwinia amylovora* with *Erwinia herbicola*
H.A.S. Epton, M. Wilson, S.L. Nicholson and D.C. Sigee 335

Index 353

Contributors

J.H. Andrews *Plant Pathology Department, University of Wisconsin–Madison, 1630 Linden Drive, Madison, WI 53706, USA.*

E. Bravo-Velasquez *Department of Biology, Catholic University, Quito, Ecuador.*

A. Culham *Schools of Plant and Animal Sciences, University of Reading, 2 Earley Gate, Reading RG6 2AU, UK.*

S.G. Edwards *Department of Agriculture, University of Aberdeen, School of Agriculture Building, Aberdeen AB9 1UD, UK.*

H.A.S. Epton *School of Biological Sciences, The University of Manchester, Oxford Road, Manchester M13 9PT, UK.*

D. Fargette ORSTOM, *Laboratoire de Phytovirologie, CIRAD, BP 5035, Montpellier 34032, France.*

N.J. Fokkema *DLO Research Institute for Plant Protection (IPO-DLO), PO Box 9060, 6700 GW Wageningen, The Netherlands.*

R.T.V. Fox *Schools of Plant and Animal Sciences, University of Reading, 2 Earley Gate, Reading RG6 2AU, UK.*

G.W. Griffith *School of Biological Sciences, University College of North Wales, Bangor, Gwynedd LL57 2UW, UK.*

R.J. Gutteridge *Plant Pathology Department, AFRC Institute of Arable Crops Research, Rothamsted Experimental Station, Harpenden, Herts AL5 2JQ, UK.*

K. Hahne *Schools of Plant and Animal Sciences, University of Reading, 2 Earley Gate, Reading RG6 2AU, UK.*

J.G. Harrison *Scottish Crop Research Institute, Invergowrie, Dundee DD2 5DA, UK.*

J.N. Hedger *Department of Biological Sciences, University of Wales, Aberystwyth, Dyfed SY23 3DA, UK.*

S.S. Hirano *Plant Pathology Department, University of Wisconsin–Madison, 1630 Linden Drive, Madison, WI 53706, USA.*

D. Hornby *Plant Pathology Department, AFRC Institute of Arable Crops Research, Rothamsted Experimental Station, Harpenden, Herts AL5 2JQ, UK.*

M. Jalaluddin *Plant Pathology Department, AFRC Institute of Arable Crops Research, Rothamsted Experimental Station, Harpenden, Herts AL5 2JQ, UK.*

M.J. Jeger *Natural Resources Institute, Central Avenue, Chatham Maritime, Kent ME4 4TB, UK.*

P. Jenkinson *Crop and Environment Research Centre, Harper Adams Agricultural College, Newport, Shropshire TF10 8NB, UK.*

J.F. Jenkyn *Plant Pathology Department, AFRC Institute of Arable Crops Research, Rothamsted Experimental Station, Harpenden, Herts AL5 2JQ, UK.*

J. Köhl *DLO Research Institute for Plant Protection (IPO-DLO), PO Box 9060, 6700 GW Wageningen, The Netherlands.*

A.K. Lees *Crop and Environment Research Centre, Harper Adams Agricultural College, Newport, Shropshire TF10 8NB, UK.*

J.M. Lenné *Natural Resources Institute, Central Avenue, Chatham Maritime, Kent ME4 4TB, UK.*

D.M. Lewis *Department of Biological Services, University College Wales, Aberystwyth, Dyfed SY23 3DA, UK.*

R. Lowe *Scottish Crop Research Institute, Invergowrie, Dundee DD2 5DA, UK.*

H.M. Manley *Schools of Plant and Animal Sciences, University of Reading, 2 Earley Gate, Reading RG6 2AU, UK.*

H.A. McCartney *Plant Pathology Department, AFRC Institute of Arable Crops Research, Rothamsted Experimental Station, Harpenden, Herts AL5 2JQ, UK.*

T. McKay *Department of Agriculture, University of Aberdeen, School of Agriculture Building, Aberdeen AB9 1UD, UK.*

P.R. Mills *Department of Microbial Biotechnology, Horticulture Research International, Wellesbourne, Warwick CV35 9EF, UK.*

S. Muthumeenakshi *Biotechnology Centre for Animal and Plant Health, Queen's University of Belfast, Newforge Lane, Belfast BT9 5PX, UK.*

S.L. Nicholson *School of Biological Sciences, The University of Manchester, Oxford Road, Manchester M13 9PT, UK.*

D.W. Parry *Crop and Environment Research Centre, Harper Adams Agricultural College, Newport, Shropshire TF10 8NB, UK.*

J.C. Peters *Department of Agricultural Botany, School of Plant Sciences, University of Reading, 2 Earley Gate, Whiteknights, PO Box 239, Reading RG6 2AU, UK.*

T.R. Pettitt *Crop and Environment Research Centre, Harper Adams Agricultural College, Newport, Shropshire TF10 8NB, UK.*

B. Seddon *Department of Agriculture, University of Aberdeen, School of Agriculture Building, Aberdeen AB9 1UD, UK.*

M.W. Shaw *Department of Agricultural Botany, School of Plant Sciences, University of Reading, 2 Earley Gate, Whiteknights, PO Box 239, Reading RG6 2AU, UK.*

T. Shepherd *Scottish Crop Research Institute, Invergowrie, Dundee DD2 5DA, UK.*

D.C. Sigee *School of Biological Sciences, The University of Manchester, Oxford Road, Manchester M13 9PT, UK.*

D.M. Teverson *Natural Resources Institute, Central Avenue, Chatham Maritime, Kent ME4 4TB, UK.*

J.M. Thresh *Natural Resources Institute, Central Avenue, Chatham Maritime, Kent ME4 4TB, UK.*

A.I. Tiffin *Schools of Plant and Animal Sciences, University of Reading, 2 Earley Gate, Reading RG6 2AU, UK.*

G. Tuitert *Department of Phytopathology, Wageningen Agricultural University, PO Box 8025, 6700 EE Wageningen, The Netherlands.*

C.D. Upper *Plant Pathology Department, University of Wisconsin–Madison, 1630 Linden Drive, Madison, WI 53706, USA.*

E. Ward *Plant Pathology Department, AFRC Institute of Arable Crops Research, Rothamsted Experimental Station, Harpenden, Herts AL5 2JQ, UK.*

N.A. Williams *Scottish Crop Research Institute, Invergowrie, Dundee DD2 5DA, UK.*

B. Williamson *Scottish Crop Research Institute, Invergowrie, Dundee DD2 5DA, UK.*

F.J. Wilson *Department of Biological Sciences, University College of Wales, Aberystwyth, Dyfed SY23 3DA, UK.*

M. Wilson *Department of Plant Pathology, University of California, Berkeley, CA 94720, USA.*

Preface

This volume contains the majority of papers presented at the December 1992 meeting of the British Society for Plant Pathology held at The University of Wolverhampton, Walsall Campus. Planning of this meeting commenced in 1991 when it was decided that pathogen ecology should provide the main basis for the meeting. Subsequently contributors to this meeting were invited to participate with a large proportion opting to provide a written version of their paper – hence this volume.

The layout adopted for this volume essentially follows the format for the meeting. Papers are accordingly grouped into the following four sections: I: Ecological Determinants of Plant Pathogen Populations, II: Methodology, III: Aspects of Autecology and IV: Microbial Interactions.

The first section covers ecological principles of a more general nature, integrating growth habit, population and microenvironmental aspects of pathogen interactions with the host or with differing complexities of agricultural ecosystems.

Understanding ecological relationships between pathogens, their hosts and the associated microflora will partly depend on selection of appropriate methodology. Much interest is currently being shown in the use of rapid, specific diagnostic methods for pathogen detection and quantification in the diverse habitats in which inoculum may occur, whether it be on or in the host, or in the soil, with or without the presence of an associated microflora. Molecular-based methods are being considered increasingly in this respect because of their accuracy and specificity. Successful control of a pathogen involves an understanding of its behaviour in relation to host, environment and microflora. Nowhere is this more important than in the application of biological control where establishing appropriate protocols for selection and application of control agents is paramount for the successful use of this technique. Development of methodologies relating to detection and quantification of pathogens must be complemented by those for studying the occurrence and distribution of specific vectors. This is especially appropriate for the study of plant viruses.

A wide range of different aspects of pathogen behaviour is covered under the section on Autecology. These each contribute in varying ways to enhancing the success of pathogens for their mode of life. Production and effective dispersal of inoculum are essential if a pathogen is to increase its chances of colonizing an appropriate host, as are developing means of survival on, or within, the host until such time as optimal susceptibility is reached, which will permit effective colonization. There are particular problems in understanding the development and spread of root pathogens in the soil because, in addition to changing patterns of host susceptibility, the presence of competing microorganisms and the ways that such populations develop over several seasons of crop monoculture may have a significant influence on the ability of a pathogen to cause damaging levels of root disease. At the other extreme, the foliar environment provides its own special circumstances influencing pathogen development. Here changes in environmental conditions can have an immediate and far-reaching effect on the success or failure of a pathogen. Pathogens on foliar surfaces are particularly exposed to microclimatic changes such as the nature and extent of rainstorms, humidity and temperature changes. Foliar bacterial pathogens are amongst the most responsive to changes in their environment as large cell populations occur as epiphytes and there is a complex interaction which determines the change to pathogenesis. Pathogen ecology is more frequently studied in the temperate regions of the world. However, quite different sets of circumstances will influence the behaviour and spread of tropical pathogens, but a common feature is the reservoir of inoculum and genetic interactions between pathogens on crop plants and their wild relatives in the ecosystem. The ecology of viruses has probably been under-represented in the literature in recent years. Since viruses cause devastating disease losses in the tropics it is particularly appropriate that this area of study is represented.

The last section covers Microbial Interactions. The saprophytic microflora in the pathogen environment can interact at different stages in the pathogen life cycle and is strongly influenced by agricultural practices associated with management of the crop, such as straw incorporation. Here the influence will primarily be on persistence of pathogen inoculum, but on the surfaces of the living crop microflora-based interactions will affect multiplication of inoculum (especially with bacterial pathogens) and formation and development of infection structures together with consequent influences on the success of host entry. At a later stage, microbial interactions in necrotic tissues may influence pathogen inoculum production and hence future disease spread. Interactions may also be important between pathogens, especially where a species/strain complex exists for a particular host or taxonomically related group of hosts.

By necessity, because of the breadth of the subject area, the organizers of the December 1992 BSPP meeting, and consequently this volume which is derived from it, have had to be selective in the requesting of papers, but it is hoped that the majority of readers of this volume, both established pathologists, teachers and students alike, will benefit from the range, nature and distinctive approach of the contributions. Although the topics covered are diverse, throughout the volume the authors and editors alike have tried to ensure an ecological relevance to the subject matter wherever possible.

The editors would like to thank Philip Smith of the Scottish Crop Research Institute, Dundee for considerable help in proofreading of all the chapters.

J.P. Blakeman
B. Williamson

I ECOLOGICAL DETERMINANTS OF PLANT PATHOGEN POPULATIONS

1 All Creatures Unitary and Modular

J.H. Andrews

Plant Pathology Department, 1630 Linden Drive, University of Wisconsin, Madison, Wisconsin 53706, USA.

Introduction

This chapter considers aspects of the biology of two fundamental categories of organisms – unitary and modular – and argues that fungi and bacteria belong in the latter group. The significance is that members of either category share properties or what might be called 'solutions' to common ecological 'problems'. Therefore, research on modular macroorganisms should have relevance and conceptual value for studies on microorganisms. Likewise, fungi and bacteria could be tractable model systems for answering basic questions pertaining to the ecology of their larger modular counterparts. Only the underlying principles and general patterns are stated here and, of course, there are some exceptions.

Unitary and Modular Organisms

In unitary organisms, the zygote develops into a determinate structure that is repeated only at the start of the life cycle; in modular organisms, the product of the zygote is an indeterminate structure, units of which are iterated throughout the life cycle (Harper, 1985; Harper *et al.*, 1986). Unitary organisms follow sequential life cycle phases predictably and their morphology is usually fixed early in ontogeny. The genetic individual (genet) and the physiological or numerical individual are the same entity. Modular organisms are moulded much more by the environment which acts on the unit of

Table 1.1. Major attributes of unitary and modular organisms.[1]

Attribute	Unitary organisms	Modular organisms
Branching	Generally nonbranched	Generally branched
Mobility	Mobile; active	Nonmotile;[2] passive
Germplasm	Segregated from soma	Not segregated
Development[3]	Typically preformistic	Typically somatic embryogenesis
Growth pattern	Noniterative; determinate	Iterative; indeterminate
Internal age structure	Absent	Present
Reproductive value	Increases with age, then decreases; generalized senescence	Increases; senescence delayed or absent; directed at module
Role of environment in development	Relatively minor	Relatively major especially among sessile forms
Examples	Mobile animals generally, especially the vertebrates	Many of the sessile invertebrates such as hydroids; corals; colonial ascidians; also plants; fungi; bacteria

[1] Modified from Andrews (1991). These are generalizations. There are exceptions (see text for examples).
[2] Juvenile or dispersal phases mobile. Many bacteria and protists are somewhat mobile.
[3] Pertains to degree to which embryonic cells are irreversibly determined. Preformistic = all cell lineages so determined in early ontogeny; somatic embryogenesis = organisms capable of regenerating a new individual from some cells at any life cycle stage (cells totipotent or pluripotent). See Buss (1983).

construction (module). Modules in aggregate constitute the genet. The term module has been defined variously (Watkinson and White, 1985) but is frequently taken to be a repeated, multicellular unit of organization. A typical, easily visualized module is the leaf together with its axillary bud and internode of a stem. Modules may remain attached and contribute substantially to the organism's architecture (as in trees and corals, much of which often consists of accumulated dead modules). Alternatively, they can operate as separate physiological units or ramets (see below). Examples of unitary organisms include most mobile animals, particularly the vertebrates. Modular organisms include plants and many of the sessile invertebrates such as bryozoans, colonial ascidians and hydroids. As developed below, it can be argued that microorganisms belong to the modular group.

The major characteristics of modular and unitary organisms are summarized in Table 1.1. These characteristics are generalizations and some of the exceptions should be noted in passing. For example, not all unitary species sequester the germ line. The solitary anemones (Cnidaria) do not sequester and are not modular. Such species are typically members of phyla with lots of modular species. Likewise, the descriptor 'sessile' is not invariably a reliable discriminator between modular and unitary forms: for instance, the Siphonophores in the Cnidaria can be modular, but they are pelagic; conversely, there are numerous sessile unitary animals (e.g. mussels, oysters, barnacles, tube-dwelling polychaete worms).

At the outset it is also worthwhile to comment on the distinction among terms that have been used loosely or incorrectly. Indeterminate implies that there is no genetically fixed upper size limit to the organism, together with the capacity to increase or decrease size substantially with changing environmental conditions (broadly construed to include such factors as nutrition, parasitism, predation or competition) (Sebens, 1987). While all modular organisms display indeterminate growth, not all indeterminately growing organisms are modular; examples include the echinoderms, annelids and molluscs. Clonal growth implies production of genetically identical units which may remain aggregated or be dispersed (Hughes and Jackson, 1980). Each individual unit of a clone at the level of potential or actual independent existence is a ramet, alluded to above (Harper, 1977). Modular organisms may or may not be clonal. For instance, while all trees are modular, some are clonal (aspens) while others are not (maples). Furthermore, many animals can reproduce asexually, thus are clonal, but are not necessarily modular (e.g. aphids, starfish and earthworms).

A fungus is a population of hyphal apices just as a tree is a population of shoot and root tips (Harper, 1977, 1985). Fungal colonies are branched, sessile, and have an age structure, just as trees grow by producing cohorts of leaves. The 'hyphal growth unit' (Trinci, 1973), a standard measure of growth used for decades in the study of fungal kinetics, is quite useful conceptually in modular semantics (Andrews, 1991). For yeasts, as for bacteria (below), the module is the individual cell. Thus, fungi can be considered as modular rather than unitary organisms.

For the most part, bacteria are essentially sessile organisms that grow in microcolonies, frequently as biofilms, at surfaces (Characklis and Marshall, 1990). Indeed, there are planktonic forms (in water or, occasionally, bloodstreams) but these cells are analogous to portions of dispersed, floating ramets of *Lemna* and *Salvinia*, i.e. to genets of organisms that fall apart as they grow (Harper, 1977, Harper *et al.*, 1986). The lifespan of the clonal lineage of bacteria is unlimited and the individual cell, iterated indefinitely, can be taken to be the module in the bacterial context (Andrews, 1991).

Implications of a Modular Lifestyle

Modular organisms, being sessile, adjust to their environment in a fundamentally different way from mobile unitary creatures. Selection in the former case has been for the ability to change form (plasticity) in response to changing conditions, along with the development of life cycle phases associated with dispersal and survival; in the latter it has been for mobility and behavioural adaptation. All one has to do to confirm this is to observe the variation in plant phenotype and in allocation of resources that results from crowding or other forms of physiological stress (Palmblad, 1968; Coleman *et al.*, 1992). Modules can be added progressively by growth or acquired suddenly by fusion with compatible genotypes: bryozoans and corals add zooids, lichen colonies may converge, and fungal hyphae often anastomose. Indeed, the presence of neighbours, and contact of parts of organisms, is a common and distinctive feature of modular growth (Harper, 1985; Harper *et al.*, 1986). Conversely, shrinkage or 'degrowth' occurs locally by fragmentation or mortality resulting from dispersal or death of modules, respectively (Hughes and Jackson, 1980; Sebens, 1987). Hence, for modular organisms, size and age typically are not correlated.

The extreme case in growth potential is the clonal organism that can increase exponentially in size and fecundity (Watkinson and White, 1985). Consequently, reproductive value (Williams, 1957; Begon *et al.*, 1986) increases and generalized senescence is absent, consistent with theoretical predictions (Williams, 1957). The same genet can be exposed to very different environments and hence selection pressures. The most striking example is long distance movement of microbial clones which routinely occurs by atmospheric transport. In contrast, reproductive value in unitary organisms typically rises swiftly to a peak near sexual maturity and then declines; senescence occurs and is expected based on theory (Williams, 1957; Begon *et al.*, 1986; Andrews, 1991). Clonal organisms, especially those that grow prostrate rather than upright, escape many of the problems of large size such as the need to support a massive bulk.

Modular organisms are potentially very long lived (Hughes and Jackson, 1980; Watkinson and White, 1985). Thus, for a given genet there is more opportunity for somatic change to occur and be expressed. A related and more important issue is whether somatic mutations enter the germ line to any evolutionary significant degree. Germ cells are sequestered early in the ontogeny of most unitary organisms, and are insulated from the entry of somatic mutations. The germ line is not isolated in modular creatures (Buss, 1983, 1990). Where organisms retain cellular totipotency, and hence can reproduce clonally, somatic variants can be transmitted to progeny. In the fungi, for instance, mutants arising in any tissue can be transferred asexually and, potentially, sexually. Because asexual reproductive rates of fungi are so high, favourable mutants are rapidly increased through natural selection. Hence, through clonal growth, the fungi can evolve significantly by mutation alone in the absence of recombination. Likewise, mutation is a major evolutionary force in bacteria (Andrews, 1991). Among plants, the phenomenon of somatic variation has been exploited for centuries by horticulturists. Various characteristics of plants related to their modular growth form allow mutations to accumulate (e.g. Gill, 1986). Because of the clonal aspects of development, somatic mutations in precursors of a floral lineage can potentially be passed on through gametes. That this does happen has been shown in work with transposable elements of maize (e.g. Fedoroff, 1989). For example, if a genetic change occurs during the first embryotic cell division, a plant with genotypically and phenotypically distinct halves is created. Each half will go on to produce different gametes. Therefore, somatic changes can not only alter the fitness of the carrier but can, at least in some instances, be passed on to offspring produced sexually. However, to date there is virtually no solid evidence on the magnitude of gametic as opposed to the asexual transfer of somatic variation for plants or other modular organisms.

Analogies based on modular growth

An important prediction from Table 1.1 and the foregoing remarks is that natural selection should operate rather differently on modular and unitary organisms. Thus, assuming that we have correctly categorized species according to the unitary/modular dichotomy, then selection should result in similarities in life histories among members within each respective group. The following general issues for modular organisms are examined below: (i) formation of resource depletion zones and the consequent search strategies for resources; (ii) growth as a single unit or as multiple units; and (iii) consequences of contact.

Resource depletion zones (RDZ) and foraging

Because modular organisms occupy relatively fixed positions, resources sooner or later tend to become locally exhausted, creating RDZs (Harper, 1985). In mycological semantics, these would be called zones of substrate exhaustion or enzymic erosion. The depletion concept has been applied primarily to roots but is relevant in principle to all sessile organisms. For example, leaves form depletion zones for light by shading their neighbours, and passive suspension feeders locally deplete waters of nutrient particles.

To move themselves away from RDZs and into new terrain, modular organisms must grow. Growth includes both linear expansion and asexual 'reproduction'. For example, the fungal genet or clone expands by diffuse or directed hyphal growth as an intact unit, as well as by sporulation (Andrews, 1992). From a foraging perspective, dispersal by spores offers the opportunity to exploit relatively distant nutrient stores not accessible by hyphal spread.

The architecture of organisms, as it relates to foraging theory, has been discussed elsewhere (Andrews, 1991, 1992); here the idea of fungal dimorphism as it may relate to nutrient acquisition is considered. A fascinating case concerns the yeast *Saccharomyces cerevisiae*, which can grow either as single cells (unpolarized colonial growth) or filaments (polarized or pseudohyphal growth). The latter phase can occur only in diploid strains heterozygous at the a/α mating type locus (Gimeno *et al.*, 1992). Activation of the *RAS* pathway and resultant elevated levels of intracellular cAMP are directly or indirectly involved in the transition. Pseudohyphal growth has been interpreted adaptively (Gimeno *et al.*, 1992) as a mechanism of the fungus to forage for nutrients (in other words as a means to escape RDZs). Thus we have a good microbial analogue for a phenomenon pertaining to modular organisms generally, and the first instance where morphogenetic switching related to the nutrient level is being explained at the genetic and biochemical level.

How modules are arranged must influence strategies of resource capture by modular organisms. An open architecture (longer spacers) would promote rapid vertical or horizontal colonization of new habitat, while a closed architecture (shorter spacers) would favour consolidation of acquired habitat (Harper, 1985). The terms 'guerrilla' and 'phalanx' describe these respective growth forms which can be viewed as opposite poles of a continuum of habits (Lovett Doust, 1981). The guerrilla architecture is characterized by long internodes, infrequent branching, spaced modules, and minimum overlap of RDZs. Examples of guerrilla-type plants include creeping buttercup (*Ranunculus repens*), white clover (*Trifolium repens*) and wild strawberry (*Fragaria vesca*). In phalanx organisms, internodes are short, branching is frequent and modules are closely packed. Consequently, space is tightly occupied and RDZs overlap. Plant examples (Harper, 1985) include various tussock grasses such as *Deschampsia caespitosa*, the common goldenrod (*Solidago canadensis*) and figwort (*Scrophularia nodosa*).

The descriptors phalanx and guerrilla also fit the growth forms of other sessile organisms. Crustose (mat-forming) lichens grow as phalanxes, as do some corals, bryozoans, and colonial ascidians; other corals and the foliose (aerially branched) bryozoans and lichens have, relatively speaking, a guerrilla-type habit (Harper, 1985). There are six basic forms of colonial marine animals (Jackson, 1979), of which 'runners' and 'vines' comprise essentially a guerrilla habit, while the 'trees', 'plates', 'mounds' and 'sheets' are to varying degrees phalanx in form.

The terminology is also relevant to certain microbial patterns. For example, some fungi, particularly the Zygomycetes and Oomycetes, extend rapidly in guerilla-like fashion. The open, relatively noncompetitive colonization of fields of canola (oilseed rape) by numerous clones of the pathogen *Sclerotinia sclerotiorum* has been interpreted as proceeding in guerrilla fashion (Kohli *et al.*, 1992), whereas the dense occupation of terrain by, and clear demarcation among, the few clones of the root pathogen *Armillaria bulbosa* in woodland seems to depict localized, mutually exclusive spread in phalanx fashion (Smith *et al.*, 1992).

The guerrilla/phalanx continuum is a useful descriptor of growth form. However, it does not identify why a particular form has arisen or is maintained, and this may be for quite different reasons in different phyla. For instance, though plants can change their form conspicuously in response to environment (Harper, 1977, 1985), fungal thalli are even more plastic (Andrews, 1991, 1992). Nevertheless, within the guerrilla/phalanx concept are some testable specific questions of relevance to microbiologists (for plant analogies see Schmid, 1986; Sutherland and Stillman, 1988). How does growth form change in response to nutrient conditions (or other environmental stimuli such as antimicrobial compounds)? What are the genetic and biochemical mechanisms behind such phenotypic responses (cf. for plants the touch response; Braam and Davis, 1990)? How does growth form relate to the relative competitive or colonizing ability of parasitic or saprophytic fungi?

To divide or not to divide

Should a clonal modular organism grow as an intact, progressively expanding entity, or undergo rounds of division and dispersal? Whether selection favours one course or the other will depend ultimately on which alternative most reduces the probability of genotype death (Cook, 1979). All clonal organisms fragment at some point so the question really becomes what sets the upper size limit. This could be size-selective predators, mechanical factors, the abundance or quality of nutrients, or energetic limitations determined by the ability of an expanding unit to feed itself (Sebens, 1982).

For various marine invertebrates it has been shown that asexual reproduction can be more favourable for survival of the genotype than continual increase in body size (Sebens, 1982; McFadden, 1986). Passive suspension feeders such as sea anemones and soft corals capture food particles by exposing their feeding surfaces (polyps) to ambient water currents. Rates of food intake depend on effective surface area available. Energetic costs rise faster with size than does prey capture because polyps tend to interfere with each other, and as the colony grows there is a declining ratio of peripheral to internal polyps. The former are more effective than the latter at capturing nutrient particles. Overall, per polyp capture rates decline as colony size increases (McFadden, 1986). Thus, these organisms should reach an energetically optimum size that would be habitat-dependent because sites vary in aspects such as prey abundance, current velocity, water temperature, and feeding (immersion) time for animals in the intertidal zone.

How relevant is the lifestyle of passive suspension feeders to that of other modular organisms such as the fungi? Hyphae are absorptive surfaces analogous to polyps. The question is to what extent fungi (or other modular organisms) can increase feeding (absorptive) surface proportionately to mass. If intake is indeed proportional to biomass, an energetic limit on size or an optimum size based on nutritional criteria is not expected (Sebens, 1982). Examples include invertebrates that grow as flat sheets or as many branches (Sebens, 1982). By virtue of their multiple hyphae, filamentous fungi grow to some extent the same way. However, hyphae often are aggregated and, even in a diffuse mode, the feeding activity of one hypha of the colony tends to compete with other hyphae in the same area. As described earlier, fungal colonies form RDZs, and energetic costs would seem likely to increase with colony size faster than absorptive capacity for two reasons. First, although the shape of the volume explored and depleted by a colony will be irregular, as a first approximation it might be considered to be spherical. Clearly this will be closer to reality in some situations (e.g. aquatic fungi growing in stagnant water) than others. Since, for spherical objects, volume increases as d^3 but surface area as d^2, the rate at which the thallus can obtain nutrients will be a declining function of size. Second, feeding efficiency per unit biomass will also decline with size because of the impact on nutrient uptake of differentiation and senescence of individual hyphae (Jennings and Rayner, 1984; Zalokar, 1965). Asexual division allows the genet as a whole in clonal organisms to increase in surface area indefinitely and proportionately to mass, even though the individual ramets (e.g. fungal colonies) would be size-limited by the increasing energetic cost function.

So for fungi, as for certain marine organisms, it may be energetically better for a genotype to exist as many small units rather than one large one. Certainly this prediction is consistent with a wealth of information implicating nutrient exhaustion as a trigger for sporulation (Smith, 1978).

Competition or complementation

That a modular organism grows in indeterminate, sessile fashion, frequently as a clone, means that contact with parts of itself or other organisms is common (Buss, 1990). There are many interesting implications, among them: (i) the origin of recognition mechanisms and the extent to which they may be fundamentally similar across phyla; (ii) the types of rejection responses; and (iii) the benefits and hazards of fusion. Only the latter issue can be addressed briefly here.

Fusion offers the potential for a small organism to become instantly large, thereby acquiring powers of competitive advantage and deterrence of predators. In the microbial world the former might include increased inoculum potential, or higher enzymatic or antibiotic capability. Furthermore, fusion can confer physiological complementation such that the chimera gains phenotypic traits lacking in the donor and recipient. This has been particularly well documented for the fungi. Commonly, for instance, fungal heterokaryons outperform their component homokaryons (Jinks, 1952; Davis, 1966). Buss (1982) notes that genetic variability contributed to chimeras by somatic mutation may be especially important in clonal organisms (favourable gene combinations would not be lost in meiosis) and haploid organisms (disadvantageous recessives are masked). There are two other benefits of fusion (Buss, 1982): elimination of mate location problems (for those creatures that are both somatically and sexually compatible) should environmental conditions make sexual reproduction advantageous; and developmental synergism, whereby the component elements of a chimera collaborate to produce supporting tissue. To the extent to which fusion can result in a better 'survival machine' (Dawkins, 1989) for genes, the phenomenon should be favoured by natural selection.

Fusion does involve certain hazards. The recipient may acquire systemic pathogens or its genes may come to be under-represented in gametes produced by a chimeric individual if donor nuclei overwhelm those of the recipient (somatic cell parasitism). Chimeric hydroids have a common gastrovascular system and thus an open pathway for movement of cells between the partners (Buss, 1990). This, together with the fact that the germ line is not sequestered in cnidarians (somatic embryogenesis, Table 1.1), means that the cell line of one or the other component can be disproportionately represented in the gametes arising from the chimera (Buss, 1990). Formation of clamp connections and the close, stable association of the different nuclear types in pairs (dikaryosis), a characteristic of the basidiomycete fungi, appears to be a self-limiting mechanism controlling nuclear representation in sexual or asexual propagules (Buss, 1982). More generally, the operation in modular organisms of recognition systems, which frequently link somatic with sexual compatibility and which restrict fusions to closely related individuals (Grosberg, 1988; Marchalonis and Reinisch, 1990; Rayner, 1991) limits the impact of genomic replacement (Buss, 1982, 1990; Buss and Dick 1992) or pathogens (Caten, 1972) on members of a chimera. This implies that many if not all modular organisms compete both at the level of the colony, manifested by rejection responses such as overgrowth, zone lines, and other somatic incompatibility reactions, as well as at the level of the cell lineage, manifested by the invasion of a soma by foreign nuclei or cells (Buss, 1990).

Analogies – pro and con

One can see in nature either infinite variety at all levels of biological organization or common themes as expressed, for example, by the responses of modular versus unitary organisms to natural selection. Despite their great diversity in life histories, the organisms about us obviously are all extant: they (or at least their parents) have survived what is often called the sieve of natural selection. However, as Jacob (1982, pp. 14–15) says, selection is more than a sieve, over time integrating mutations and ordering them into adaptively coherent patterns, giving direction to change, and orienting chance. Not surprisingly, therefore, similar strategies emerge as 'solutions' to common ecological 'problems'. Therefore, research on a particular system has implications that transcend it. Analogies provide intellectual stimulation and the impetus to probe further. They also are the way towards insight from a new vantage point, to broader application of knowledge, and to a common ground for discussions among biologists representing diverse specialties. For elaboration of these aspects see Andrews (1991), Price (1992) and Lawton (1992).

This is not to imply that analogies are without limitations. Like models they are simplifications. Accuracy is sacrificed for generality. Many analogues may well be incorrect, misleading, or cast at such a general level as to be bland and trivial. They do not provide solutions to practical problems.

Table 1.2. Some testable predictions emerging from the three analogies or general principles pertaining to modular organisms (see text).

	Analogy or principle		Relevant testable hypotheses
1.	Creation of RDZs; evolution of foraging mechanisms	A.	Nutrient levels are the trigger for fungal dimorphism
		B.	Fungal morphological responses are relatable to localized nutrient reservoirs
2.	Habitat-dependent size optima for aclonal modular organisms; ramets of clonal forms are habitat-dependent in size	A.	Energetic costs for a fungal colony increase with size faster than rate of nutrient absorption
		B.	Nutrient thresholds trigger asexual sporulation
3.	Reaction of parts of an organism to parts of another	A.	Fusion results in increased fitness to members of a chimera
		B.	Ancient evolutionary origin of gene sequences governing recognition mechanisms are homologous in unitary and modular organisms

I cannot and do not pretend to tell you how to control any plant disease based on my ecological analogies, although obviously the growth form of modular organisms establishes basic limitations as well as opportunities for microorganisms. It dictates how pathogens compete within the microbial community, how they colonize plants, and how they disperse to new hosts. Analogies typically fall into the realm of science called concept formation and, couched in general terms, are not testable. Nevertheless, analogies can form the basis for fruitful hypotheses (Table 1.2); they also should play a role in inductive inference for generalizing the specifics of research findings.

Synopsis and a Concluding Perspective

The largest and smallest of organisms differ by some 20 orders of magnitude in size (McMahon and Bonner, 1983) and clearly must experience quite different worlds (Andrews, 1991). Unfortunately, this difference historically has overshadowed commonalities between micro- and macroorganisms and has been reinforced by the specialization of biologists, disciplinary isolation, and distinctive approaches to the study of various phyla. It seems more biologically defendable to integrate with respect to size and to split the world along an axis of modular versus unitary design. This would bring into sharper focus the meaningful commonalities and distinctions among creatures and would open vistas for fruitful analogies.

This is not to suggest that it is a simple matter to classify every organism as modular or unitary, or that this demarcation is the only sensible one. Natural groupings along conventional phylogenetic lines obviously convey much information; these delineations are being confirmed or reordered based on unfolding evidence from molecular systematics. Other useful alignments of organisms, depending on one's objective, are clonal versus aclonal, determinate versus indeterminate, colonial versus solitary, and parasitic versus free-living. Pathologists, being for the most part microbiologists, have tended to be reductionists. Our thinking can be parochial. More can be gained by blending the rigour of mechanistic interpretation possible at the genetic/biochemical level in microbiology with the theory and perspective of macroecology. The reductionist approach, based on the specifics of individual pathogens and disease cycles, has often served plant pathology well. Nevertheless, if our discipline is to advance as a science the search for depth must be matched by a search for breadth characterized by generalization, synthesis, and unifying biological theories.

Acknowledgements

This is a contribution from the College of Agricultural and Life Sciences, University of Wisconsin, Madison, USA. Grant support during the time these ideas were formulated was provided by the National Science Foundation (Grant No. DEB-9119476), the USDA (Grant No. 62-016-02934), and the Environmental Protection Agency (Grant No. R819377-01-0). I thank D. Padilla and R. Lindroth for comments on the manuscript.

References

Andrews, J.H. (1991) *Comparative Ecology of Microorganisms and Macroorganisms.* Springer-Verlag, New York.

Andrews, J.H. (1992) Fungal life-history strategies. In: Carroll, G.C. and Wicklow, D.T. (eds), *The Fungal Community.* 2nd edn. Marcel Dekker, New York, pp. 119–145.

Begon, M., Harper, J.L. and Townsend, C.R. (1986) *Ecology: Individuals, Populations and Communities.* Blackwell Scientific Publications, Oxford.

Braam, J. and Davis, R.W. (1990) Rain-, wind-, and touch-induced expression of calmodulin and calmodulin-related genes in *Arabidopsis. Cell* 60, 357–364.

Buss, L.W. (1982) Somatic cell parasitism and the evolution of somatic tissue compatibility. *Proceedings of the National Academy of Sciences, USA* 79, 5337–5341.

Buss, L.W. (1983) Evolution, development, and the units of selection. *Proceedings of the National Academy of Sciences, USA* 80, 1387–1391.

Buss, L.W. (1990) Competition within and between encrusting clonal invertebrates. *Trends in Ecology and Evolution* 5, 352–356.

Buss, L.W. and Dick, M. (1992) The middle ground of biology: themes in the evolution of development. In: Grant, P.R. and Horn, H.S. (eds), *Molds, Molecules and Metazoa.* Princeton University Press, Princeton, New Jersey, pp. 77–97.

Caten, C.E. (1972) Vegetative incompatibility and cytoplasmic infection in fungi. *Journal of General Microbiology* 72, 221–229.

Characklis, W.G. and Marshall, K.C. (1990) *Biofilms.* Wiley, New York.

Coleman, J.S., Jones, C.G. and Krischik, V.A. (1992) Phytocentric and exploiter perspectives of phytopathology. *Advances in Plant Pathology* 8, 149–195.

Cook, R.E. (1979) Asexual reproduction: a further consideration. *The American Naturalist* 113, 769–772.

Davis, R.H. (1966) Mechanisms of inheritance. 2. Heterokaryosis. In: Ainsworth G.C. and Sussman, A.S. (eds), *The Fungi. An Advanced Treatise. Volume II. The Fungal Organism.* Academic Press, London, pp. 567–589.

Dawkins, R. (1989) *The Selfish Gene*, 2nd edn. Oxford University Press, Oxford.

Fedoroff, N.V. (1989) Maize transposable elements. In: Berg, D.E. and Howe, M.M.

(eds), *Mobile DNA*. American Society for Microbiology, Washington, D.C., pp. 375–411.

Gill, D.E. (1986) Individual plants as genetic mosaics: ecological organisms versus evolutionary individuals. In: Crawley, M.J. (ed.), *Plant Ecology*. Blackwell Scientific Publications, London, pp. 321–343.

Gimeno, C.J., Ljungdahl, P.O., Styles, C.A. and Fink, G.R. (1992) Unipolar cell divisions in the yeast *S. cerevisiae* lead to filamentous growth: regulation by starvation and *RAS*. *Cell* 68, 1077–1090.

Grosberg, R.K. (1988) The evolution of allorecognition specificity in clonal invertebrates. *Quarterly Review of Biology* 63, 377–412.

Harper, J.L. (1977) *Population Biology of Plants*. Academic Press, New York.

Harper, J.L. (1985) Modules, branches, and the capture of resources. In: Jackson, J.B.C., Buss, L.W. and Cook, R.E. (eds), *Population Biology and Evolution of Colonial Organisms*. Yale University Press, New Haven, Connecticut, pp. 1–33.

Harper, J.L., Rosen, B.R. and White, J. (1986) The growth and form of modular organisms. *Philosophical Transactions of the Royal Society of London* B 313, 1–25.

Hughes, T.P. and Jackson, J.B.C. (1980) Do corals lie about their age? Some demographic consequences of partial mortality, fission, and fusion. *Science (USA)* 209, 713–715.

Jackson, J.B.C. (1979) Morphological strategies of sessile animals. In: Larwood, G. and Rosen, B.R. (eds), *Biology and Systematics of Colonial Organisms*. Academic Press, New York, pp. 499–555.

Jacob, F. (1982) *The Possible and the Actual*. Pantheon, New York.

Jennings, D.H. and Rayner, A.D.M. (1984) *The Ecology and Physiology of the Fungal Mycelium*. Cambridge University Press, Cambridge.

Jinks, J.L. (1952) Heterokaryosis: a system of adaptation in wild fungi. *Proceedings of the Royal Society of London* B 140, 83–99.

Kohli, Y., Morrall, R.A.A., Anderson, J.B. and Kohn, L.M. (1992) Local and trans-Canadian clonal distribution of *Sclerotinia sclerotiorum* on canola. *Phytopathology* 82, 875–880.

Lawton, J. (1992) There are not 10 million kinds of population dynamics. *Oikos* 63, 337–338.

Lovett Doust, L. (1981) Population dynamics and local specialization in a clonal perennial (*Ranunculus repens*). I. The dynamics of ramets in contrasting habitats. *Journal of Ecology* 69, 743–755.

Marchalonis, J.J. and Reinisch, C.L. (1990) *Defense Molecules*. Wiley-Liss, New York.

McFadden, C.S. (1986) Colony fission increases particle capture rates of a soft coral: advantages of being a small colony. *Journal of Experimental Marine Biology and Ecology* 103, 1–20.

McMahon, T.A. and Bonner, J.T. (1983) *On Size and Life*. Scientific American Books, New York.

Palmblad, I.G. (1968) Competition in experimental populations of weeds with emphasis on the regulation of population size. *Ecology* 49, 27–34.

Price, P.W. (1992) Evolutionary perspectives on host plants and their parasites. *Advances in Plant Pathology* 8, 1–30.

Rayner, A.D.M. (1991) The challenge of the individualistic mycelium. *Mycologia* 83, 48–71.

Schmid, B. (1986) Spatial dynamics and integration within clones of grassland perennials with different growth form. *Proceedings of the Royal Society of London* B 228, 173–186.

Sebens, K.P. (1982) The limits to indeterminate growth: an optimal size model applied to passive suspension feeders. *Ecology* 63, 209–222.

Sebens, K.P. (1987) The ecology of indeterminate growth in animals. *Annual Review of Ecology and Systematics* 18, 371–407.

Smith, J.E. (1978) Asexual sporulation in filamentous fungi. In: Smith, J.E. and Berry, D.R. (eds), *The Filamentous Fungi. Vol. 3. Developmental Mycology.* Wiley, New York, pp. 214–239.

Smith, M.L., Bruhn, J.N. and Anderson, J.B. (1992) The fungus *Armillaria bulbosa* is among the largest and oldest living organisms. *Nature* 356, 428–431.

Sutherland, W.J. and Stillman, R.A. (1988) The foraging tactics of plants. *Oikos* 52, 239–244.

Trinci, A.P.J. (1973) The hyphal growth unit of wild type and spreading colonial mutants of *Neurospora crassa. Archiv für Mikrobiologie* 91, 127–136.

Watkinson, A.R. and White, J. (1985) Some life-history consequences of modular construction in plants. *Philosophical Transactions of the Royal Society of London* B 313, 31–51.

Williams, G.C. (1957) Pleiotrophy, natural selection, and the evolution of senescence. *Evolution* 11, 398–411.

Zalokar, M. (1965) Integration of cellular metabolism. In: Ainsworth, G.C. and Sussman, A.S. (eds), *The Fungi. An Advanced Treatise. Vol. I. The Fungal Cell.* Academic Press, New York, pp. 377–426.

2 The Biological Environment and Pathogen Population Dynamics: Uncertainty, Coexistence and Competition

M.W. Shaw and J.C. Peters

Department of Agricultural Botany, School of Plant Sciences, University of Reading, 2 Earley Gate, Whiteknights, PO Box 239, Reading RG6 2AU, UK.

Introduction

For ecological purposes, organisms are often categorized by the position they occupy in a food chain, their trophic level. This is a convenient way to classify the biological environment of a pathogen. For the purposes of this chapter, it is useful to distinguish three levels, in a sequence. Firstly, an ecologically obligate plant pathogen needs a host, on the trophic level below. Secondly, any pathogen population coexists with many other organisms on the same trophic level, dependent on carbon fixed at the host trophic level; some of these species interact with some or all of the pathogen individuals, and thereby alter the population. Lastly, there are species which parasitize or consume pathogens, either as a sole diet, or as one component in a diet. The effects of these biological interactions on pathogen populations have been relatively neglected, compared to the influences of physical factors such as the weather. Biological factors differ crucially from physical factors in one respect: that they can differ in their effects according to the size of the pathogen population, and may therefore regulate or destabilize the population. The physical environment sets the stage on which biotic influences perform; in what follows, the values of population parameters are determined inevitably by the physical environment, and any assertion made about a biological influence will only be useful if it remains true even though the parameters of the system vary about a mean value. Aspects of the physical environment can interact very powerfully with biological influences to produce surprising effects on population size and variation; an aspect we shall stress in particular is the consequence of annual cycles in the weather.

Population dynamics is the study of changes in populations and the determinants of population size. Insect ecology was for a long time influenced by the question 'why is the world green?' (Hairston *et al.*, 1960): if insect populations possess the capacity to increase rapidly, why do they not continue to increase until limited by lack of food? In the most simplistic view this should occur only when individuals are obviously competing for food, which does not appear to be the case in the dense green foliage of many productive ecosystems. Neither pathogens nor insects usually defoliate an ecosystem; the spectacular exceptions are rare enough to be the subject of intensive study: Dutch elm disease, potato blight, chestnut blight or southern corn leaf blight. In fact, in neither agricultural nor natural systems do pathogens parasitize all the available plant tissue.

A system of dynamics which predicts 100% severity of disease as the equilibrium state of an ecosystem obviously requires revision, but it is reasonable to ask as a preliminary to modelling, how common are pathogens? To provide a basis for discussion, therefore, we quote data from a few well-studied systems. Table 2.1 refers to results from the winter cereals agroecosystem in the UK in the period 1970 to 1977 (King, 1977a,b; Polley and Thomas, 1991). After this time the use of fungicides became too common to make any comment about the intrinsic disease levels in the system. Kranz (1990) quoted data from several natural plant communities in Germany, which showed that about half of all species in the communities had disease incidences greater than 50%, while about a third of all species had severities greater than 10%. Some results from our own studies in successional grassland in Berkshire, southeast England, are shown in Fig. 2.1. In our data, the mean disease severity averaged over species at any one time was *c.* 3%, in 2 years of markedly different weather. These data are not dissimilar to those reported by Kranz (1990).

Table 2.1. Severity of all foliar diseases combined on winter wheat and barley during the 1970s in England and Wales, based on data in Polley and Thomas (1991) and King (1977a, b). Data are the average of severity on the flag leaf and leaf 2 over all crops surveyed. The interquartile range refers to variation in this figure over years.

Crop	Mean	Interquartile range
Barley	12.8	10.4–15.4
Wheat	9.2	3.8–13.2

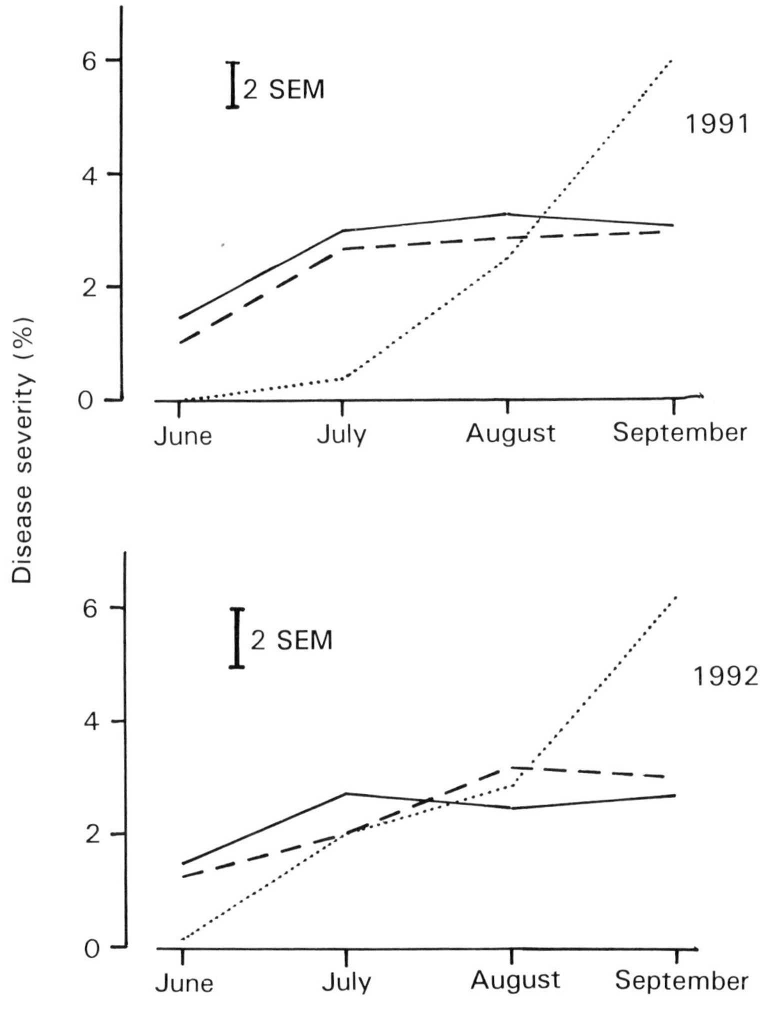

Fig. 2.1. The mean severity of disease caused by fungal pathogens on above ground plant parts throughout the summers of 1991 and 1992 in experiments on naturally regenerating pasture set up by Dr V.K. Brown at Silwood Park, Berkshire, UK. The data are averaged over all plant species present at a sampling date, weighted by the abundance of each species. (····) plot cleared of vegetation with herbicide and cultivated in spring of the year of assessment, (– – – –) plot cleared and cultivated in 1988 and (———) plot cleared and cultivated in 1985.

The cereal data refer to a single sampling time, late in the season, and, for much of the life of the crop, a random sample of foliage would have had much less disease. The cereal data also concern a single host species, whereas the data from natural populations are averaged for many species, with substantial turnover of the species included in the samples. Nonetheless, the disparity between the unmanaged, varied vegetation and the farmed monoculture is not marked. A theory which suggested disease severities on aerial plant parts of the order of magnitude of 5% at equilibrium would not be falsified by these data. Very little is known of the average severity of underground pathogens; it would involve a large effort and destructive sampling to collect such data.

How can this relative rarity of pathogens be explained? Many factors could be involved, but it may be profitable to start with simple models and see how much complexity is needed to explain the disease patterns we commonly record. Regulation of pathogen populations could be mediated directly through interactions with hosts; by interactions with other organisms living in, on or around the hosts, whether competitive or mutualistic; or by predation or parasitism. What levels of disease might be expected from regulation at each level, and what disease dynamics would result?

Interactions with the Host

Consider first the trophic level below pathogens: the plant hosts. These can usefully be regarded, both above and below ground, as populations of modules of varying lifetimes (Harper, 1977). The resource lifetime, the lifetime of an individual module, is usually short – only a few pathogen life cycles. The pathogen generation time is usually matched to the module lifetime, because long-lived modules, such as the trunk and major branches of trees, are very strongly defended, and pathogens must grow slowly or exploit rare events to invade the tissue. Thus, the host population is dynamic and its size is set by the balance between birth and death of modules. The pathogen population is similarly dynamic. In our first model we start with a benign environment: a monoculture in a continuously favourable environment.

First, we need to describe the population dynamics of the plant. We will denote the host population by h, measured in whatever way is suitable for the population under consideration. Then we suppose that the population changes according to the equation:

$$\frac{dh}{dt} = \lambda h - \mu h^2 \tag{2.1}$$

This has an equilibrium at $\hat{h} = \lambda/\mu$. The birth rate of plant organs is represented by λ. These organs may be leaves for a tree or a grass, or entire ramets if we are interested in systemic diseases such as wilts. (A ramet is a physiological individual of a plant, as distinguished from a genet, or genetical individual (Harper, 1977).) The death rate of modules is represented by μ. Because of shading and other crowding effects, the death rate increases with increasing density of plant organs. The simplest way to capture this density dependence is to assume that the death rate, μ, is directly proportional to density; this produces equation (2.1). Structurally equation (2.1) is a logistic, and the plant population will return to equilibrium from a perturbation in a logistic way. This approach is similar to that taken by Jeger (1987), but differs in the way the equation is written and in the interpretation of the parameters.

A pathogen can now be added to the initial model. A closed system is used so that the density of propagules produced by the pathogen should be related to the pathogen density. The number of propagules that survive to give rise to new individuals also depends on the density of healthy host tissue. Any realistic model should also incorporate a death rate for the pathogen. This suggests the following equations:

$$\frac{dh}{dt} = \lambda h - (\mu h + k(p/h))h = \lambda h - \mu h^2 - kp$$

$$\frac{dp}{dt} = rp(h-p) - \mu_p p = (rh - \mu_p)p - rp^2 \tag{2.2}$$

Here the *per capita* death rate of the host is increased by a factor which depends proportionately on the pathogen load per host individual, with a constant of proportionality k; this links host and pathogen dynamics in a plausible way.

The equilibria in this model are relatively simple to plot (Figs 2.2, 2.3 and 2.4). The equilibrium population is much lower for pathogens with more deleterious effects, even when the pathogen has only a small impact on the host population at equilibrium; thus the detrimental effect of pathogenicity on the pathogen population is greater than the effect of pathogenicity on the host population.

By itself, the introduction of a cyclically varying environment has little effect in this model. Although an analytical solution to equations (2.2) is not possible under these circumstances, the consequences can be explored by numerical simulation of the equations. This shows that if lifetimes of host and pathogen are approximately equal to or longer than the annual cycle, the model can be used directly, provided average values are used. If the lifetimes of host modules and pathogen individuals are much shorter than a year, populations will track one another as they vary, and equilibria will not be dramatically altered.

In a diverse community, such a model must apply to each species

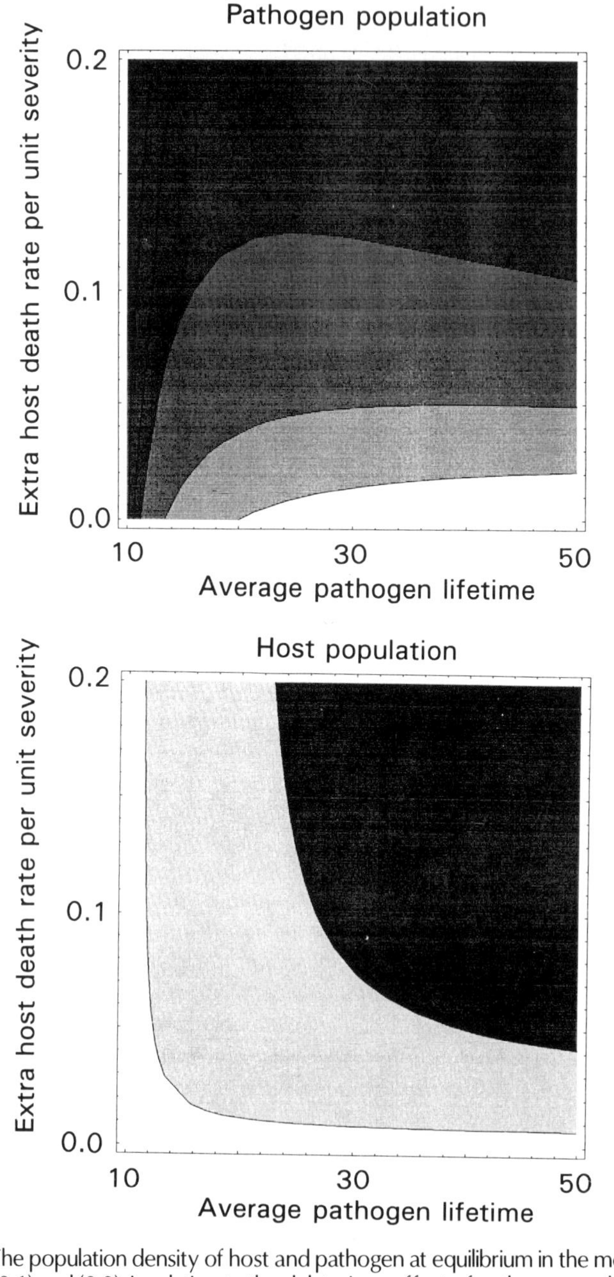

Fig. 2.2. The population density of host and pathogen at equilibrium in the model in equations (2.1) and (2.2), in relation to the deleterious effect of pathogen on the host and the lifetime of the pathogen. The birth and death rates of the host modules (leaves, roots, stems etc.), λ and μ, are both 0.05 per unit time; the intrinsic growth rate of the disease (apparent infection rate), *r*, is 0.1 per unit time.

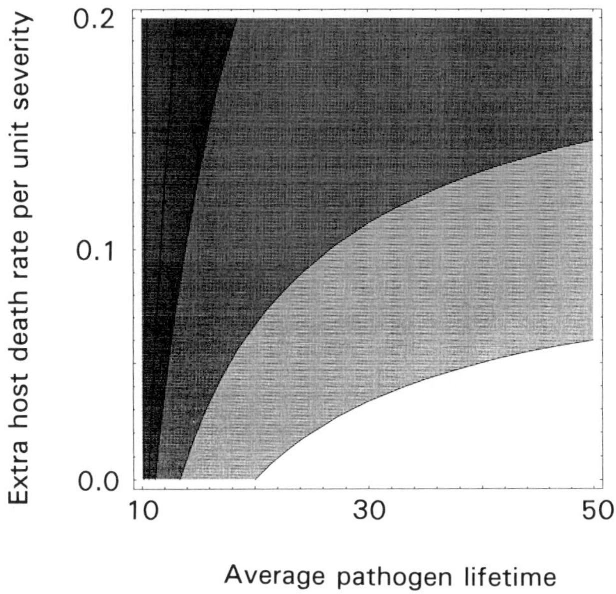

Fig. 2.3. The data in Fig. 2.2 represented as the ratio of pathogen population to host population ('severity').

separately. The main effect of diversity is presumably to reduce the birth rate of each pathogen present, since fewer propagules will succeed. This is an effect similar to some features of the physical environment; the birth rates of all pathogens on all species will be reduced greatly or slightly, and the overall severity of disease will be reduced (compare Figs 2.3 and 2.4). This would accord with the common perception that disease is commoner in agricultural than in natural systems. By this theory, one should find a correlation between host density and disease incidence. In our data from naturally regenerating vegetation at Silwood Park in Berkshire, southeast England (Fig. 2.5), we found a weak correlation between disease incidence and host density for necrotrophic pathogens, most of which probably have splash-dispersed spores. This fits with other work, for example the relationship between disease progress rate and host density found by Burdon and Chilvers (1976). However, the correlation scarcely seemed to hold for the rusts. Although the rust sample size in our data was small, it is worth considering why the correlation might be poor.

Consider a plant species whose individuals are not in a monospecific community, and have an average spacing, for example, 1 m^{-2} or 0.1 m^{-2}. Now suppose the density of spores deposited, perhaps by wind, falls off proportionally to:

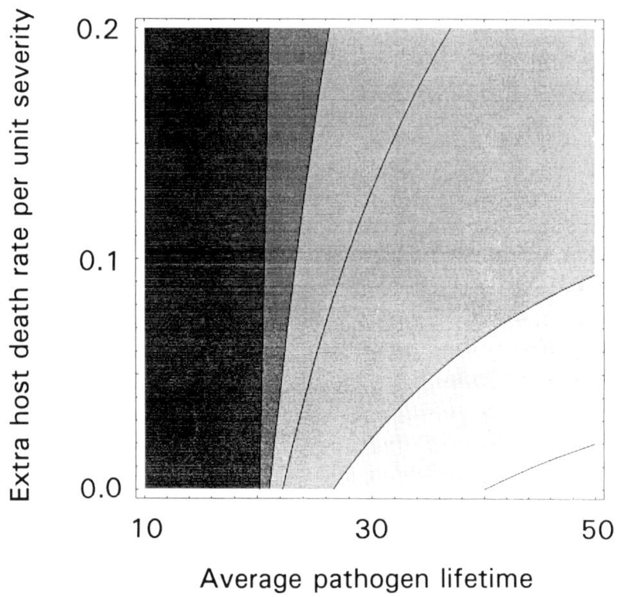

Fig. 2.4. As Fig. 2.3, but with the intrinsic growth rate of the disease decreased to 0.05 per unit time.

$$\frac{1}{(1 + ax)^b} \qquad (2.3)$$

where a and b are constants and x is the distance. Empirically, $1 < b < 3$ (Fitt *et al.*, 1987), although the average distance moved by a spore is infinite if $b > 2$, so that the interpretation of the equation needs care (Shlesinger *et al.*, 1993). Suppose the probability of a spore infecting is i. Then while the density of spores landing is greater than the leaf area index of a plant species, disease incidence will be independent of inoculum density, especially if propagule production very close to the first infection on a plant is abundant, so that all that matters is the arrival on a plant of enough spores to start a single infection. In a denser plant population more plants will be infected, but the ratio of infected to healthy plants may not differ. In the context of pollen-transmitted diseases Antonovics and Alexander (1992) have introduced the idea of frequency-dependent disease dynamics. Similarly, for a widely dispersed pathogen, plant density, over a wide range of densities, does not determine disease progress in terms of the ratio p/h, although it does in terms of p as an absolute density. These ideas are closely related to the ratio-dependent models which, it is argued, predict properties of real ecosystems

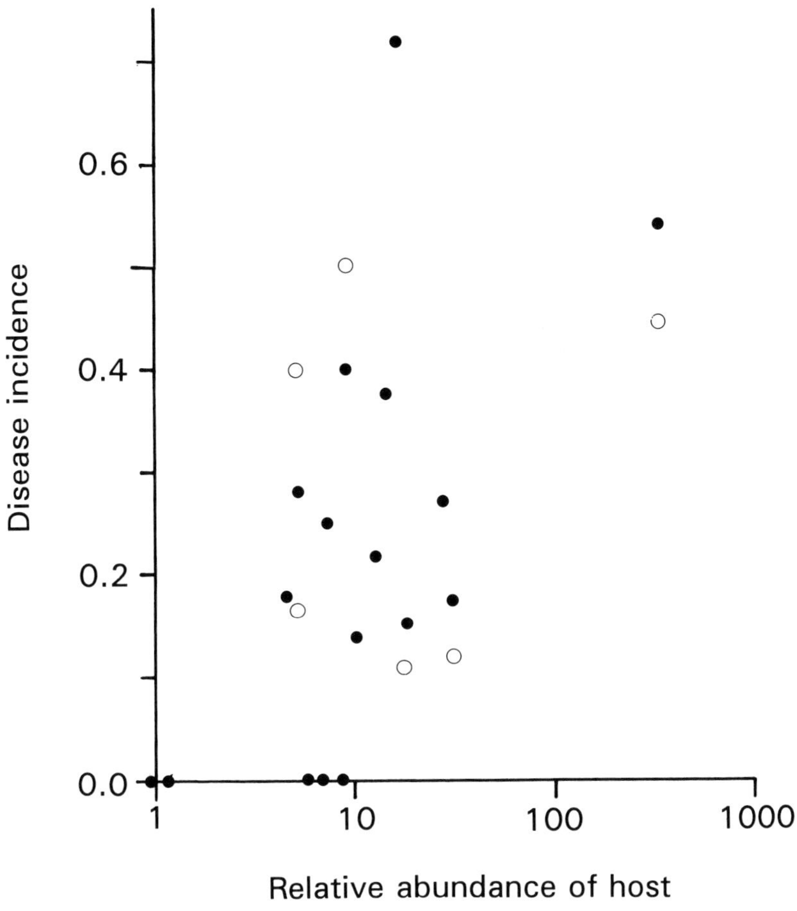

Fig. 2.5. The relationship between foliar pathogen incidence and host abundance in natural vegetation at Silwood Park, Berkshire, UK, for twenty host species. Species were ranked in order of abundance and the most common and least common sampled so as to achieve similar sampling patterns on both lists. Incidence of necrotrophic pathogens, probably mostly splashborne, ●; incidence of biotrophic pathogens (rusts and mildews), ○.

better than more traditional formulations (Arditi and Ginzburg, 1989; Ginzburg and Akcakaya, 1992).

The simple model outlined, of a set of two interacting populations of modules seems useful as an approximate mathematical description of plant pathogen communities. Pathogen incidence of about 10% emerges with weak dependence on the exact parameters chosen, and without the need for any further assumptions.

The stable dynamics described above are in contrast with some descriptions of host–pathogen dynamics. For example, May (1985) pointed out that an equation which apparently describes host pathogen dynamics is intrinsically chaotic, with no stable or periodic solutions. This conclusion depends on close coupling of pathogen load to death rate in an annual plant. The final number of hosts infected depends on how much host density exceeds the threshold density for disease to increase in the population. If the host density is high in one year, few individuals will set seed, and the pathogen may produce little mortality the next year. Because of the threshold, the effect of disease on seed-set depends delicately on the starting population of plants. The outcome is chaotic, unpredictable, variation in population size from year to year. Moderating the assumption that the disease is fatal – for example, if plants infected late in their lives produce some seed, rather than none – could dramatically alter the dynamics, possibly leading to stable equilibria.

Interactions on the Pathogen Trophic Level

Competition affects pathogens most obviously at the point of infection, where large populations of epiphytic microorganisms can substantially reduce the infection and establishment of pathogens (Fokkema and van der Meulen, 1976; Williamson and Fokkema, 1985; Gowdu and Balasubramanian, 1988). In some cases this would affect pathogen communities in the same way as a direct effect of the physical environment. For example, an environment in which epiphytes did not thrive would be more favourable to pathogens in just the same way as an environment in which temperatures were optimal for pathogen growth. One possible advantage of the pathogenic way of life is that it enables the pathogen to avoid competition in the phylloplane. Interactions could, however, be more complicated. For example, pathogens which damaged tissue and so increased the availability of tissue exudates might well improve conditions for saprophytes; these abundant populations would provide an abundance of inoculum to colonize newly-appearing healthy tissue, which would strongly antagonize pathogen propagules arriving on this tissue. The net effect would be a density-dependent regulation of the pathogen population, with competition becoming a more effective factor in denser pathogen populations.

There is evidence for direct interactions between pathogen species. We quote two examples from wheat in Europe. *Rhizoctonia cerealis* increases in incidence and severity when *Pseudocercosporella herpotrichoides* is controlled by triazole fungicides which are relatively inactive against *R. cerealis* (Daamen, 1990; Polley and Thomas, 1991). The increased incidence of *Septoria tritici* coincides with the decreased incidence of *Septoria nodorum*

(Royle *et al.*, 1986). In several instances, necrotrophic pathogens have been shown to reduce the growth rate or final population sizes of biotrophic pathogens; interactions between biotrophic pathogens have also been recorded (Brokenshire, 1974; Simkin and Wheeler, 1974; Adee *et al.*, 1990; Weber, 1992).

The strength of competition has not always been contrasted with the strength of intraspecific competition. In Adee *et al.*'s (1990) experiments with the wheat pathogens *Pyrenophora tritici-repentis* and *Septoria nodorum*, a replacement series (Harper, 1977) was used in which each pathogen was inoculated both alone and in mixtures. The mixtures were such that complementary proportions of each pathogen were used, i.e. if one component was at y% of the inoculum dose used for the pure inoculation, the other was at $(100 - y)$%. The area of leaf showing symptoms of each disease was noted. As an example of the results, the area diseased with *S. nodorum* when 50% of spores of *P. tritici-repentis* were replaced with *S. nodorum* was much less than half that when twice the concentration of *S. nodorum* spores were applied alone. But this experiment gives no information about the effect of the two fungi on each other, unless we know their response to the inoculum changes in the absence of the other. It is often implicitly assumed that responses to spore dose will be linear, provided much less than 100% of plant tissue is infected. This is not so; for example in the experiments of Jeger *et al.* (1985) changes in dose of five orders of magnitude only changed the area diseased by one order of magnitude. To know that two species compete is insufficient: do they compete with each other more or less strongly than with themselves?

Competition is not the only common interaction among pathogens. Biotrophic pathogens often facilitate the infection of hosts by necrotrophs, while necrotrophs antagonize the biotrophs. For example, the infection and multiplication of *Botrytis cinerea* on *Senecio vulgaris* was enormously increased in plants already infected with *Puccinia lagenophorae* (Hallet and

Table 2.2. Co-occurrence of the rust *Puccinia coronata* and a leaf spot complex of *Ascochyta leptospora* and *Colletotrichum holci* on *Holcus lanatus* leaves in natural mixed vegetation at Shinfield, Reading in England (J.C. Peters and M.W. Shaw, unpublished). Assessments were made in September 1992. Assessments made earlier in 1992 were consistent with these. Data are averaged over plots, but the result is statistically significant using a test based on combining measures of association within plots.

	Without rust		With rust	
	Actual	Expected	Actual	Expected
Without leaf spot	**213**	181	**159**	190
With leaf spot	**23**	55	**90**	58

Ayres, 1992; Hallet *et al.*, 1990). In a similar interaction in our own studies on natural vegetation, infection of *Holcus lanatus* by a leaf spot complex was greater on plants also infected with *Puccinia coronata* (Table 2.2). Other cases include a greater susceptibility to *Septoria nodorum* in plants infected with *Erysiphe graminis* (Weber, 1992) or *Pseudocercosporella herpotrichoides* (Gareth Jones and Jenkins, 1978). In none of these examples have the longer-term population dynamics been studied.

Thus, although there is evidence that interactions exist within plants, as well as immediately around them in the phyllosphere and rhizosphere, there is little evidence as to the consequences of these interactions for community structure or the population dynamics of species. Models may make predictions which will suggest fruitful experiments and observations.

One way to construct a model would be to develop the model in equation (2.2) by adding more pathogen species. This would include the consequences of any increased deleterious effect on the host; it would however introduce many parameters. It seems better to use, at least initially, the information derived from standard competitive systems (May, 1974), but recognize that this work does not explicitly include the impact of the competing species on their host and its growth rate.

The work of which May (1974) represents an example is almost all based on the Lotka-Volterra equations or variations of them. These equations have the form

$$\frac{dp_1}{dt} = r_1 p_1 \left(1 - \frac{(p_1 + \alpha_{12} p_2)}{K_1}\right)$$

$$\frac{dp_2}{dt} = r_2 p_2 \left(1 - \frac{(p_2 + \alpha_{21} p_1)}{K_2}\right) \qquad (2.4)$$

Here p_1 and p_2 are the two pathogens. If each is alone each will increase until it reaches K_1 or K_2, limited by competition within the species. There is less detail than in equation (2.1) as to how this competition operates; this compromise is necessary to reduce the number of equations and parameters. When both pathogens grow together, each inhibits the other, with a strength relative to an individual unit of the other of α_{12} (the second species effect on the first) or α_{21} (the first species effect on the second). If α_{ij} is negative, this represents an interaction like that described above: the growth rate of species i is increased by the presence of species j. This old model is simplistic but well understood, and can serve to focus as a basis for further work. There are three possible outcomes: (i) extinction of whichever species is initially rarest; (ii) extinction always of one of the species; and (iii) stable coexistence (Fig. 2.5).

Both species can coexist if $\alpha_{12}K_2 < K_1$ and $\alpha_{21}K_1 < K_2$: that is if the strength of interspecific competition at its most extreme is less than that of

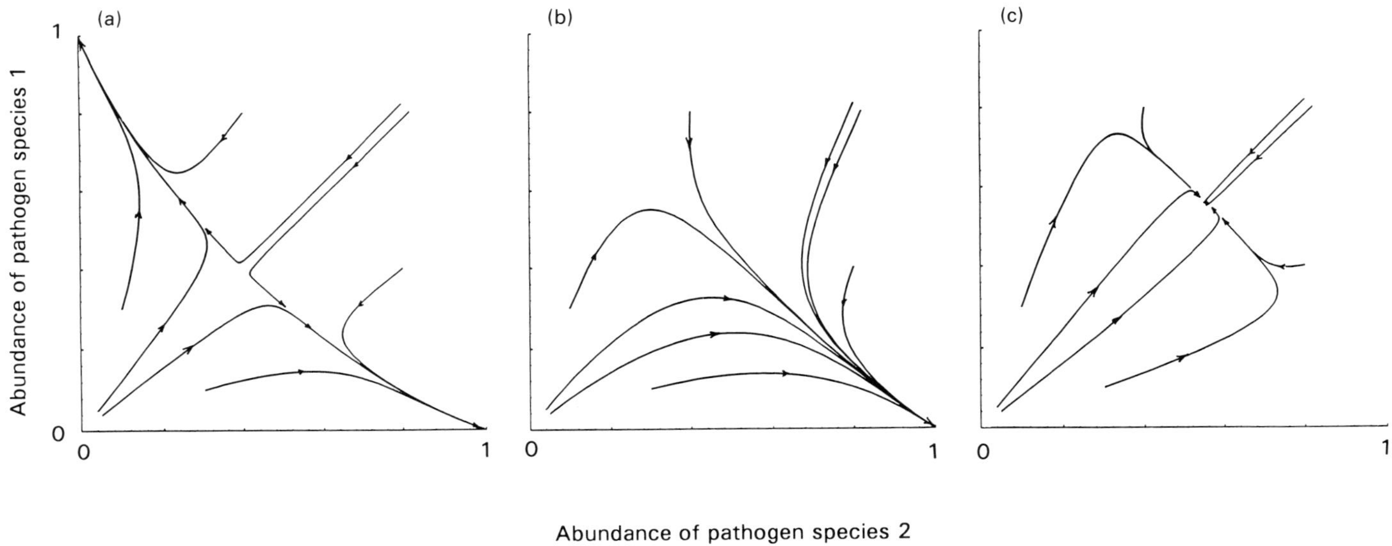

Fig. 2.6. The possible outcomes of competition among two pathogens. (a) Competitive exclusion of either pathogen by the other. (b) Competitive exclusion of the stronger pathogen by the weaker. (c) Coexistence. In each case the successive populations of each pathogen are joined by a line, with time increasing in the direction of the arrow; the trajectory followed depends on the proportion of each species at the start.

intraspecific competition, for both species. The equilibrium levels are:

$$\begin{aligned} p_1 &= (K_1 - \alpha_{12}K_2)/(1 - \alpha_{12}\alpha_{21}) \\ p_2 &= (K_2 - \alpha_{21}K_1)/(1 - \alpha_{12}\alpha_{21}) \end{aligned} \tag{2.5}$$

If the species did not compete the total pathogen load at equilibrium would be $K_1 + K_2$; if they competed equivalently, presumably with $K_1 = K_2 = K$, the total population would be K. So the effect of competition is to reduce the load of each pathogen, and the overall load, since the sum of equations (2.5) must be less than $K_1 + K_2$.

We have ignored the plant hosts in this model, except as mediators of competition. This role of mediator is important for it is entirely possible, even likely, that competition coefficients α might be greater than one. That is, competition is occurring not by simple competition for resources, but on an active substrate, which may be made dramatically more or less favourable for another species by the presence of a first. For example, if a pathogen triggers any kind of systemic response in a plant, then there will be very strong competition between it and another pathogen. Almost regardless of what the other pathogen does, this will lead to competitive exclusion: that is, one or other pathogen will, eventually, die out in any given patch.

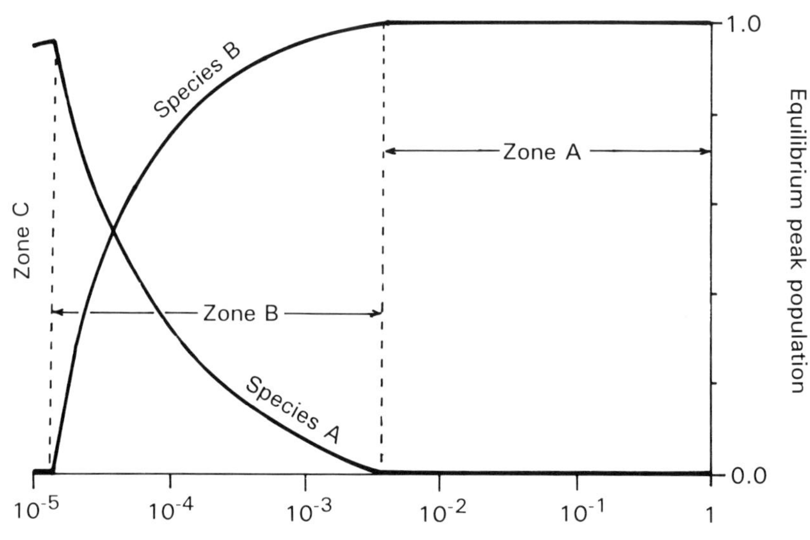

Fig. 2.7. The effect of seasonality in a competition model. The variation in the outcome of competition with varying proportions of the populations overseasoning is shown. Zone A: species B excludes species A; Zone B: both species coexist; Zone C: species A excludes species B. The parameters were: $r_1 = 0.04\ \text{day}^{-1}$; $r_2 = 0.035\ \text{day}^{-1}$; $\alpha_{12} = 1.1$; $\alpha_{21} = 0.5$; $K_1 = K_2 = 1$.

The analysis above ignores factors such as annual cycles in the environment, patchiness, and the effect of the pathogen on the availability of host tissue. Here we will explore only cyclic variations in the environment. This factor has the interesting effect of altering and, generally, broadening the conditions under which coexistence between pathogens can occur. The easiest way to model this is by assuming that each autumn, a large proportion of both host and pathogen disappear either by being harvested or through adverse weather. A consequence of this is that pathogens repeatedly compete at all densities. A fast growing pathogen may out-compete a slower but more competitive form, simply because there is more of it around earlier in the season, so it makes up in weight of numbers what it lacks in aggressiveness. This may lead to a reversal of competitive ranking, or to coexistence where none was possible before. Figure 2.7 shows an example.

What are the patterns of coexistence seen in nature? Most plants are host to many pathogens on each organ, and pathogens often occur together. In a crop, there are usually only a few major pathogens, and fewer still that are serious in any given portion of a particular field. This generalization is interesting if put in an evolutionary context: a coevolutionary race between a host and one pathogen may have coincidentally much greater effects on other pathogens and lead to reduction in the total number of pathogens coexisting on a single host. It is impossible to extrapolate further without models which incorporate the effects of the pathogens on the host.

Interactions with Higher Trophic Levels

The trophic layer above pathogens includes fungal hyperparasites, animal predators, and bacterial or viral diseases of pathogens. In insect ecology, this trophic level is commonly credited with regulating herbivore numbers and preventing unsustainable consumption of plants (Hairston *et al.*, 1960). Although this conclusion is disputed by others (Wratten *et al.*, 1988), the phenomenon of secondary resurgence of insect pests following pesticide application shows that predators and hyperparasites undoubtedly regulate some insect herbivore populations (Debach, 1974). The failure of most attempts to harness hyperparasites as biological control agents suggests that the role of parasitism in controlling plant disease may be small. However, diseases like oak powdery mildew (*Microsphaera alphitoides*) or graminaceous rusts represent a substantial standing biomass and it would be surprising if no organism specialized in eating them. Many organisms are known to parasitize plant pathogenic fungi (Hijwegen and Buchenaker, 1984; Puzanova, 1984; Sachan and Sharma, 1988; Turman, 1990). Fleming (1980) argued that there may be substantial but unrecognized regulation of rusts by generalist predators and hyperparasites. He argued that populations of typical predators and hyperparasites of rusts grew much more slowly than the rust populations; epidemics occurred when the rust population became so large that the depredations of the slowly responding predatory populations were negligible. But there are many examples of organisms that live on pathogens which have generation times comparable with the pathogens they destroy (Sztejnberg *et al.*, 1989), so it is of interest to consider what can happen if the hyperparasite population changes at rates comparable to the pathogen population.

Once again, it is helpful to make a simplified but precise argument with a model. Suppose we have a pathogen that increases according to a logistic model but is also subject to parasitism by a hyperparasite, which increases in frequency at a rate which increases the more abundant the pathogen becomes. Both populations are dramatically reduced by the onset of autumn, but the hyperparasite has a small alternative host which allows it to persist when the pathogen is rare. The hyperparasite decreases the growth rate of the pathogen according to the number of individuals infected. The equations for pathogen and hyperparasite increase within a year are:

$$\begin{aligned} \frac{dp}{dt} &= r_p p\,(1-p) - b_1 p v \\ \frac{dv}{dt} &= r_v v\,(c+p) - b_2 v \end{aligned} \tag{2.6}$$

in which p is the pathogen population density, scaled from 0 to a maximum of 1, v is the hyperparasite population density, c is a constant which specifies how much hyperparasite the alternative hosts will allow to exist in the absence of any host, r_p and r_v determine the birth rates of pathogen and hyperparasite respectively, b_1 determines the impact of the hyperparasite on the pathogen growth rate, and b_2 is the death rate of the hyperparasite. Once each year, both populations are reduced by a factor a, representing the loss in population caused by seasonal fluctuations in host availability.

What are the patterns of behaviour possible in this model? Without the alternative host, the hyperparasite cannot persist if pathogen density is not high enough to permit the hyperparasite population to increase for long enough to recover from the devastating effects of autumn. The hyperparasite can only persist in a relatively restricted set of circumstances in which it does not damage the host too much, yet increases rapidly. With the alternative host, there are various possible outcomes: extinction of one or other component; a stable equilibrium or repeating cycle of abundant and scarce disease years; or a chaotic, apparently random, fluctuation in numbers from year to year (Fig. 2.8). The stable equilibrium does not necessarily imply that

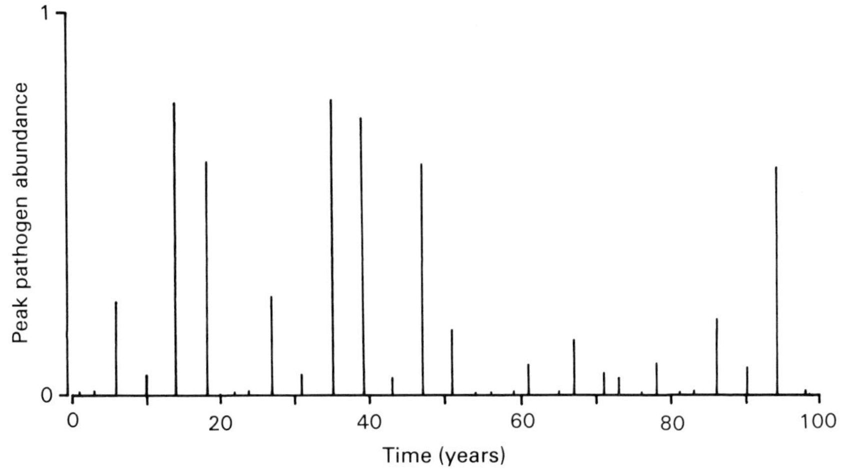

Fig. 2.8. A time series showing the peak pathogen population each year for 100 years, using the model in equations (2.6). All parameters are constant throughout the interval shown; the population fluctuations are caused solely by the interaction of the pathogen and its hyperparasite. The parameters are: r_p, 0.04 day^{-1}; r_v, 0.6 day^{-1}; b_1, 1.0; b_2, 0.04; c, 0.01; a, 0.01.

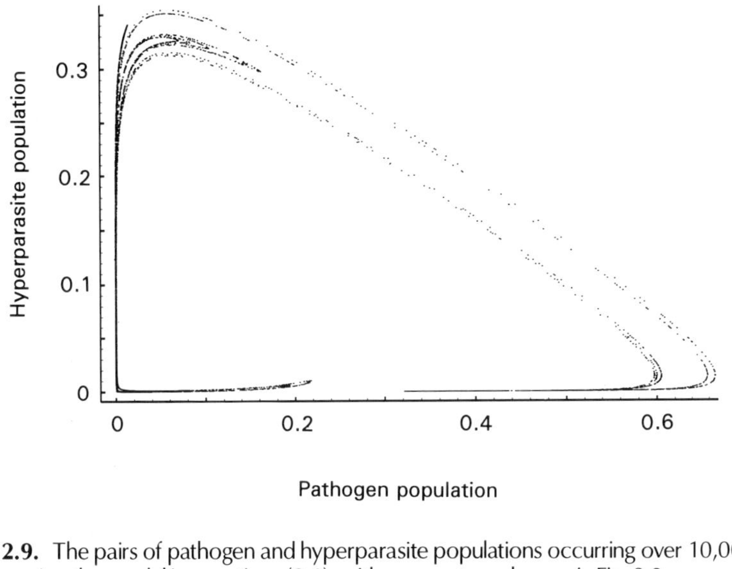

Fig. 2.9. The pairs of pathogen and hyperparasite populations occurring over 10,000 years, using the model in equations (2.6), with parameter values as in Fig. 2.8.

the hyperparasite is controlling the level of disease to 'acceptable' levels, nor that the level of disease is constant within a year. When numbers fluctuate apparently randomly, some order is nonetheless maintained in this apparent random fluctuation (Fig. 2.9).

Such fluctuations take place in this model in an absolutely constant environment. Faced with a time series like that shown in Fig. 2.8 it would be tempting to look for an environmental factor correlated with the outbreak years, and with a short series of data we could probably find one. The result would however have no long-term predictive value.

Conclusions

The biological environment of a pathogen population could in principle induce a wide variety of behaviour in the population. The host trophic level may exert, through the effect of pathogens on the individual host modules, a determining influence on average size of the pathogen population. This is not the same as the effect of host resistance, which enters equation (2.3) directly as the birthrate of the pathogen; obviously such resistance is critical, but our aim here is to focus attention on the dynamic aspects of regulation of pathogen populations.

A variety of interactions on the pathogen trophic level could occur, with important consequences for both pathogen and host populations. In the main, these interactions seem to lead to relatively stable long-term patterns of population abundance before environmental perturbations are taken into account. Competition can reduce the overall abundance of pathogens, but not necessarily to any great extent. The argument also suggests that synergistic interactions may be common and important in pathogen communities.

The trophic level above pathogens seems most likely to give rise to extreme or unstable population dynamics. Hyperparasites could very considerably reduce peak pathogen populations; they might also cause apparently patternless fluctuations in pathogen abundance from year to year, even when no environmental feature changes. It is not obvious how environmental effects could be distinguished from intrinsic population instability in real data, but it is important to be aware of the possibility.

Acknowledgements

Parts of this work were supported by the University of Reading Research Endowment Fund and the Natural Resources Institute (Contract X0128).

We are extremely grateful to Dr V.K. Brown for encouraging us to work in the successional sites she has developed at Silwood Park. We are grateful to Colin Birch, Carl Pijls and David Royle for commenting on the manuscript.

References

Adee, S.R., Pfender, W.F. and Hartnett, D.C. (1990) Competition between *Pyrenophora tritici-repentis* and *Septoria nodorum* in the wheat leaf as measured with de Wit replacement series. *Phytopathology* 80, 1177–1181.

Antonovics, J. and Alexander, H.M. (1992) Epidemiology of anther-smut infection of *Silene alba* (= *S. latifolia*) caused by *Ustilago violacea*: patterns of spore deposition in experimental poulations. *Proceedings of the Royal Society of London* B 250, 157–163.

Arditi, R. and Ginzburg, L.R. (1989) Coupling in predator–prey dynamics: ratio-dependence. *Journal of Theoretical Biology* 139, 311–326.

Brokenshire, T. (1974) Predisposition of wheat to *Septoria* infection following attack by *Erysiphe*. *Transactions of the British Mycological Society* 63, 393–397.

Burdon, J.J. and Chilvers, G.A. (1976) Epidemiology of *Pythium* induced damping-off in mixed species seedling stands. *Annals of Applied Biology* 82, 233–240.

Daamen, R.A. (1990) Surveys of cereal disease and pests in the Netherlands. 2. Stem-base diseases of winter wheat. *Netherlands Journal of Plant Pathology* 96, 251–260.

Debach, P. (1974) *Biological Control by Natural Enemies.* Cambridge University Press, Cambridge.

Fitt, B.D.L., Gregory, P.H., Todd, A.D., McCartney, H.A. and MacDonald, O.C. (1987) Spore dispersal and plant disease gradients: a comparison between two empirical models. *Journal of Phytopathology* 118, 227–242.

Fleming, R.A. (1980) The potential for control of cereal rust by natural enemies. *Theoretical Population Biology* 18, 374–395.

Fokkema, N.J. and van der Meulen, F. (1976) Antagonism of yeastlike phyllosphere fungi against *Septoria nodorum* on wheat leaves. *Netherlands Journal of Plant Pathology* 82, 13–16.

Gareth Jones, D. and Jenkins, P.D. (1978) Predisposing effects of eyespot (*Pseudocercosporella herpotrichoides*) on *Septoria nodorum* infection of winter wheat. *Annals of Applied Biology* 90, 45–49.

Ginzburg, L.R. and Akcakaya (1992) Consequences of ratio-dependent predation for steady-state properties of ecosystems. *Ecology* 73, 1536–1543.

Gowdu, B.J. and Balasubramanian, R. (1988) Role of phylloplane micro-organisms in the biological control of foliar plant diseases. *Zeitschrift für Pflanzenkrankheiten und Pflanzenschutz* 95, 310–331.

Hairston, N.G., Smith, F.E. and Slobodkin, L.B. (1960) Community structure, population control, and competition. *American Naturalist* 94, 421–425.

Hallet, S.G. and Ayres, P.G. (1992) Invasion of rust (*Puccinia lagenophorae*) aecia on groundsel by secondary pathogens – death of the host. *Mycological Research* 96, 142–145.

Hallet, S.G., Paul, N.D. and Ayres, P.G. (1990) *Botrytis cinerea* kills groundsel (*Senecio vulgaris*) infected by rust (*Puccinia lagenophorae*). *New Phytologist* 114, 105–109.

Harper, J.L. (1977) *Population Biology of Plants.* Academic Press, London.

Hijwegen, T. and Buchenaker, H. (1984) Isolation and identification of hyperparasitic fungi associated with Erysiphaceae. *Netherlands Journal of Plant Pathology* 90, 79–87.

Jeger, M.J. (1987) The potential of analytic compared with simulation approaches to modeling in plant disease epidemiology. In: Leonard, K.J. and Fry, W.E. (eds), *Plant Disease Epidemiology Volume 1: Population Dynamics and Management.* Macmillan, New York, pp. 255–281.

Jeger, M.J., Griffiths, E. and Jones, D.G. (1985) The effects of post-inoculation wet and dry periods, and inoculum concentration, on lesion numbers of *Septoria nodorum* in spring wheat seedlings. *Annals of Applied Biology* 106, 55–63.

King, J.E. (1977a) Surveys of diseases of winter wheat in England and Wales 1970–75. *Plant Pathology* 26, 8–20.

King, J.E. (1977b) Surveys of foliar diseases of spring barley in England and Wales, 1972–75. *Plant Pathology* 26, 21–29.

Kranz, J. (1990) Fungal diseases in multispecies plant communities. *New Phytologist* 116, 383–405.

May, R.M. (1974) *Stability and Complexity in Model Ecosystems*, 2nd edn Princeton University Press, Princeton, New Jersey.

May, R.M. (1985) Regulation of populations with non-overlapping generations by microparasites: a purely chaotic system. *American Naturalist* 125, 573–584.

Polley, R.W. and Thomas, M.R. (1991) Surveys of diseases of winter wheat in England and Wales, 1976–1988. *Annals of Applied Biology* 119, 1–20.

Puzanova, L.A. (1984) Hyperparasites of *Ampelomyces* Ces ex Schlecht and their possible application to biological control of powdery mildew. *Mikologya i Fitopatologiya* 18, 333–339.

Royle, D.J., Shaw, M.W. and Cook, R.J. (1986) The natural development of *Septoria nodorum* and *S. tritici* in some winter wheat crops in Western Europe, 1981–83. *Plant Pathology* 35, 466–476.

Sachan, S.A. and Sharma, M.R. (1988) *Cicinnobolus cesatii* de Bary as a hyperparasite on some powdery mildew fungi from Doon Valley. *Indian Journal of Forestry* 11, 74–76.

Shlesinger, M.F., Zaslavsky, G.M. and Klafter, J. (1993) Strange kinetics. *Nature* 363, 31–37.

Simkin, M.B. and Wheeler, B.E.J. (1974) Effects of dual infections of *Puccinia hordei* and *Erysiphe graminis* on barley cv. Zephyr. *Annals of Applied Biology* 78, 237–250.

Sztejnberg, A., Galper, S., Mazor, S. and Lisker, N. (1989) *Ampelomyces quisqualis* for biological and integrated control of powdery mildew in Israel. *Journal of Phytopathology* 124, 285–295.

Turman, G. (1990) Further hyperparasites of *Rhizoctonia solani* Kuhn as promising candidates for biological control. *Zeitschrift für Pflanzenkrankheiten und Pflanzenschutz* 97, 208–216.

Weber, G.E. (1992) Interaktionen zwischen *Erysiphe graminis* f.sp. *tritici*, *Septoria nodorum* sowie *Pseudocercosporella herpotrichoides* im Agroökosystem Winterweizen und deren Modellierung. Unpublished Ph.D. thesis, University of Giessen, Germany.

Williamson, M.A. and Fokkema, N.J. (1985) Phyllosphere yeasts antagonise penetration from appressoria and subsequent infection of maize leaves by *Colletotrichum graminicola*. *Netherlands Journal of Plant Pathology* 91, 265–276.

Wratten, S.D., Edwards, P.J. and Winder, L. (1988) Insect herbivory in relation to dynamic changes in host plant quality. *Biological Journal of the Linnean Society* 35, 339–350.

3 The Microchemistry of Plant/Microorganism Interactions

T. Shepherd

Scottish Crop Research Institute, Invergowrie, Dundee DD2 5DA, UK.

Introduction

Living plants release organic chemicals from their root systems into the rhizosphere, via active excretion, leakage, abrasion and wounding, and by cell death (Rovira, 1969; Hale and Moore, 1979). These mechanisms will hereafter be referred to collectively as root exudation. The chemicals released range from simple low molecular weight amino acids, organic acids, carbohydrates, phenolic acids and various secondary metabolites, to more complex polypeptides, polysaccharides, enzymes and other proteins (Rovira, 1969; Hale and Moore, 1979; Whipps, 1990).

A relatively high proportion of net fixed carbon (> 20% in some species) is released into the root-zone as organic carbon compounds via root exudation (Helal and Sauerbeck, 1989; Whipps, 1990; Shepherd and Davies, 1993). Such losses represent a major drain of photoassimilate theoretically available for plant growth and development. The question therefore arises as to what effect such inputs of carbon have on the rooting environment and its complex ecology and, as a consequence, on the growth of the source plant itself.

The most profound effects of root exudates on growth and development probably arise from interaction of the chemicals with root-zone microflora. It is becoming evident that root-derived chemicals have a significant, and possibly a controlling effect, on the ordering and maintenance of the rhizosphere microfloral populations. Root exudates interact with micro-organisms principally by induction of chemotaxis (Bowen and Rovira, 1976),

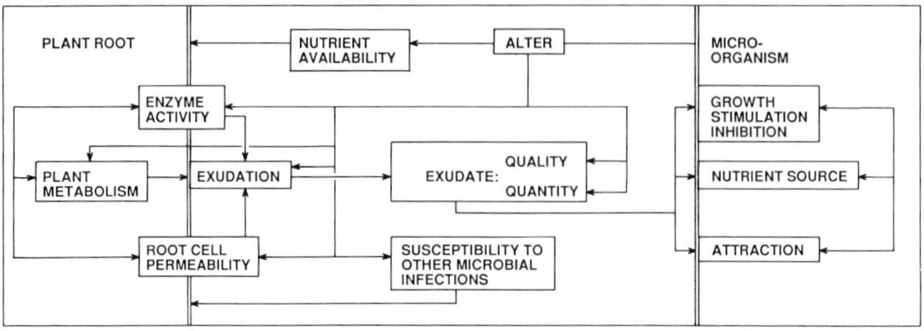

Fig. 3.1. Schematic representation of interactions between plant roots, root exudate and rhizosphere microorganisms.

stimulation or inhibition of germination of propagules, and microbial growth (Rovira, 1969; Hale and Moore 1979; Nelson, 1990; Sundin *et al.*, 1990). In turn, microorganisms can interact with plant roots by altering cell permeability, enzyme activity and other elements of plant metabolism (Hale and Moore, 1979). These processes, together with differential microbial assimilation of specific exudate components, can result in changes in the quantity and quality of chemicals released from roots. Microorganisms can also affect the susceptibility of plants to infection by other root-zone microorganisms, particularly pathogens (Hale and Moore, 1979), and the availability of inorganic nutrients to the source plant (Hale and Moore, 1979; Lynch, 1990). A special case of the involvement of both plant- and microbially-produced metabolites is the initiation of symbiotic or parasitic interrelationships between plants and microorganisms (Keen and Staskawicz, 1988; Dénarié *et al.*, 1992). These various interactions and their interrelationships are shown in Fig. 3.1.

Chemicals released from roots must be present in a suitable physical state for interaction with microorganisms. The diffusion of dissolved non-volatile materials away from roots is affected by soil porosity and moisture content, adsorption of the chemicals on to the soil matrix, and their assimilation and metabolism by microorganisms (Bowen and Rovira, 1976). Non-volatile soluble root-derived chemicals can be detected more than 3 cm from roots (Helal and Sauerbeck, 1984). Volatile chemicals may penetrate the soil to greater distances, but they must have a suitable partition coefficient in the aqueous phase for interaction with microorganisms.

Chemotaxis Induction by Exuded Compounds

Root-zone bacteria and fungal zoospores require a continuous water film for movement through soil pores towards roots in response to chemoattractants (Bowen and Rovira, 1976). Many instances of root exudate-induced chemo-

Table 3.1. Induction of chemotaxis by root exudates and root exudate components.

Plant	Microorganisms	Chemotaxis inducing chemicals (Ref.)
A	Fungal zoospores	
Pea	*Pythium aphanidermatum*	Amino acids (ammonium glutamate, Gln, Ser); sugars (Glc, Suc); (Chang-Ho and Hickman, 1970)
Beet	*Aphanomycetes cochlioides*	Organic acids (including gluconic acids); sugars; (Chang-Ho and Hickman, 1970)
Pea	*A. cochlioides*	Gluconic acids; (Chang-Ho and Hickman, 1970)
–	*Phytophthora cinnamomi*	Amino acids (4-NH_2-butyric acid, Asp, Glu); (Chang-Ho and Hickman, 1970; Zentmyer, 1970)
–	*P. cactorum*	Amino acids (4-NH_2-butyric acid, Asp, Asn, Glu, Gln); sugars (Glc, Suc); organic acids (e.g. citric acid); (Zentmyer, 1970)
–	*P. capsici*	
–	*P. citrophthora*	
–	*P. palmivora*	Amino acids (4-NH_2-butyric acid, Asp, Glu); (Zentmyer, 1970)
B	Rhizobia	
Soyabean	*Bradyrhizobium japonicum*	Amino acids (Asp, Glu); TCA-cycle acids; cinnamic acids (caffeic, cinnamic, *p*-coumaric, ferulic, sinapic); (Barbour *et al.*, 1991; Kape *et al.*, 1991, 1992)
Alfalfa	*Rhizobium meliloti*	Flavone (luteolin); (Caetano-Anolles *et al.*, 1988)
–	*R. leguminosarum*	Sugars (Raf, Suc, Xyl); monocyclic phenols (vanillyl alcohol, 4-OH-benzoic acid, 3, 4-$(OH)_2$-benzoic acid; flavonoids (apigenin, luteolin); coumarin (umbelliferone); acetophenone (acetosyringone); (Aguilar *et al.*, 1988)

taxis have been described, and in general, the microorganism is attracted to the elongation zone between the root tip and emerging root hairs. This region has been identified as one of the principal sites of exudation from roots, along with sites of lateral root emergence (Hale and Moore, 1979).

Chemotaxis to roots of various plant species may be non-specific, as is the case with the bacterium *Pseudomonas lachrymans*, and zoospores of many phycomycetes (Misaghi, 1982). However, there are instances of specificity to susceptible varieties of a species, as illustrated by *Phytophthora megasperma*, which is attracted to susceptible alfalfa seedlings, but not to non-susceptible ones (Misaghi, 1982). Amino acids, carbohydrates and organic acids, the ubiquitous constituents of root exudates, are the most common inducers of chemotaxis, both *in vitro* and in soil. However, there are instances of attraction to specific chemicals; *Erwinia amylovora* is attracted to aspartate and to various organic acids, but not to simple sugars (Misaghi, 1982). Attraction is often stronger if the attracting chemical is also a suitable nutrient for the organism. Such chemicals are unlikely to persist within the rhizosphere, and other types of chemical may be involved in chemotactic induction. Examples of this are the attraction of the rhizobia *Bradyrhizobium japonicum*, *Rhizobium meliloti* and *R. leguminosarum* to flavonoids and other phenolic compounds present in root exudates of their legume hosts, in addition to amino acids, sugars and organic acids (Aguilar *et al.*, 1988; Caetano-Anolles *et al.*, 1988; Barbour *et al.*, 1991; Kape *et al.*, 1991). These phenolic compounds are involved in specific initiation or inhibition of the symbiotic relationship between rhizobia and their host plants by triggering transcription of various bacterial nodulation (*nod*) genes (Dénarié *et al.*, 1992). The same *nod* genes are required for chemotaxis induction towards the specific *nod*-inducing phenolics (Caetano-Anolles *et al.*, 1988; Dénarié *et al.*, 1992), suggesting that, in some instances, chemotaxis may be specifically induced by chemicals involved in some form of signal exchange process between a particular microorganism and its prospective host(s). A list of chemotaxis inducing compounds is given in Table 3.1.

Germination of Propagules

Germination of propagules of soil-borne pathogenic fungi requires a source of carbon and nitrogen, primarily amino acids and sugars.[1] In general,

---

[1]The following abbreviations apply to amino acids and sugars. Amino acids: Alanine (Ala); Arginine (Arg); Asparagine (Asn); Aspartic acid (Asp); γ-Aminobutyric acid (Gaba); Glutamine (Gln); Glutamic acid (Glu); Glycine (Gly); Histidine (His); Isoleucine (Ile); Leucine (Leu); Lysine (Lys); Methionine (Met); Phenylalanine (Phe); Proline (Pro); Serine (Ser); Threonine (Thr); Tryptophan (Trp); Tyrosine (Tyr); Valine (Val). Sugars: Fructose (Fru); Galactose (Gal); Glucose (Glc); Maltose (Mal); Raffinose (Raf); Rhamnose (Rha); Ribose (Rib); Sucrose (Suc); Xylose (Xyl).

propagules of *Macrophominia phaseolina*, *Phytophthora* spp. and *Verticillium dahliae* respond to amino acids, but not to sugars, while propagules of *Fusarium* spp. and *Pythium* spp. respond to amino acids and sugars (Nelson, 1990). Chlamydospores of *Phytophthora cinnamomi* and *P. parasitica* are stimulated to germinate *in vitro* equally by the amino acids Ala, Asn, Asp, Glu, Gly, Leu, Lys, Met, Phe, Ser and Thr, and also by casein hydrolysates. Citric acid is also an active inducer, whereas the sugars Glc, Suc and Fru, and inorganic nitrogen sources do not induce germination (Mircetich and Zentmyer, 1970). Various combinations of glucose, inorganic nitrogen and mineral compounds, with or without thiamine, fail to induce more germination of chlamydospores of *P. cinnamomi* than asparagine and citrate alone, indicating the specificity of the nutrient requirements for induction of germination. Strong vegetative growth of fungal hyphae following germination is supported by a mixture of glucose, fructose and ammonium buffered with fumaric acid and nitrate, showing that there can be different nutritional requirements for propagule germination and fungal growth (Mircetich and Zentmyer, 1970).

Germination of chlamydospores of *Fusarium solani* f.sp. *phaseoli* and *F. solani* f.sp. *pisi* can be stimulated by the amino acids Asn, Asp, Gln, Glu, Gly and Phe, and by the sugars Glc, Suc and Fru, both in soil and *in vitro* (Toussoun, 1970; Nelson, 1990). Germination of chlamydospores of *F. solani* f.sp. *phaseoli* in soil is stimulated more effectively by amino acids and ammonium glutamate than by glucose, and not by inorganic nitrogen. Following germination, fungal hyphae are subject to lysis at higher nutrient levels than required for germination, asparagine having the most pronounced effect. For combined chlamydospore germination and hyphal growth, glucose together with natural soil nitrogen were the most effective stimulators (Toussoun, 1970). Mixtures of amino acids and sugars stimulate the germination of sporangia of *Pythium ultimum* and *P. irregulare*, while mixtures of organic acids and glucose stimulate germination of *P. aphanidermatum* (Nelson, 1990).

Germination of spores can be stimulated by other classes of compound. Endoconidia and chlamydospores of *Thielaviopsis basicola* are germinated by C_{16}–C_{18} fatty acids (palmitic, stearic, oleic, linoleic and palmitoleic) released by several plant species grown in sand culture, and not by amino acids or sugars (Nelson, 1990). Sclerotia of *Sclerotium cepivorum* (which has a narrow host range, only infecting members of the genus *Allium*) are induced to germinate *in vitro* by volatile alkyl sulphides, and not by amino acids and sugars. These compounds are formed in the soil by microbial degradation of alkyl cysteins and cysteine sulphoxides, released from roots of *Allium* (Coley-Smith and King, 1970; Nelson, 1990). There have been various other reports of propagule germination by root or seed-derived volatile chemicals. Germination of sporangia of *Pythium ultimum*, *P. irregulare* and *P. sylvaticum* is stimulated by volatiles released by germinating seeds of cotton and other plants (Nelson,

Table 3.2. Activation of propagules of plant pathogenic fungi by root and seed exudates. (Adapted from Nelson, 1990.)

Organism	Propagule	Exudate source root (R), seed (S)	Stimulant molecules (*in-vitro* (*i-v*) or soil (*s*) bioassay)
Aspergillus flavus	conidium	*Arachis hypogaea* (R)	Glucose + unknown N source (*s*)
Fusarium oxysporum	chlamydospore, macroconidium	*Arachis hypogaea* (R) *Lens esculenta* (S)	Glucose + unknown N source (*s*) ethanol (*i-v*)
F. solani f.sp. *phaseoli*	chlamydospore, microconidium	various (R,S)	Amino acids, sugars, lipids (*i-v, s*)
F. solani f.sp. *pisi*	chlamydospore; microconidium	*Pisum sativum* (R,S)	Sugars + unknown volatiles (*s*)
Macrophominia phaseolina	sclerotium	*Pinus lambertiana* (R)	Amino acids (*s*)
Phytophthora cinnamomi	chlamydospore	*Persea americana* (R)	Amino acids, organic acids (*i-v*)
P. megasperma f.sp. *medicaginis*	oospore	various (R)	Amino acids
Puccinia punctiformis	teliospore	*Cirsium arvense* (R)	Aplotaxene (*i-v*)
Pythium aphanidermatum	zoospore cyst	*Pisum sativum* (R) *Phaseolus aureus* (R)	Amino acids, sugars (*i-v*)
P. irregulare	sporangium	*Pinus resinosa* (S)	Amino acids, sugars (*s*)
P. ultimum	sporangium	various (S)	Ethanol, glucose (*s*)
Sclerotium cepivorum	sclerotium	*Allium* spp. (R)	Alkyl sulphides (*s*)
Thielaviopsis basicola	endoconidium, chlamydospore	various (R)	Fatty acids (*s*)
Ustilago maydis	chlamydospore	*Zea mays* (S)	Ethanol (*i-v*)
Verticillium albo-atrum, *V. dahliae*	microsclerotium	various (R)	Amino acids (*i-v, s*)

1990). These volatiles contain ethanol and acetaldehyde which are active at nanomolar and sub-nanomolar concentrations, and are inhibitory to germination of sporangia at higher concentrations. Ethanol and acetaldehyde can increase microbial respiration in the soil without changing the size of soil populations, and they can also be metabolized by quiescent macroconidia of *F. solani* without inducing germination (Nelson, 1990).

Other components of the rhizosphere microflora can also influence propagule germination and growth of pathogenic fungi. The germination of *P. cinnamomi* chlamydospores in natural soils and soils amended with asparagine is greater in the presence of certain antibiotics (vancomycin and nystatin) (Mircetich and Zentmyer, 1970). This suggests that competition between microflora for available nutrients can have a significant effect on propagule germination. Mycorrhiza or mycorrhizal roots can release antibiotic substances which inhibit root pathogens. Volatile mono- and sesquiterpenes released from the roots of *Pinus sylvestris* infected with the mycorrhizal fungus *Boletus variegatus* inhibit vegetative growth of *P. cinnamomi* and *Heterobasidion annosum*, both in soil and *in vitro*. Other volatile chemicals produced by *B. variegatus* in culture such as isobutanol, isoamyl alcohol, acetoin, and isobutyric acid also inhibit *P. cinnamomi* and *H. annosum* (Krupa and Fries, 1971; Krupa and Nylund, 1972). A list of chemicals in root and seed exudates which promote germination of fungal propagules is given in Table 3.2.

Microbial Growth

The results of several studies of interactions between plants and microorganisms suggest the differential utilization of organic substrates in root exudates by microorganisms. Subsequent to inoculation of wheat roots with *Helminthosporium sativum*, there was a reduction in the exudation of total carbohydrate, with an increase in exudation of the individual sugars Rib, Mal, Raf and Suc, and a reduction in Glc, Fru, Gal, Xyl and Rha. Exudation of the amino acids Gly, Phe and Tyr was increased, while no Asp, Asn, Glu, Gln, Trp and Gaba could be detected. Exudation of organic acids was negligible, although more glycolic and succinic acid was released (Hale and Moore, 1979). Inoculation of oilseed rape with bacterial isolates from rape roots reduced the level of total carbohydrate and amino acids in root exudates by 75%. Individually, exudation of the sugars Glu and Suc was reduced and Fru was not detected, whereas there was a slight increase in release of *myo*-inositol. Exudation of the amino acids Ala, Asp, Glu, Gly, Ile, Lys, Pro, Tyr, Trp, Thr and Orn was reduced, while there was an increase in loss of Ser (Sundin *et al.*, 1990).

The substantial reductions in the root exudation of specific molecules, as described above, is indicative of their preferential assimilation by rhizosphere microorganisms. Further evidence for this comes from studies of the dynamics of amino acid loss from forage rape. Amino acids released from roots were accumulated in nutrient media over periods of 4.5–6 h and the accumulation patterns of individual amino acids in the root-zones of axenic and non-sterile plants were compared. This revealed that the absolute rates of amino acid consumption were highest for those acids present in the greatest concentrations (Ser, Gly, Thr, Leu, His, Ala, Asp, Val, Gaba, Gln) i.e. the major constituents (2–25%) of the exudate. This was considered to be evidence for concentration-dependent differential assimilation of amino acids (Shepherd and Davies, 1994a,b). However, when the rates of amino acid assimilation (R_a) were related to the rates of amino acid supply through exudation (R_e), then a relative affinity factor (R_a/R_e) for microbial assimilation of specific amino acids could be derived. For most acids, including Gly, Thr, Ser, Ala, etc. the ratio (R_a/R_e) was close to unity (1.05–1.15). However, for Asn, Glu, Gln, Arg and Lys the ratio was larger (1.3–1.7), which suggests a degree of selectivity by the bacteria for glutamate and nitrogen-rich amino acids (Asn, Gln, Arg and Lys). This was taken to be evidence for structure-dependent differential assimilation of amino acids (Shepherd and Davies, 1994a,b).

Effects of Microorganisms on Plants and Root Exudation

In general, microorganisms increase root exudation; the levels of total carbon in root exudates of maize, wheat and barley were 2–2.5 times greater in the presence of microorganisms, compared to axenically grown plants (Barber and Martin, 1976). Amino acid release from 7- and 21-day-old liquid-cultured non-sterile rape seedlings was 3–4.5 times greater than from axenic plants (Shepherd and Davies, 1994a). An increase in overall exudation can be attributed mainly to changes in membrane permeability of root cells. Microorganisms increase root lysis; more than 70% of seminal root cortical cells of 4-week-old wheat seedlings, and 60% of root cells of marigold plants infected with *Penicillium simplicissimum* are dead (Hale and Moore, 1979). Metabolites secreted by fungal pathogens (penicillin, victorin, pectic enzymes, etc.) and soil bacteria (e.g. the polypeptide antibiotic polymyxin from *Bacillus polymyxa*) may also change root cell permeability (Rovira, 1969).

Changes in the quantity of specific chemicals exuded from roots can arise through changes in membrane permeability, coupled with changes in plant metabolism or differential utilization of exudate components by microorganisms. Pisatin, a phytoalexin flavonoid, is found in root exudate from non-sterile pea plants, but not in exudate of axenic plants (Hale and Moore, 1979). This is clearly an example of a microbially-induced change in plant metabolism, in this case initiation of the plant's defence mechanisms. Inoculation of roots of marigold with *P. simplicissimum* resulted in exudation of more proteins, total amino acids, water-soluble carbohydrate and reducing sugars, and enhanced shoot growth compared with non-inoculated controls (Hale and Moore, 1979). This may also be indicative of microbially-induced changes in plant metabolism. In contrast, inoculation of tobacco with *Trichoderma harzianum* resulted in a 50% fall in exudation of reducing sugars (Hale and Moore, 1979), probably due to microbial utilization of the sugars.

Changes in the pattern of root exudation can also arise following infection with virus pathogens. Increased exudation of amino acids, nucleotides and carbohydrates followed infection of pea with bean yellow mosaic virus (Hale and Moore, 1979); infection also resulted in increased root rot by *F. solani*. In contrast, infection of squash with squash mosaic virus causes increased resistance to *F. oxysporum*. An increase in amino acid exudation also followed infection of red pepper (*Capsicum annuum*) with tobacco mosaic virus, and infection of hyacinth bean by the foliar saprophyte *Beijerinckia* sp. (Hale and Moore, 1979).

Some of the effects of microorganisms on plant growth have been attributed to the action of microbial metabolites, including plant growth regulators, phytotoxins and antibiotics (Lynch, 1990). The release of such substances after a microorganism has penetrated the cortex would undoubtedly have a significant effect on plant growth, and will not be considered further here. However, certain microbial metabolites, particularly those that can diffuse through the soil effectively, may affect both plant roots and microorganisms in the rhizosphere. Such chemicals include aliphatic and phenolic acids, which are the products of microbial transformations on plant residues. The toxic gases hydrogen sulphide (H_2S) and hydrogen cyanide (HCN) are microbial metabolites, and being volatile they are potentially effective phytotoxins. Hydrogen sulphide is produced by the anaerobic bacterial genera *Desulfotomaculum*, *Desulfovibrio*, *Desulfomonas* and *Desulfuromonas*, but the gas is usually inactivated in soil and may only be a significant toxin in wet acidic soils. Cyanide is formed by a number of organisms, including the bacteria *Chromobacterium violaceum*, *Pseudomonas aeruginosa*, *P. fluorescens*, *P. chlorapsis* and *P. aureofaciensis*, the fungus, *Marasmius oreades* and the green alga *Chlorella vulgaris* (Lynch, 1990). Microbial antibiotic metabolites such as streptomycin from *Streptomyces* spp. may affect plant growth by changing the population distribution of other rhizosphere microorganisms, particularly pathogens. Several bacterial antibiotics have been implicated in control of fungal diseases. *Gaeumannomyces graminis* (take-all) is one of several diseases controlled by 6n-pentyl-

2H-pyran-2-one, a volatile antibiotic produced by *T. harzianum* and *T. koningii* (Lynch, 1990).

Rhizosphere microorganisms can have various effects on the nutritional status of plants. Mycorrhiza associated with plant roots help to increase the uptake of phosphorus by roots by exploiting nutrient sources removed from the vicinity of the root system. Microorganisms have a central role in the cycling of nitrogen within the rhizosphere. Root exudates may stimulate heterotrophic bacteria, which in turn are consumed by bacterial-feeding nematodes and protozoa. The digestion of the bacteria results in the release of a proportion of the bacterial nitrogen as ammonium at the root surface (mineralization), which is available to the plant. Other soil bacteria can oxidize ammonium to nitrate (nitrification), and reduce nitrate to molecular nitrogen (denitrification). The ammonium and nitrate can be incorporated into the microbial biomass (immobilization) and the molecular nitrogen is available to non-symbiotic nitrogen-fixing bacteria. Microorganisms can also stimulate the uptake of metals such as manganese (Mn), iron (Fe) and zinc (Zn) (Hale and Moore, 1979). The solubility of Fe^{3+} in soil is raised by chelation with special binding agents released by fungi and bacteria (siderophores), and by plants (phytosiderophores) (Neilands and Leong, 1986; Lynch, 1990). Two mechanisms for iron acquisition in plants have been suggested, characterized by (i) an inducible plasma membrane-bound reductase (in most graminaceous species) and (ii) enhanced phytosiderophore release, coupled with a highly specific uptake process for Fe^{3+} phytosiderophores (in grass species). The extent of siderophore involvement in iron uptake by plants therefore depends on factors such as the type and specificity of the uptake mechanisms in roots, and the relative stability constants of the Fe^{3+} siderophore and Fe^{3+} phytosiderophore complexes. For example, the uptake of Fe^{3+} phytosiderophores by barley is 100–1000 times faster than uptake of microbial Fe^{3+} siderophores (e.g. ferrichrome), or synthetic Fe^{3+} chelates (e.g. FeEDTA). Pseudobactin, the siderophore from *Pseudomonas aureofaciens* ATCC 15926, inhibits uptake of Fe^{3+} ion by roots of pea and maize, and this was attributed to competitive Fe binding. In contrast, aerobactin from *Klebsiella pneumoniae* has no effect on Fe^{3+} uptake by maize, and agrobactin from *Agrobacterium tumefaciens* increases the uptake of Fe^{3+} by young bean and pea plants (Neilands and Leong, 1986; Lynch, 1990).

Chemical Signal Exchange

There is a growing body of evidence to suggest that interactions between microorganisms and plants proceed through a number of stages initiated or inhibited by exchange of chemical signals. Two systems have been subject to detailed study, the development of the symbiotic relationship between rhizobia and their host legumes, and the infection of plants with the pathogenic bacterium *Agrobacterium tumefaciens*. Although symbiosis between rhizobia and legumes is not itself a pathogenic interaction, the mechanism of symbiosis is similar to that of agrobacterium pathogenicity. An understanding of the molecular mechanisms of these interactions may provide a model for other pathogenic systems.

Rhizobium–legume symbiosis

There are three distinct genera of rhizobia, characterized by their ability to establish N-fixing symbiosis with legumes; fast-growing *Rhizobium*, slow-growing *Bradyrhizobium* and *Azorhizobium* (characterized by N_2-dependent growth in free-living conditions). Bacterial infection of root hairs of host plants initiates a complex series of events leading to formation of nodules, specialized structures incorporating elements of both plant and bacterial origin, in which N_2 is reduced to ammonia (Dénarié *et al.*, 1992).

Rhizobia generally have a narrow, well-defined host range. The specificity of rhizobia varies greatly: *Rhizobium meliloti* on alfalfa (*Medicago* spp.), *Melilotus* spp. and *Trigonella* spp.; *R. leguminosarum* biovar *viciae* on pea (*Pisum sativum*) and vetch (*Vicia* spp.); *R. leguminosarum* biovar *trifolii* on clover (*Trifolium* spp.); *R. leguminosarum* biovar *phaseoli* on bean (*Phaseolus* spp.); *Rhizobium loti* on lotus; *Rhizobium fredii* on soyabean (*Glycine*); *Rhizobium* sp. NGR234 on more than 35 different genera of tropical legumes and *Parasponia* (a non-legume); *Rhizobium tropici* on bean (*Phaseolus* spp.) and *Leucaena*; *Bradyrhizobium japonicum* on soyabean (*Glycine*); tropical *Bradyrhizobium* sp. on many tropical legumes; *Azorhizobium caulinodans* on *Sesbania* (stem-nodulating) (Dénarié *et al.*, 1992).

Rhizobia have large symbiosis (Sym) plasmids containing several genes essential for host infection and nodulation, and genes for nitrogen fixation in nodules. The host infection genes are responsible for the synthesis, transport and assembly of surface components (*exo*-lipopolysaccharides, β-1,2-glucans) involved in recognition and binding of the bacterium by host plant root lectins. The nodulation (*nod*) genes include a highly conserved operon of three genes, *nodABC*, which are required for the initial stages of infection (root hair curling). Expression of these genes is regulated by a *nodD* gene, closely linked to *nodABC*. The *nodD* gene product (NodD protein) activates transcription of *nodABC* and other *nod* genes by binding to promoter regions containing extended conserved sequences, referred to as *nod* boxes (Dénarié *et al.*, 1992). The activation of *nod* gene transcription by the NodD protein requires the presence of plant-derived signal compounds, flavonoids or closely related phenolic compounds (Redmond *et al.*, 1986; Zaat *et al.*,

Table 3.3. Induction of *Rhizobium* NodD proteins by phenolic compounds.

	Substitution pattern							NodD protein					
	3	5	7	2'	3'	4'	5'	R_t	R_l	R_m		NGR	B_j
Compounds										D_1	D_2		
Flavones (1)													
Luteolin[a]		OH	OH		OH	OH		S	S	S	W	S	W
Luteolin-7-OGlc		OH	OGlc		OH	OH		nd	nd	W		nd	nd
Apigenin[a]		OH	OH			OH		S	S	M	W	S	M
–[c]			OH		OH	OH		nd	S	S	W	nd	nd
Chrysoeriol		OH	OH		OCH_3	OH		nd	nd	S	W	nd	nd
–		OCH_3	OH		OH	OH		nd	nd	W		nd	nd
–		OCH_3	OH		OCH_3	OH		nd	nd	W		nd	nd
–					OH	OH		nd	nd	W	W	nd	nd
–[a,b,c]			OH			OH		S	nd	M	nd	S	M
Chrysin		OH	OH					M	I	W	W	S	W
Flavanols (3-OH flavones)													
Myricetin	OH	OH	OH		OH	OH	OH	W	W	W	W	M	nd
Quercetin	OH	OH	OH		OH	OH		W	W	W	W	S	W

Kaempferol	OH	OH	OH			OH	W	I	W	W	S	M
Morin	OH	OH	OH	OH		OH	nd	nd	I		nd	nd
Flavanones (2)												
–[a]			OH			OH	nd	nd	W		nd	nd
Eriodictoyl[d]		OH	OH		OH	OH	M	S	M	W	nd	W
Naringenin		OH	OH			OH	S	S	I		S	W
Hesperetin		OH	OH		OH	OCH_3	W	S	W	W	S	nd
Isoflavones (3)												
Genistein		OH	OH			OH	W	W	W	W	S	S
Daidzein[e]			OH			OH	W	W	W	W	S	S
Coumarins (5)												
Umbelliferone			OH				nd	nd	I		nd	nd
Chalcones (4)	4			2'		4'						
–[a]	OH			OCH_3		OH	nd	nd	S	S	nd	nd
Isoliquiritigenin[e]	OH			OH		OH	nd	nd	S		nd	nd

Compounds detected in root exudates of: [a]alfalfa, [b]clover, [c]lentil, [d]pea, [e]soyabean. *nod*-Inducing activities as determined with rhizobial strains carrying the *nodD* genes of *Rhizobium trifolii* (R_t); *R. leguminosarum* (R_l); *R. meliloti* (R_m); *Rhizobium* NGR234 (NGR) and *Bradyrhizobium japonicum* (B_j). Induction activity, (percentage of maximum induction): S (strong, 50–100%); M (medium, 10–50%); W (weak, 0–10%); I (inhibition); nd (not determined). Numbers associated with different classes of compound refer to structures (1)–(6).

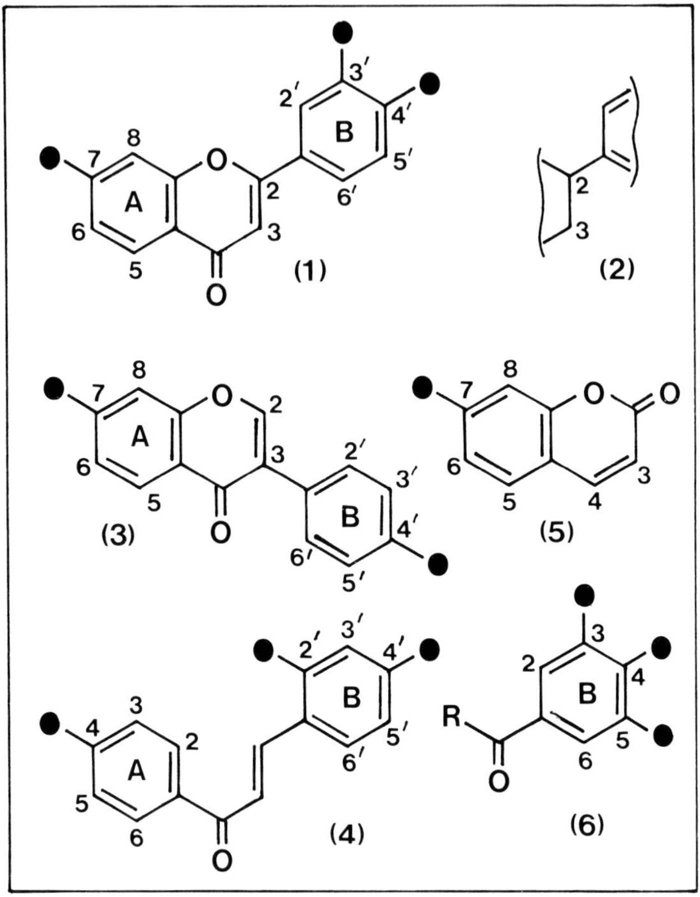

1987; Peters and Long, 1988). Host range is determined mainly by the regulatory (*nodD*) and structural (*nodABC* etc.) genes (Spaink *et al.*, 1987; Hartwig *et al.*, 1990; Dénarié *et al.*, 1992). The level of induction of the NodD protein from a given rhizobial strain is highest in the presence of root exudates from plants nodulated by that particular rhizobium species.

A range of phenolic compounds, both natural and synthetic, have been used to define the structural features of these compounds required for induction of different *nod* genes by different NodD proteins. The results of these studies are listed in Table 3.3 for induction of *nod* genes by NodD proteins from *R. trifolii* (Redmond *et al.*, 1986), *R. leguminosarum* (Firmin *et al.*, 1986; Zaat *et al.*, 1987), *R. meliloti* (Peters *et al.*, 1986; Peters and Long, 1988; Maxwell *et al.*, 1989; Hartwig *et al.*, 1990), *Rhizobium* sp. NGR234 (Bassam *et al.*, 1988) and *B. japonicum* (Kape *et al.*, 1991, 1992). These different chemical species are numbered as shown in the above box. In general, inducing molecules are flavones (1), flavanols (3-OH flavones) and flavanones (2). A hydroxyl group at C-7, C-3' and/or C-4' is essential for activity. A hydroxyl group at C-2' or C-3 (flavanols) abolishes or greatly reduces activity. Alkylation (CH_3) at positions 3' and 5 reduces induction activity. Inclusion of a sugar at position 7 (7-*O*-Glc derivatives) reduces activity, while 8-*C*-glucosides are inactive. Unsaturation at C-2,3 reduces activity, i.e. flavanones are generally more active than flavones. Isoflavones (3) are inactive or weakly active with *Rhizobium* NodD proteins, whereas they are strong activators with *B. japonicum* NodD proteins. For induction activity, isoflavones must have hydroxyl groups at C-4' and C-7. Chalcones (4) are strong activators of *R. meliloti* NodD proteins, and hydroxyl groups at C-4 and C-4' are required. It can be seen from Table 3.3 that for the narrow host-range rhizobia (*R. trifolii*, *R. leguminosarum*, and *R. meliloti*) NodD proteins are activated by only a few flavonoids, while NodD from the wide host-range *Rhizobium* NGR234 is activated by a wider group of compounds. In particular, NodD from *Rhizobium* NGR234 is activated by flavanols, which are almost inactive with other NodD proteins. A number of these NodD-inducing compounds have been identified in the root exudates of the legumes nodulated by the same rhizobial strains from which the NodD proteins were derived, and are indicated in Table 3.3 (D'Arcy-Lameta, 1986; Firmin *et al.*, 1986; Redmond *et al.*, 1986; D'Arcy-Lameta and Jay, 1987; Maxwell *et al.*, 1989; Maxwell and Phillips, 1990). Root exudates of a given legume may contain several flavonoids, of which only one, or a few, induce NodD activity in the specific symbiont. Flavonoids may be released as aglycones, or as the more soluble glycosides which are hydrolysed to the aglycones by bacterial glycosidases.

A number of the flavonoids tested inhibit *nod* activation by active inducers (Firmin *et al.*, 1986; Peters and Long, 1988). These compounds include some flavones (1), flavanones (2), isoflavones (3), coumarins (5) and acetophenones (6); they have structures similar to inducers, and act like competitive inhibitors, since the inhibition can be overcome by increasing the inducer concentration. Synergistic interactions between inducer molecules have also been observed, which may be explained if *nodD* exists as a multimer. *R. meliloti* carries three *nodD* genes, $nodD_1$, D_2, D_3; $nodD_1$ is activated by the flavones luteolin and chrysoeriol, and 4,4'-$(OH)_2$-2'-(OCH_3)-chalcone. $nodD_2$ is activated by the chalcone alone (Hartwig *et al.*, 1990).

The *nod* genes induced by the NodD proteins are involved in production of extracellular Nod factors, chemical signals which stimulate root hair curling, the next step in nodule formation (Dénarié *et al.*, 1992). The *nodABC* genes (e.g. *nodAB* of *R. meliloti*) specify the backbone of the structure, and the other *nod* genes (e.g. *nodH*, *nodQ* of *R. meliloti*) cause structural changes, altering the host specificity of the NodABC factor which is also a determinant of host-range. A number of these Nod factors have been

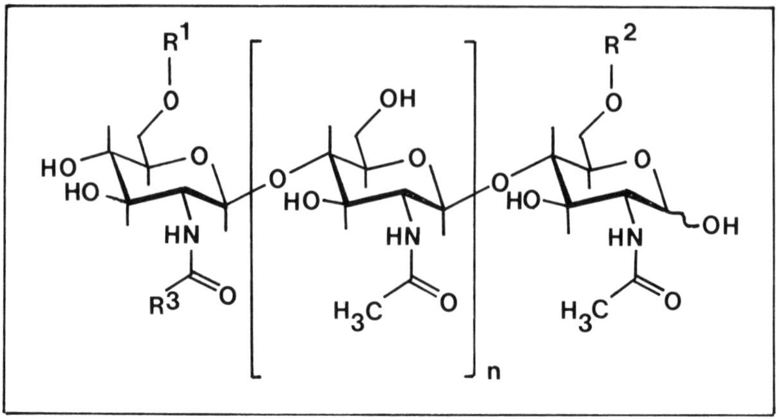

R^1 = CH_3O (Ac), H
R^2 = SO_3H
R^3 = C16:2 (2,9)
n = 2,3

NodRm-IV(Ac, S) n = 2
NodRm-IV(S) n = 2
NodRm-V(Ac, S) n = 3
NodRm-V(S) n = 3

R. meliloti

R^1 = CH_3O (Ac)
R^2 = H
R^3 = C18:4(2,4,6,11), C18:1 (11)
n = 2,3

NodRlv-IV(Ac, C18:4) n = 2
NodRlv-IV(Ac, C18:1) n = 2
NodRlv-V(Ac, C18:4) n = 3
NodRlv-V(Ac, C18:1) n = 3

R. leguminosarum

Fig. 3.2. Nod factors secreted by *Rhizobium meliloti* and *R. leguminosarum*.

isolated, and the structures of the *R. meliloti* and *R. leguminosarum* Nod factors are shown in Fig. 3.2 (Roche *et al.*, 1991; Spaink, *et al.*, 1991). The molecules consist of a backbone of four or five β-1,4-linked D-glucosamine residues, which are *N*-acetylated except on the non-reducing terminal sugar, which is *N*-acylated by (*R. meliloti*) a C16:2 fatty acid with double bonds at C-2, and C-9; (*R. leguminosarum*) a C11:1 fatty acid with a *cis* double bond at C-11, or a C18:4 fatty acid with *trans* double bonds conjugated to the acyl carbonyl, and a fourth double bond at C-11. The non-reducing terminal sugar is *O*-acetylated at C-6 in some compounds, and the reducing sugar may carry a sulphate group on C-6 (*R. meliloti*). Purified Nod factors elicit deformation of root-hairs (at pM concentrations), and cortical cell division and nodule formation (at nM concentrations). Nod Rlv factors also cause changes in the pattern of flavonoid release from root exudates (at 50 nM), a process which results in the release of more effective NodD inducers and is referred to as increased *nod* gene inducing activity (INI) (van Brussel *et al.*, 1990). Modifications to the structures of these factors result in alteration of their ability to induce nodulation in hosts. For example, if the *R. meliloti* Nod Rm factors are transformed by reduction of the anomeric carbon of the reducing sugar, removal of sulphate, or reductive resaturation of the *N*-aryl chain, their induction activity is strongly decreased. Nod factors isolated from other rhizobia have similar structures (Dénarié *et al.*, 1992).

Plant–*Agrobacterium* pathogenic interaction

Agrobacterium tumefaciens and *A. rhizogenes* cause crown gall disease and hairy root respectively. *A. tumefaciens*, although a root colonizer, may not cause disease unless the root is wounded. The oncogenic factor is carried on an extrachromosomal Ti plasmid, a portion of which is transferred to the host plant (Keen and Staskawicz, 1988). Gene expression causes excessive formation of zeatin (a cytokinin) and IAA (an auxin). The products of bacterial virulence (*vir*) genes (located on the Ti plasmid and bacterial chromosomes) are required by plant cells for transfer of T-DNA from the bacterium to the plant. Other genes may be involved in modification of extracellular polysaccharides (possibly involving synthesis of a cyclic β-1,2-linked glucan), a process required for attachment of *A. tumefaciens* to wounded cells (Keen and Staskawicz, 1988). The *vir* genes are normally repressed, but are activated by phenolic compounds present in root exudate of the host plant. *A. tumefaciens* generally has a wide host range, including most dicotyledonous species. The levels of these phenolic compounds are very low in several monocotyledons, which may account for their low susceptibility to this pathogen. The gene products of the *virD* locus excise the T-DNA from the Ti plasmid. Transfer of the T-DNA involves the *virC* gene product, and possibly other *vir* proteins. Expression of these *vir* genes is activated by the gene products of the regulatory *virA* and *virG* genes. The VirA protein (membrane-associated) interacts with the phenolic compounds in root exudates, and then alters the intracellular VirG protein such that it binds and activates the other gene products (Keen and Staskawicz, 1988).

Studies of the *vir* gene-activating properties of various plant phenolic compounds have revealed details about the structural features required for activity (Stachel *et al.*, 1985; Bolton *et al.*, 1986; Spencer and Towers, 1988). The active compounds fall into two groups (Table 3.4):

Table 3.4. Induction of *vir* expression in *Agrobacterium tumefaciens* by phenolics and flavonoids.

R^2	R^1	Acetophenones (6)	Activity	R^2	R^1	Cinnamic acids (8)	Activity
H	H	Vanillin	M	Me	H	Vanillylacetone	M
H	OMe	Syringaldehyde	S	OH	H	Ferulic acid	W
Me	H	Acetovanillone	W	OH	OMe	Sinapic acid	M
Me	OMe	Acetosyringone[a]	S	OMe	H	Ferulic acid Me ester	M
OH	OMe	Syringic acid	W	OMe	OH	5-OH-Ferulic acid Me ester	M
OMe	OMe	Syringic acid Me ester	M	OMe	OMe	Sinapic acid Me ester	S
R^1		Monolignols (7)		R^1		Chalcones (9)	
H		Coniferyl alcohol	S	H		2',4',4-$(OH)_3$-3-(OMe)-chalcone	M
OMe		Sinapyl alcohol	M	OMe		2',4',4-$(OH)_3$-3,5-$(OMe)_2$-chalcone	S
Flavanol glycosides							
5,7,4'-$(OH)_3$-Flavone-3-*O*-glucosylgalactoside			S	5,7,3',4'-$(OH)_4$-Flavone-3-*O*-glucosylgalactoside			M
5,7,3',4'-$(OH)_4$-Flavone-3-*O*-rutinoside			S	5,7,3',4',5'-$(OH)_5$-Flavone-3-*O*-galactoside			M
5,7,4'-$(OH)_3$-3'-(OMe)-Flavone-3-*O*-rutinoside			S				

Induction activity: S (strong); M (medium); W (weak). [a]Detected in root exudates of tobacco.
The positions of substituents R^1 and R^2 in the different classes of compound are shown in structures (6)–(9).

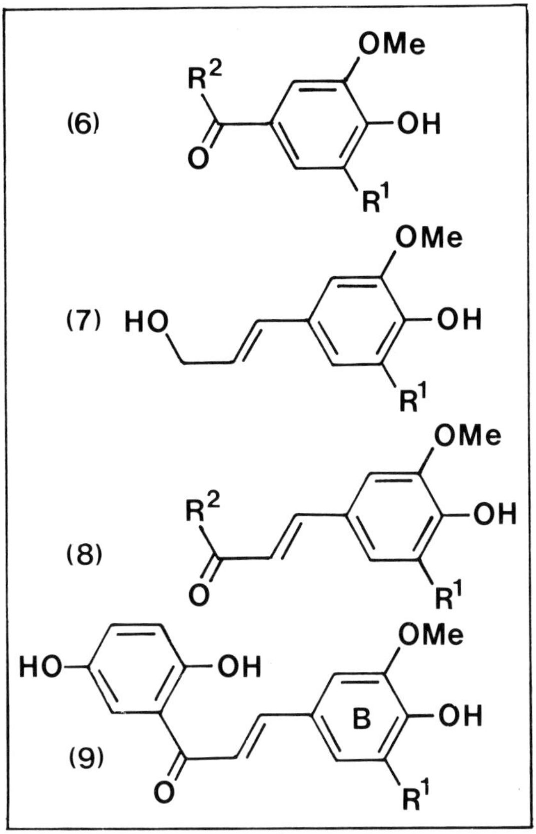

Acetophenones (6), monolignols (7), cinnamic acids (8) and chalcones (4, 9)

In general, compounds with the syringyl substitution pattern (R^1 = OMe) are more active than those with the guaiacyl substitution pattern (R^1 = H), except for monolignols where the reactivity order is reversed. Methyl esters of acids (R^2 = OMe) are more active than acids (R^2 = H) which are the least active inducers. Ethyl esters are less active than methyl esters. The double bond in cinnamic acids (8) enhances activity compared to acetophenones (6). Chalcones (4, 9) are active at lower concentrations than the other inducers. Monolignols excepted, a carbonyl group is apparently required on a substituent *para* to the ring hydroxyl. This must be one carbon atom removed (6), or three carbon atoms removed, with an intervening double bond (8), which allows conjugation of electronic charge from the hydroxyl oxygen to the carbonyl oxygen (Spencer and Towers, 1988).

Flavones (1) and flavanols

The structural features of these compounds required for strong *vir* induction are different from those required for rhizobial *nodD* induction, in that flavanols are strong inducers. In that respect *Agrobacterium* spp. are similar to *Rhizobium* NGR234, otherwise structural requirements seem to be similar (Zerback *et al.*, 1989).

It is therefore evident that plant/microorganism interactions, including those involving pathogens, may proceed via a series of steps involving exchange of small molecules between both participants in the interaction, and that the structures of the molecules determine the specificity of the interactions at each step.

The question arises as to whether such signal transduction mechanisms might have a more general role in interactions between plants and microorganisms. There is an interesting parallel between nodulation induction in rhizobia and infection of host plants by the parasitic plant *Striga* spp. (Lynn and Chang, 1990). A hydroquinone released from roots of the *Striga* spp. host induces germination of the parasite, and a root grows towards the host plant, apparently along a concentration gradient of the hydroquinone. The parasitic root exudes an oxidase enzyme which releases a second signal molecule, a benzoquinone, from the host root. The quinone induces formation of a haustorium, a special attachment organ. The benzoquinone is derived from syringic acid and a number of related compounds by oxidation (Lynn and Chang, 1990). The phenolic compounds that induce *Agrobacterium vir* gene expression, and those that are converted into benzoquinone in the *Striga* spp./host relationship are also remarkably similar. The interaction between *Striga* and its host has been described as a 'chemical radar' detection system, the enzyme released from the parasite removes quinones from the host root surface as signals of a viable host. In support of such a concept is the secretion by *F. solani* of cutinase following induction with host wall-derived hydroxy acids, which are themselves the products of cutinase digestion (Lynn and Chang, 1990). Undoubtedly, more examples of the involvement of phenolic compounds in initiation of plant/microorganism interactions will be uncovered in due course.

Conclusions

A number of factors will determine the spatial distibution of rhizosphere microorganisms. Root exudation is not uniform along the length of the root. Individual organisms probably occupy ecological microniches, where prevailing conditions meet their particular requirements for growth. As plants age, the quantity and quality of root exudate changes, reflecting the different metabolic state of the plant, which can change the nature and distribution of the associated microflora. Plants release chemicals which can initiate and sustain a variety of microbial activities, including chemoattraction, germination of propagules and growth. Interactions with exudate components may be general, or highly specific, and may involve exchange of chemical signals. In their turn, microorganisms can modify the chemical composition of the rhizosphere by changing plant root metabolism, and by exuding organic chemicals themselves.

Acknowledgement

The author thanks the Scottish Office Agriculture and Fisheries Department for financial support.

References

Aguilar, J.M.M., Ashby, A.M., Richards, A.J.M., Loake, G.J., Watson, M.D. and Shaw, C.H. (1988) Chemotaxis of *Rhizobium leguminosarum* biovar *phaseoli* towards flavonoid inducers of the symbiotic nodulation genes. *Journal of General Microbiology* 134, 2741–2746.

Barber, D.A. and Martin, J.K. (1976) The release of organic substance by cereal roots into soil. *New Phytologist* 76, 69–80.

Barbour, W.M., Hattermann, R. and Stacey, G. (1991) Chemotaxis of *Bradyrhizobium japonicum* to soybean exudates. *Applied and Environmental Microbiology* 57, 2635–2639.

Bassam, B.L., Djordjevic, M.A., Redmond, J.W., Batley, M. and Rolfe, B.G. (1988) Identification of a *nodD*-dependent locus in the *Rhizobium* strain NGR234 activated by phenolic factors secreted by soybeans and other legumes. *Molecular Plant–Microbe Interactions* 1, 161–168.

Bolton, G.W., Nester, E.W. and Gordon, M.P. (1986) Plant phenolic compounds induce expression of the *Agrobacterium tumefaciens* loci needed for virulence. *Science* 232, 983–985.

Bowen, G.D. and Rovira, A.D. (1976) Microbial colonization of plant roots. *Annual Review of Phytopathology* 14, 121–144.

Caetano-Anolles, G., Crist-Estes, D.K. and Bauer, W.D. (1988) Chemotaxis of *Rhizobium meliloti* to the plant flavanone luteolin requires functional nodulation genes. *Journal of Bacteriology* 170, 3164–3169.

Chang-Ho, Y. and Hickman, C.J. (1970) Some factors involved in the accumulation of Phycomycete zoospores on plant roots. In: Toussoun, T.A., Bega, R.V. and Nelson, P.E. (eds), *Root Diseases and Soil-Borne Pathogens*. University of California Press, Berkeley, pp. 103–108.

Coley-Smith, J.R. and King, J.E. (1970) Response of resting structures of root-infecting fungi to host exudates: an example of specificity. In: Toussoun, T.A., Bega, R.V. and Nelson, P.E. (eds), *Root Diseases and Soil-Borne Pathogens.* University of California Press, Berkeley, pp. 130–133.

D'Arcy-Lameta, A. (1986) Study of soybean and lentil root exudates II. Identification of some polyphenolic compounds, relation with plantlet physiology. *Plant and Soil* 92, 113–123.

D'Arcy-Lameta, A. and Jay, M. (1987) Study of soybean and lentil root exudates III. Influence of soybean isoflavonoids on the growth of rhizobia and some rhizospheric microorganisms. *Plant and Soil* 101, 267–272.

Dénarié, J., Debellé, F. and Rosenberg, C. (1992) Signalling and host range variation in nodulation. *Annual Review of Microbiology* 46, 497–531.

Firmin, J.L., Wilson, K.E., Rossen, L. and Johnston, A.W.B. (1986) Flavonoid activation of nodulation genes in *Rhizobium* reversed by other compounds present in plants. *Nature* 324, 90–92.

Hale, M.G. and Moore, L.D. (1979) Factors affecting root exudation II: 1970–1978. *Advances in Agronomy* 31, 93–124.

Hartwig, U.A., Maxwell, C.A., Joseph, C.M. and Phillips, D.A. (1990) Effects of alfalfa *nod* gene-inducing flavonoids on *nod*ABC transcription in *Rhizobium meliloti* strains containing different *nod*D genes. *Journal of Bacteriology* 172, 2769–2773.

Helal, H.M. and Sauerbeck, D.R. (1984) Influence of plant roots on carbon and phosphorus metabolism in soil. *Plant and Soil* 76, 175–182.

Helal, H.M. and Sauerbeck, D.R. (1989) Carbon turnover in the rhizosphere. *Zeitschrift für Pflanzenernahrung und Bodenkunde* 152, 211–216.

Kape, R., Parniske, M. and Werner, D. (1991) Chemotaxis and *nod* gene activity of *Bradyrhizobium japonicum* in response to hydroxycinnamic acids and isoflavonoids. *Applied and Environmental Microbiology* 57, 316–319.

Kape, R., Parniske, M., Brandt, S. and Werner, D. (1992) Isoliquiritigenin, a strong *nod* gene and glyceollin resistance inducing flavonoid from soybean root exudate. *Applied and Environmental Microbiology* 58, 1705–1710.

Keen, N.T. and Staskawicz, B. (1988) Host range determinants in plant pathogens and symbionts. *Annual Review of Microbiology* 42, 421–440.

Krupa, S. and Fries, N. (1971) Studies on ectomycorrhizae of pine. I. Production of volatile organic compounds. *Canadian Journal of Botany* 9, 1425–1431.

Krupa, S. and Nylund, J.E. (1972) Studies on ectomycorrhizae of pine. III. Growth inhibition of two root pathogenic fungi by volatile organic constituents of ectomycorrhizal root systems of *Pinus sylvestris* L. *European Journal of Forest Pathology* 2, 88–94.

Lynch, J.M. (1990) Microbial metabolites. In: Lynch, J.M. (ed.), *The Rhizosphere.* Wiley, Chichester, pp. 177–206.

Lynn, D.G. and Chang, M. (1990) Phenolic signals in cohabitation: Implications for plant development. *Annual Review of Plant Physiology and Plant Molecular Biology* 41, 497–526.

Maxwell, C.A. and Phillips, D.A. (1990) Concurrent synthesis and release of *nod*-gene-inducing flavonoids from alfalfa roots. *Plant Physiology* 93, 1552–1558.

Maxwell, C.A., Hartwig, U.A., Joseph, C.M. and Phillips, D.A. (1989) A chalcone and two related flavonoids released from alfalfa roots induce *nod* genes of *Rhizobium meliloti. Plant Physiology* 91, 842–847.

Mircetich, S.M. and Zentmyer, G.A. (1970) Germination of chlamydospores of *Phytophthora*. In: Toussoun, T.A., Bega, R.V. and Nelson, P.E. (eds), *Root Diseases and Soil-Borne Pathogens*. University of California Press, Berkeley, pp. 112–115.

Misaghi, I.J. (1982) *Physiology and Biochemistry of Plant–Pathogen Interactions*. Plenum Press, New York and London.

Neilands, J.B. and Leong, S.A. (1986) Siderophores in relation to plant growth and disease. *Annual Review of Plant Physiology* 37, 187–208.

Nelson, E.B. (1990) Exudate molecules initiating fungal responses to seeds and roots. *Plant and Soil* 129, 61–73.

Peters, N.K. and Long, S.R. (1988) Alfalfa root exudates and compounds which promote or inhibit induction of *Rhizobium meliloti* nodulation genes. *Plant Physiology* 88, 396–400.

Peters, N.K., Frost, J.W. and Long, S.A. (1986) A plant flavone, luteolin, induces expression of *Rhizobium meliloti* nodulation genes. *Science* 233, 977–980.

Redmond, J.W., Batley, M., Djordjevic, M.A., Innes, R.W., Kuempel, P.L. and Rolfe, B.G. (1986) Flavones induce expression of nodulation genes in *Rhizobium*. *Nature* 323, 632–635.

Roche, P., Lerouge, P., Ponthus, C. and Promé, J.C. (1991) Structural determination of bacterial nodulation factors involved in the *Rhizobium meliloti*–alfalfa symbiosis. *Journal of Biological Chemistry* 266, 10933–10940.

Rovira, A.D. (1969) Plant root exudates. *The Botanical Review* 35, 35–57.

Shepherd, T. and Davies, H.V. (1993) Carbon loss from the roots of forage rape (*Brassica napus* L.) seedlings following pulse-labelling with $^{14}CO_2$. *Annals of Botany* 72, 155–163.

Shepherd, T. and Davies, H.V. (1994a) Pattern of short-term amino acid accumulation and loss in the root-zone of liquid-cultured forage rape (*Brassica napus* L.). *Plant and Soil* 158, 99–109.

Shepherd, T. and Davies, H.V. (1994b) Effect of amino acids, glucose and citric acid on the patterns of short-term accumulation and loss of amino acids in the root-zone of sand-cultured forage rape (*Brassica napus* L.). *Plant and Soil* 158, 111–118.

Spaink, H.P., Sheeley, D.M., van Brussel, A.A.N., Glushka, J., York, W.S., Tak, T., Geiger, O., Kennedy, E.P., Reinhold, V.N. and Lugtenberg, B.J.J. (1991) A novel highly unsaturated fatty acid moiety of lipo-oligosaccharide signals determines host specificity of *Rhizobium*. *Nature* 354, 125–130.

Spaink, H.P., Wijffelman, C.A., Pees, E., Okker, R.J.H. and Lugtenberg B.J.J. (1987) *Rhizobium* nodulation gene *nod*D as a determinant of host specificity. *Nature* 328, 337–340.

Spencer, P.A. and Towers, G.H.N. (1988) Specificity of signal compounds detected by *Agrobacterium tumefaciens*. *Phytochemistry* 27, 2781–2785.

Stachel, S.E., Messens, E., van Montagu, M. and Zambryski, P. (1985) Identification of the signal molecules produced by wounded plant cells that activate T-DNA transfer in *Agrobacterium tumefaciens*. *Nature* 318, 924–929.

Sundin, P., Valeur, A., Olsson, S. and Odham, G. (1990) Interactions between bacteria-feeding nematodes and bacteria in the rape rhizosphere: effects on root exudation and distribution of bacteria. *FEMS Microbiology Ecology* 73, 13–22.

Toussoun, T.A. (1970) Nutrition and pathogenesis of *Fusarium solani* f.sp. *phaseoli*.

In: Toussoun, T.A., Bega, R.V. and Nelson, P.E. (eds), *Root Diseases and Soil-Borne Pathogens.* University of California Press, Berkeley, pp. 95–98.

van Brussel, A.A.N., Recourt, K., Pees, E., Spaink, H.P., Tak, T., Wijffelman, C.A., Kijne, J.W. and Lugtenberg, B.J.J. (1990) A biovar-specific signal of *Rhizobium leguminosarum* bv. *viciae* induces increased nodulation gene-inducing activity in root exudate of *Vicia sativa* subsp. *nigra*. *Journal of Bacteriology* 172, 5394–5401.

Whipps, J.M. (1990) Carbon economy. In: Lynch J.M. (ed.), *The Rhizosphere.* Wiley, Chichester, pp. 59–97.

Zaat, S.A.J., Wijffelman, C.A., Spaink, H.P., van Brussel, A.A.N., Okker, R.J.H. and Lugtenberg B.J.J. (1987) Induction of the *nod*A promoter of *Rhizobium leguminosarum* Sym plasmid pRL1JI by plant flavanones and flavones. *Journal of Bacteriology* 169, 198–204.

Zentmyer, G.A. (1970) Tactic responses of zoospores of *Phytophthora*. In: Toussoun, T.A., Bega, R.V. and Nelson, P.E. (eds), *Root Diseases and Soil-Borne Pathogens.* University of California Press, Berkeley, pp. 109–111.

Zerback, R., Dressler, K. and Hess, D. (1989) Flavonoid compounds from pollen and stigma of *Petunia hybrida*: inducers of the *vir* region of the *Agrobacterium tumefaciens* Ti plasmid. *Plant Science* 62, 83–91.

4 Evaluation of Plant Pathogens in Complex Ecosystems

J.M. Lenné, D.M. Teverson and M.J. Jeger

Natural Resources Institute, Central Avenue, Chatham Maritime, Kent ME4 4TB, UK.

Introduction

Many plants with utilitarian value commonly grow in complex ecosystems and not in the monocultures typified by modern agriculture. Approximately 60% of global agriculture is cultivated by subsistence farmers using traditional methods (Francis, 1986). For example, varietal mixtures, intercrops and polycultures constitute at least 80% of the cultivated area of West Africa, while much of the production of staple crops in Latin America occurs in polycultures (Francis, 1986). More than 1050 million ha in the tropics and subtropics is suitable only for grazing and much is covered by natural and improved mixed perennial pastures (Lenné and Sonoda, 1990). Natural mixed forests contain useful multi-purpose woody species used for timber, fuel, and forage and commonly used in agroforestry systems (Nair, 1989). All important food crops originated in natural plant communities, many of which still exist and contain wild relatives of crop and pasture species of use in crop improvement, especially for disease resistance (Lenné and Wood, 1991).

Diseases caused by many pathogens occur in most complex ecosystems where climatic, edaphic and host factors can have profound effects on disease development. Effects of pathogens may be restricted to locations or seasons in which the environment is favourable for their growth and development. Pathogens may affect the outcome of inter- and intraspecific competition, the distribution of plant species, the genetic structure of populations, and the diversity of plant communities (Burdon, 1987). The proceedings of two conferences have brought together valuable information about pathogen and pest interactions in plant communities (Thresh, 1981; Burdon and Leather, 1990), yet information is still lacking on the function of diversity for disease resistance in complex ecosystems and how to evaluate the multiplicity of interactions which occur between hosts and pathogens. This chapter reviews the information on evaluation of pathogens in a range of complex ecosystems and identifies areas where further work is needed. Emphasis is given to complex ecosystems in the tropics.

Main Characteristics of Complex Ecosystems

Complex plant ecosystems are usually characterized by one or more of the following parameters:

1. Spatial and temporal heterogeneity which creates patchiness and an often complex age structure (Jeger, 1989; Alexander, 1989).
2. Intra- and interspecific diversity. Such diversity encompasses variation within landraces and varietal mixtures of single crops; intercrops which may vary from cereal/legume associations to grass/legume perennial pastures; polycultures typified by home gardens; and natural plant communities.
3. Multiple host–pathogen interactions.

Plant pathology developed as a science through the necessity to solve disease problems of food crops, and existing methodology to study plant pathogens is firmly linked with the development of modern agriculture. Modern agriculture is largely breeder-dominated with most emphasis being placed on yield. Assessment of the effect of disease on yield is generally the principal determining factor as to whether a disease merits control. Survival, competitiveness and persistence, especially important for perennials and self-seeding annuals in mixed populations, have been ignored. As modern highly-bred varieties tend to be affected by single major diseases, methods for evaluating effects of pathogens on useful plants over time in mixed populations is rudimentary. Appreciation of the wider range of potential control strategies available in complex ecosystems, such as cultural and biological control and strategic association (deliberate heterogeneity), is lacking.

Complex ecosystems to be discussed in this chapter include: traditional farming systems such as varietal mixtures and intercropping; perennial mixed tropical pastures; and natural populations of wild relatives of crops and pasture plants.

Subsistence Farming Systems

Subsistence farmers in developing countries use traditional practices of mixed cropping, varietal and within-varietal or landrace diversity. Intercropping of cereals and legumes is common and includes associations such as maize/bean, maize/groundnut (peanut), sorghum/cowpea, millet/cowpea, etc. (Allen, 1990). Within-crop diversity is characteristic of traditional landraces selected by farmers and is especially complex where mixtures of landraces are used such as *Phaseolus vulgaris* beans in East Africa, rice in West Africa, sweet potato in the Pacific and cassava in South America (Wood and Lenné, 1993).

Intercropping

More studies have been done on the effects of intercropping on pathogens than any other complex cropping system and most work has been done on cereal–legume associations (Allen, 1990; Thurston, 1992). The most commonly reported effect of intercropping is the reduced incidence and severity of disease in the intercrop, relative to the pure stand (Allen, 1990). The severity of anthracnose (*Colletotrichum lindemuthianum*), rust (*Uromyces appendiculatus*), halo-blight (*Pseudomonas syringae* pv. *phaseolicola*), and bean common mosaic potyvirus (BCMV) on *Phaseolus vulgaris* beans and the severity of Ascochyta blight (*Ascochyta phaseolorum*) and several viruses on cowpeas were reduced up to 47% and 63% respectively in association with maize compared to monocropping (Allen, 1990).

The main factors postulated as responsible for these effects include physical barriers to spread of aerial pathogens and their vectors (including trapping aphid vectors of viruses); altered microclimates – changes in shading and relative humidity have been found to affect disease severity depending on the pathogen; and specific host–pathogen interactions such as induced resistance, for example, in rusts (Allen, 1990). These factors also operate in other complex ecosystems. However, there are also cases where diseases are more severe in association with other crops relative to monoculture. Severity of angular leaf spot (*Phaeoisariopsis griseola*) of *Phaseolus vulgaris* beans showed increases of 11 to 37% in association with maize, cassava and sweet potato over the monoculture (Allen, 1990).

Intercropping shows potential to provide protection from pathogens. But whether diseases increase or decrease in intercrops depends on numerous factors affecting the crops, the pathogens, and the environment (Thurston, 1992). Too little is known of the relative importance of the mechanisms that underlie the observed effects on diseases (Allen, 1990). More evaluation of pathogens in intercrops is needed to provide the information to effectively harness this potential in the development of stable strategies for crop protection.

Varietal mixtures

Mixing of individuals of differing resistances sets in motion a complex series of interconnected changes which affect the ability of pathogens to survive and reproduce and consequent disease development (Burdon, 1987). Replacement of susceptible plants by resistant ones reduces the amount of susceptible tissue which should also reduce the amount of inoculum available for dispersal. Increased distance between susceptible individuals increases the average distance that inoculum has to travel, which is likely to reduce effective spread. Resistant plants interfere with the movement of inoculum between susceptible individuals and induced resistance or cross-protection may also reduce disease development.

Development of pathogens in cereal mixtures in temperate environments has been studied extensively, especially in developed countries (Browning, 1974, 1979, 1988; Wolfe and Barrett, 1980; Jeger *et al.*, 1981a,b; Chin and Wolfe, 1984; Wolfe, 1985, 1989; Mundt, 1989). Use of wheat and barley mixtures with different proportions of resistant varieties have resulted in reduced disease and increased yield varying up to 10% increase over single varieties but rarely over that of the best single variety. In Africa, replacement of traditional bean mixtures by 25% or more of a variety resistant to angular leaf spot (Pyndji and Trutmann, 1992) and halo-blight (Teverson, 1991) reduced disease and, in the former case, increased yield. During the last 10 years, there has been substantial theoretical clarification of the factors that affect plant pathogens in mixtures of single crops (Marshall, 1989; Mundt, 1989). No work has been done, however, to quantify the inherent contribution of varietal mixtures in subsistence farming systems to disease management.

Phaseolus vulgaris bean is the most important food legume in East Africa (Allen, 1983). Mixtures of landraces of beans are commonly grown by small farmers in East Africa. Farmers handle an array of bean genotypes (seed colours, patterns, shapes and sizes), mixtures being selected for different seasons; for different cropping systems; and for different soil conditions (Ayeh, 1985; Martin and Adams, 1987a,b; Hardman and Lamb, 1988; Allen *et al.*, 1989; Pyndji and Trutmann, 1992; Voss, 1992). Diseases are major production constraints in East Africa. The most widespread and serious bean diseases include: angular leaf spot; common bacterial blight (*Xanthomonas campestris* pv. *phaseoli*); halo-blight; anthracnose; rust; and necrotic strains of BCMV. Beans in traditional mixtures in East Africa are an excellent system for investigations of interacting heterogeneous populations of hosts and pathogens.

Mixtures potentially provide farmers with more reliable production under unpredictable pathological and environmental stresses. Yet diversity *per se* has been assumed to contribute to improved disease control and increased yield. No reliable estimates of the function of diversity inherent in traditional bean mixtures have been made. A project has been established between the Natural Resources Institute in collaboration with Horticultural Research International, Wellesbourne, UK and the National Bean Programmes in East Africa and the Regional Bean Programmes of the Centro Internacional de Agricultura Tropical (CIAT) to investigate whether different morphological and physiological phenotypes of beans have been indirectly selected by farmers as markers for different disease resistances.

Three mixtures from the Southern Highlands of Tanzania have been selected for detailed study. Different components were separated on the basis of component phenotype. Both between-component and within-component reaction to the important bean pathogens mentioned above is being evaluated *ex situ*. Studies by Teverson (1991) and Trutmann (unpublished data) have shown that both between- and within-component variability for reaction to halo-blight and anthracnose, respectively, exists in bean mixtures and could contribute to disease control. The inherent contribution of mixtures to disease control will be quantified *in situ*; mixtures will then be supplemented with the resistances lacking and their contribution to disease control and increased yield will be measured.

The usefulness of disease control by mixtures is somewhat dependent on the pathogen and its means of dispersal. Diseases for which existing levels of resistance are currently inadequate (e.g. common bacterial blight and angular leaf spot); diseases whose pathogens are known to be variable (e.g. anthracnose, rust and halo-blight); and diseases with powdery spores which are largely wind-dispersed (e.g. rust) have more potential to be controlled through mixtures. Priority is being given to these diseases to ensure the introduction of satisfactory control strategies. Knowledge of the types and range of resistances in traditional mixtures and their contribution to disease control under field conditions, as well as quantification of the value of supplementing mixtures with additional sources of resistance, will facilitate assessment of the contribution of diversity for pathogen resistance to disease control in bean mixtures.

The importance of leguminous weeds as alternative hosts of bean pathogens in production systems in East Africa is also of concern. The indigenous legume *Neonotonia wightii* is a common alternative host of several races of the halo-blight pathogen *P. syringae* pv. *phaseolicola* (Teverson, 1991). Other bean diseases including indigenous strains of BCMV (Spence and Walkey, 1991) have been found on other legumes in Uganda and Malawi. Recent work in Tanzania has demonstrated that the halo-blight pathogen can spread naturally from *N. wightii* to susceptible beans (Teverson *et al.*, 1993). As *N. wightii* and other legumes are common components of the weedy flora of farmers' bean fields in East Africa, they could be important sources of infection of several diseases to beans.

Tropical Perennial Pastures

Perennial mixed pastures are diverse, heterogeneous communities of various species of different ages supporting continuous pathogen associations (Lenné, 1989; Lenné and Sonoda, 1990). Use of inter- and intra-specific mixtures of both grasses and legumes creates heterogeneity which can stimulate pathogen diversity at both specific and sub-specific levels. Single or multiple diseases may build up over seasons. A succession of chronic attacks may result in progressive changes in botanical composition, reduced productivity and reduced persistence (Carr, 1975; O'Rourke, 1976; Thomas, 1985). The diversity of tropical environments from semi-arid savannas to humid tropical woodlands; the diversity of farming systems from intensive dual-purpose enterprises to extensive cattle production operations; within-pasture temporal and spatial diversity; and the dynamic nature of the pasture ecosystem create a diversity in which epidemiological studies of pathogens are difficult (Lenné, 1989).

Various ecological and demographical techniques can be applied when evaluating diseases of heterogeneous perennial pastures. Surveys, transects and quadrats are commonly used to evaluate pathogens in perennial pasture ecosystems (Lenné, 1989). Other methods include marked and sampled plants. Periodic evaluations take account of dynamic age structure and multilocational evaluations identify host–environment interactions. Periodic surveys provide valuable information on disease incidence and distribution on a field, farm or regional basis over time. Caution should be taken, however, in estimating severity and yield loss on the basis of subjective data especially as interactions between biotic factors may make accurate evaluations of one factor impossible.

Transects, often with marked plants, are used to evaluate plant–pathogen associations and have direct application to perennial pastures. Fixed and random quadrats may be used to complement transect data. Anthracnose (*Colletotrichum gloeosporioides*) was evaluated in 3000 marked plants of *Stylosanthes guianensis* under grazing in Colombia. Survival of plants, strongly determined by anthracnose severity, was also markedly influenced by stocking rate (Lenné and Sonoda, 1990). Studies of the incidence and severity of pasture legume diseases under grazing in South America and Australia have confirmed that disease incidence and severity is higher at low grazing intensities and under long rotations (Lenné, 1989; Davis, 1991). Burdon and Chilvers (1982) stated that 'manipulation of host density is the underlying basis for disease control strategies based on spatio-temporal patterning of crops and the use of intimate mixtures'.

Investigation of biotic problems in mixed perennial pastures using random assessments may sometimes be inadequate because of patchy distribution and variable plant density. Strategic placement of quadrats (for example, in contrasting patches of diseased versus non-diseased pasture) or stratified random sampling may give more relevant information (Lenné, 1989). The association of wart disease (*Synchytrium desmodii*) of *Desmodium ovalifolium* with low-lying intermittently flooded areas of the pasture necessitated strategic, rather than random assessments, for more realistic disease evaluation (Lenné *et al.*, 1990). Under non-flooded conditions, wart disease did not reduce yield of adult plants but under intermittently flooded conditions, yield was reduced by 72.5%. Similarly, the patchy distribution of rhizoctonia foliar blight (RFB) (*Rhizoctonia* spp.) in *Centrosema brasilianum*-based pasture merited the strategic approach (Lenné and Sonoda, 1990).

Evaluating the effect of diseases and pests on seedling emergence, survival and soil seed reserves in pastures is important to the understanding of pasture population dynamics, especially in relation to persistence of the pasture. The devastating effect of fungal diseases on seedlings has been shown for several pasture legumes. *Synchytrium desmodii* severely reduced soil seed reserves, seedling survival and survival to adult plants of *D. ovalifolium* to the pasture sward (Lenné *et al.*, 1990) and *Rhizoctonia* spp. greatly reduced seedling survival and survival to adult plants of *Centrosema brasilianum* in Colombia (Lenné *et al.*, 1989).

Pasture plants are often affected simultaneously by more than one disease. Sampling plants and their progenies facilitates the evaluation of reactions to individual diseases and specific biotypes. Plants, randomly sampled from existing swards, were successfully used to evaluate the resistance structure of a white clover population to several foliar diseases in Wales (Burdon, 1987) and populations of *Stylosanthes* spp. to anthracnose in South America (Miles and Lenné, 1984; Lenné, 1988).

In mixed perennial pastures, if disease affects the growth of one component, the species balance is altered. In a ryegrass/white clover pasture, crown rust (*Puccinia coronata* f.sp. *lolii*) caused 84% yield reduction in ryegrass and a corresponding 87% increase in white clover (Latch and Lancashire, 1970). Diseases can cause undesirable changes in species composition with long-term effects on pasture productivity, which also may result in different diseases becoming problems (Lenné, 1989). More studies of this phenomenon in mixed plant populations are needed.

Grazing is always an important source of heterogeneity in a pasture (Watkin and Clements, 1978). Animals usually select plant parts and species that are palatable, accessible and available. Grazing animals sit, lie, scratch, paw, walk, run, jump, trample and excrete on the pasture. Animal activities influence species composition, productivity, and stress responses of plants, all of which can affect disease incidence and severity. No matter how severe a disease, it will only affect animal production when the availability of forage is below the critical level. Of the few studies made, Anderson *et al.* (1982) found that animal production was reduced only at higher stocking rates in scorch-affected subterranean clover pastures. At lower stocking rates, there was sufficient forage on offer for continued high animal production in spite of dry matter losses due to scorch. Losses in the production and persistence of pastures in relation to animal production must be estimated to give any meaningful evaluation of the economic significance of pathogen damage.

Natural Plant Communities Including Wild Relatives of Crops and Pasture Species

The main use of wild relatives of crops has been as sources of resistance to diseases (Burdon and Jarosz, 1989; Lenné and Wood, 1991). Resistances from wild species have played a key role in the improvement of wheat, potato and tomato and many pasture cultivars are selections from wild populations. Only a limited number of wild germplasm–pathogen systems have been extensively studied and few have been studied spatially and temporally (Burdon *et al.*, 1990). The best documented studies are of wild relatives of barley, oats and wheat and their associated rusts and mildews in Israel (Dinoor, 1970, 1977; Wahl, 1970; Browning, 1974; Manisterski *et al.*, 1991) and wild *Glycine* spp. and soyabean rust (*Phakopsora pachyrhizi*) (Burdon and Marshall, 1981; Burdon, 1987) and wild *Linum marginale* and flax rust (*Melampsora lini*) in Australia (Burdon *et al.*, 1990). Populations of tropical pasture legumes *Stylosanthes* spp. and anthracnose (*Colletotrichum gloeosporioides*) have been studied in South America (Miles and Lenné, 1984; Lenné, 1988). Wild host–pathogen systems have not been given adequate attention by plant pathologists (Lenné and Wood, 1991).

With the notable objectives of sustainable utilization and conservation of invaluable genetic resources for improved agriculture and food production, the Ammiad Project has evaluated an important wild relative of wheat *in situ* (Anon., 1991). A site for intensive genetic, ecological and pathological studies of the wild wheat *Triticum turgidum* var. *dicoccoides* was established at Ammiad, Eastern Galilee, Israel, in 1984 (Anikster and Noy-Meir, 1991). The aim of the study was to gain basic information on a variable population of a selfing wild cereal, the progenitor of durum and bread wheats, in the centre of its local distribution. *In situ* ecological studies were coupled with documentation of wheat demography, phenotypic variation and disease incidence, and with *ex situ* progeny studies of genetic variation, including reaction to diseases. From 1984 to 1989, plants were monitored at closely-spaced permanent sampling points along four topographically diverse transects. Single-spike seed collections were made at each point annually for studies of their reaction to various fungal diseases (Anikster and Noy-Meir, 1991).

The levels of microenvironmental changes and zonation of different population clusters documented by the Ammiad study cast doubt on previous conservation designs based on theories of island geography (Namkoong, 1991). Assumptions of the random distribution of genes and genotypes, constant selection pressure and uniform demographic or density effects were not well-founded. Phytopathological studies showed various types of resistances to a number of pathogens (Dinoor *et al.*, 1991). Most progenies from single-spike seed collections were susceptible to isolates of leaf (*Puccinia recondita* f.sp. *tritici*) and stem rust (*Puccinia graminis* f.sp. *tritici*) and only 22 of 225 accessions tested were resistant to mildew (*Erysiphe graminis*). As a pilot undertaking for developing methods for evaluating the structure and dynamics of populations of wild relatives of crops and their pathogens, the Ammiad project provided valuable data on the diversity of populations of wild plants in their natural habitat on which to base future strategies for *in situ* genetic conservation (Hawkes, 1991) and emphasized the need for detailed evaluation and monitoring of wild populations prior to collection, including for disease resistances.

It is still widely believed that wild germplasm in crop centres of origin and diversity is a concentrated source of disease resistance genes because of the long association of hosts and pathogens (Lenné and Wood, 1991). Most studies have shown that, although useful sources of resistance are available in wild populations, albeit at low frequency, susceptible plants are more common than resistant plants (Dinoor, 1970; Wahl, 1970; Browning, 1974; Burdon, 1987; Lenné, 1988; Dinoor *et al.*, 1991). Population defence mechanisms such as escape and avoidance supported by the heterogeneous character of natural plant communities and lack of linkages between competitive ability and resistance/susceptibility are responsible for the survival of susceptible plants (Burdon, 1987). Random sampling of wild germplasm, even from areas of host–pathogen coevolution, will collect a considerable proportion of susceptible plants.

In an evaluation of a collection of 1500 accessions of *Stylosanthes guianensis* in Colombia which had been randomly sampled from natural populations of the host from tropical America, only two accessions were sufficiently resistant to anthracnose (*C. gloeosporioides*) to be used in a breeding programme. The cost of the collection and evaluation process was estimated to be £150,000 in 1983 (Toledo *et al.*, 1989). A subsequent exploratory study on natural populations of *Stylosanthes capitata* in Central Brazil also showed that most accessions were susceptible to anthracnose but sites with higher proportions of resistant plants were readily identified which merited further attention (Lenné, 1988). In a study of crown rust of oats in Israel, Dinoor (1970) clearly showed that the most efficient single method of capturing adult *Avena sterilis* plants resistant to *Puccinia coronata* was by collection of seed from plants resistant *in situ*. Alternative methods failed to detect at least 40% of the locations where resistant plants were found by the *in situ* method.

It is recommended that *in situ* evaluation of diseases followed by targeted collection of resistant germplasm should be a more efficient strategy for utilization of wild germplasm that *ex situ* evaluation of large random collections (Lenné and Wood, 1991). As far as possible, collecting of wild germplasm for disease resistances should concentrate on areas most favourable for disease development. Studies of *Phlox* spp. and populations of associated powdery mildew (*Erysiphe cichoracearum*) in the USA strongly support this (Jarosz, 1984). Monitoring disease incidence and severity in the target area for several seasons should ensure that collections are made under high disease pressure. Periodic visits to collecting sites to tag and survey apparently resistant plants would allow collection of samples more relevant as sources of disease resistance. This could reduce the high costs in time and funds involved in evaluating, processing and maintaining large collections (Lenné and Wood, 1991).

In natural communities, environmental conditions such as periods of low rainfall and humidity are not always conducive to disease development. This makes identification of resistant plants *in situ* difficult. Brown (1991) recommended the adoption of intervention measures which speed up divergence *in situ*, such as encouraging disease development. To increase disease selection pressure, inoculum of common local pathogenic races could be introduced into wild populations in the season before sampling. High relative humidity is an important factor in the multiplication and spread of many pathogens. By increasing relative humidity in microsites in natural populations, the development of many diseases would be favoured and resistant plants more readily identified. Studies of such intervention measures are needed to assess their effects on the efficiency of sampling disease resistant germplasm from the wild.

Limited studies have been made of the virulence structure of populations of pathogens in complex ecosystems. Studies which have been done show a range of different virulence biotypes occur in pathogen populations. Determination of the virulence structure of tropical pasture pathogens has been described as a 'daunting task' because the virulence characteristics of isolates cannot be determined until suitable host lines, each with differing resistance, have been identified (Burdon, 1987). Assessment of race-specific resistance where little information on host genetics exists may result in confusion with non-specific resistance which gives a continuum of resistance reactions as found by Miles and Lenné (1984) in evaluation of genetic variation within a native population of *Stylosanthes guianensis* infected by *C. gloeosporioides* in

Colombia. Incomplete, simplistic pictures of the resistance structure of wild germplasm populations may be obtained if the races used do not reflect the spectrum of existing variability (Burdon, 1987). This will reduce the effective evaluation and utilization of disease resistances from wild germplasm. Provided methodological problems are addressed, race specific resistance can be evaluated in wild germplasm under controlled conditions. Burdon (1987) has stated, however, that race non-specific resistance is difficult to evaluate in wild germplasm populations, and consequently little is known about its frequency of occurrence or genetic control. Therefore, a proportion of the genetic defence mechanisms present in wild germplasm of potential use in crop improvement is not being utilized at present.

Conclusions

The spatial and temporal complexity of many diverse ecosystems which contain useful plants and the potential for multiple interactions between pathogens and hosts as well as environmental factors necessitates the development of ecological approaches to evaluation of pathogens in such ecosystems. It is clear that the evaluation of pathogens in complex ecosystems has been neglected due to the immense difficulties of characterizing the multitude of components and interactions occurring in such systems. Nevertheless, complex ecosystems commonly contain plants of high utilitarian value and are the major source of food for subsistence farming communities in the tropics. They clearly merit more attention from plant pathologists and ecologists. Research on intercropping and varietal mixtures has emphasized the value of such systems in providing protection from pathogens, indicated their potential for the development of widely applicable and stable strategies for crop protection, but paradoxically uncovered the lack of understanding of the bases of their effects on diseases. Techniques for evaluating interacting species populations should be widely applied to studies of host–pathogen associations in complex ecosystems. Existing methods generated from research on pathogens in intercropping systems, varietal mixtures and wild populations in temperate regions can be used and modified as necessary to evaluate pathogens in complex ecosystems in the tropics. The Natural Resources Institute is presently developing a research programme in this field, including evaluation of diseases of useful plants in subsistence farming systems and wild populations of useful plants.

References

Alexander, H.M. (1989) Spatial heterogeneity and disease in natural populations. In: Jeger, M.J. (ed.), *Spatial Components of Plant Disease Epidemics*. Prentice Hall, Englewood Cliffs, New Jersey, pp. 144–164.

Allen, D.J. (1983) *The Pathology of Tropical Food Legumes*. John Wiley and Sons, Chichester.

Allen, D.J. (1990) The influence of intercropping with cereals on disease development in legumes. In: Waddington, S.R., Palmer, A.F.E. and Edje, O.T. (eds), *Workshop on Research Methods for Cereal/Legume Intercropping in Eastern and Southern Africa*. CIAT, CIMMYT and Government of Malawi, CIMMYT Eastern and Southern Africa On-farm Research Network Report No. 17, pp. 62–67.

Allen, D.J., Dessert, M., Trutmann, P. and Voss, J. (1989) Common beans in Africa and their constraints. In: Schwartz, H. and Pastor-Corrales, M.A. (eds), *Bean Production Problems in the Tropics*. CIAT, Cali, Colombia, pp. 9–31.

Anderson, W.K., Parkin, R.J. and Dovey, M.D. (1982) Relations between stocking rate, environment and scorch disease on grazed subterranean clover pastures in Western Australia. *Australian Journal of Experimental Agriculture and Animal Husbandry* 22, 182–189.

Anikster, Y. and Noy-Meir, I. (1991) The wild-wheat field laboratory at Ammiad. *Israel Journal of Botany* 40, 351–362.

Anon. (1991) Population dynamics of the wheat progenitor, *Triticum turgidum* var. *dicoccoides*, in a natural habitat in Eastern Galilee. *Israel Journal of Botany* 40, Parts 5–6, pp. 349–536.

Ayeh, E. (1985) Seed mixtures of *Phaseolus vulgaris* in Malawi. *Bunda College Agricultural Journal* 1, 28–37.

Brown, A.H.D. (1991) Population divergence in wild crop relatives. *Israel Journal of Botany* 40, 512 (Abstract).

Browning, J.A. (1974) Relevance of knowledge about natural ecosystems to development of pest management programs for agro-ecosystems. *Proceedings of the American Phytopathological Society* 1, 191–199.

Browning, J.A. (1979) Genetic protective mechanisms of plant-pathogen populations: coevolution and use in breeding for resistance. In: Harris, M.K. (ed.), *Biology and Breeding for Resistance to Arthropods and Pathogens in Agricultural Plants*. College Station, Texas A & M University, Texas, USA, pp. 52–75.

Browning, J.A. (1988) Current thinking on the use of diversity to buffer small grains against highly epidemic and variable foliar pathogens: problems and future prospects. In: Simmonds, N.W. and Rajaram, S. (eds), *Breeding Strategies for Resistance to Rusts of Wheat*. CIMMYT, Mexico, DF, pp. 76–90.

Burdon, J.J. (1987) *Diseases and Plant Population Biology*. Cambridge University Press, Cambridge.

Burdon, J.J. and Chilvers, G.A. (1982) Host density as a factor in plant disease ecology. *Annual Review of Phytopathology* 20, 143–166.

Burdon, J.J. and Jarosz, A.M. (1989) Wild relatives as sources of disease resistance. In: Brown, A.H.D., Frankel, O.H., Marshall, D.R. and Williams, J.T. (eds), *The Use of Plant Genetic Resources*. Cambridge University Press, Cambridge, pp. 280–296.

Burdon, J.J. and Leather, S.R. (eds) (1990) *Pests, Pathogens and Plant Communities.* Blackwell Scientific Publications, Oxford.

Burdon, J.J. and Marshall, D.R. (1981) Inter- and intra-specific diversity in the disease response of *Glycine* species to the leaf-rust fungus *Phakopsora pachyrhizi. Journal of Ecology* 69, 381–390.

Burdon, J.J., Brown, A.D.H. and Jarosz, A. (1990) The spatial scale of genetic interactions in host–pathogen coevolved systems. In: Burdon, J.J. and Leather, S.R. (eds), *Pests, Pathogens and Plant Communities.* Blackwell Scientific Publications, Oxford, pp. 233–248.

Carr, A.J.H. (1975) Diseases of herbage crops – some problems and progress. *Annals of Applied Biology* 81, 235–279.

Chin, K.M. and Wolfe, M.S. (1984) The spread of *Erysiphe graminis hordei* in mixtures and varieties. *Plant Pathology* 33, 89–100.

Davis, R.D. (1991) Anthracnose (*Colletotrichum gloeosporioides*) development in a *Stylosanthes* spp. based pasture in response to fire and rain. *Tropical Grasslands* 25, 365–370.

Dinoor, A. (1970) Sources of oat crown rust resistance in hexaploid and tetraploid wild oats in Israel. *Canadian Journal of Botany* 48, 153–161.

Dinoor, A. (1977) Oat crown rust resistance in Israel. *Annals of the New York Academy of Sciences* 287, 357–366.

Dinoor, A., Eshed, N., Ecker, R., Gerechter-Amitai, Z., Solel, Z., Manisterski, J. and Anikster, Y. (1991) Fungal diseases of wild tetraploid wheat in a natural stand in northern Israel. *Israel Journal of Botany* 40, 481–500.

Francis, C.A. (1986) Introduction: distribution and importance of multiple cropping. In: Francis, C.A. (ed.), *Multiple Cropping Systems.* Macmillan, New York, pp. 1–19.

Hardman, L.L. and Lamb, E.M. (1988) Results of a survey of bean genotypes and production practices in Rwanda. *Bean Improvement Cooperative Annual Report* 31, 79–80.

Hawkes, J.G. (1991) International workshop on dynamic *in-situ* conservation of wild relatives of major cultivated plants: summary of the final discussion and recommendations. *Israel Journal of Botany* 40, 529–536.

Jarosz, A.M. (1984) *Ecological and Evolutionary Dynamics of Phlox-Erysiphe cichoracearum interactions.* Unpublished PhD thesis, Purdue University.

Jeger, M.J. (ed.) (1989) *Spatial Components of Plant Disease Epidemics.* Prentice Hall, Englewood Cliffs, New Jersey.

Jeger, M.J., Jones, D.G. and Griffiths, E. (1981a) Disease progress of non-specialised fungal pathogens in intraspecific mixed stands of cereal cultivars. I. Models. *Annals of Applied Biology* 98, 187–198.

Jeger, M.J., Jones, D.G. and Griffiths, E. (1981b) Disease progress of non-specialised fungal pathogens in intraspecific mixed stands of cereal cultivars. II. Field experiments. *Annals of Applied Biology* 98, 199–210.

Latch, G.C.M. and Lancashire, J.A. (1970) Importance of some effects of fungal diseases on pasture yield. *Proceedings of the XI International Grassland Congress*, Brisbane, Australia, pp. 688–691.

Lenné, J.M. (1988) Variation in reaction to anthracnose within native *Stylosanthes capitata* populations in Minas Gerais, Brazil. *Phytopathology* 78, 131–134.

Lenné, J.M. (1989) Evaluation of biotic factors affecting grassland production –

history and prospects. *Proceedings of the XVI International Grasslands Congress*, Nice, France, pp. 1811–1815.

Lenné, J.M. and Sonoda, R.M. (1990) Tropical pasture pathology: a pioneering and challenging endeavor. *Plant Disease* 74, 945–951.

Lenné, J.M. and Wood, D.G. (1991) Plant diseases and the use of wild germplasm. *Annual Review of Phytopathology* 29, 35–63.

Lenné, J.M., Olaya, G. and Miles, J.W. (1989) Importance of *Rhizoctonia* foliar blight of the promising tropical pasture legume genus *Centrosema*. *Proceedings of the XVI International Grasslands Congress*, Nice, France, pp. 697–699.

Lenné, J.M., Torres, C. and Garcia, C.A. (1990) Effect of wart disease (*Synchytrium desmodii*) on survival and yield of the tropical pasture legume *Desmodium ovalifolium*. *Plant Disease* 74, 676–679.

Manisterski, J., Anikster, Y., Brodny, U. and Wahl, I. (1991) The fertile crescent as a source of genes for disease resistance. *Israel Journal of Botany* 40, 515 (Abstract).

Marshall, D.R. (1989) Modelling the effects of multiline varieties on the population genetics of plant pathogens. In: Leonard, K.J. and Fry, W.E. (eds), *Plant Disease Epidemiology*, Vol. 2. McGraw-Hill, New York, pp. 284–317.

Martin, G.B. and Adams, M.W. (1987a) Landraces of *Phaseolus vulgaris* (Fabaceae) in Northern Malawi. 1. Regional variation. *Economic Botany* 41, 190–203.

Martin, G.B. and Adams, M.W. (1987b) Landraces of *Phaseolus vulgaris* (Fabaceae) in Northern Malawi. II. Generation and maintenance of variability. *Economic Botany* 41, 204–215.

Miles, J.W. and Lenné, J.M. (1984) Genetic variation within a natural *Stylosanthes guianensis*, *Colletotrichum gloeosporioides* host–pathogen population. *Australian Journal of Agricultural Research* 35, 211–218.

Mundt, C.C. (1989) Modelling disease increase in host mixtures. In: Leonard, K.J. and Fry, W.E. (eds), *Plant Disease Epidemiology*, Vol. 2. McGraw-Hill, New York, pp. 150–184.

Nair, P.K.R. (1989) *Agroforestry Systems in the Tropics*. Kluwer Academic Publishers, Dordrecht, Netherlands.

Namkoong, G. (1991) Dynamics of *in-situ* conservation: can fragmentation be useful? *Israel Journal of Botany* 40, 518 (Abstract).

O'Rourke, C.J. (1976) *Diseases of Grasses and Forage Legumes in Ireland*. An Foras Taluntais, Carlow, Ireland.

Pyndji, M.M. and Trutmann, P. (1992) Managing angular leaf spot on common bean in Africa supplementing farmer mixtures with resistant varieties. *Plant Disease* 76, 1144–1147.

Spence, N.J. and Walkey, D.G.A. (1991) The occurrence of bean common mosaic virus in legume weed species and other non-*Phaseolus* hosts in Africa. *Bean Improvement Cooperative Annual Report* 34, 7–8.

Teverson, D.M. (1991) *Genetics of Pathogenicity and Resistance in the Haloblight Disease of Beans in Africa*. Unpublished PhD Thesis, University of Birmingham, UK.

Teverson, D.M., Allen, D.J. and Massomo, S.M. (1993) *Neonotonia wightii* as an alternative host for halo-blight of beans in Africa: assessment of disease spread under semi-natural conditions. *Bean Improvement Cooperative Annual Report* 36, 148–149.

Thomas, M.R. (1985) Assessment of disease in grassland. In: Brockman, J.S. (ed.), *Weeds, Pests and Diseases of Grassland and Herbage Legumes.* British Crop Protection Council, University of Nottingham, UK, pp. 188–194.

Thresh, J.M. (ed.) (1981) *Pests, Pathogens and Vegetation.* Pitman, London.

Thurston, H.D. (1992) *Sustainable Practices for Plant Disease Management in Traditional Farming Systems.* Westview Press, Boulder.

Toledo, J.M., Lenné, J.M. and Schultze-Kraft, R. (1989) Effective utilization of tropical pasture germplasm. In: *FAO Plant Production and Protection Paper No. 94, Utilization of Genetic Resources: Suitable Approaches, Agronomical Evaluation and Use*, FAO, Rome, pp. 27–57.

Voss, J. (1992) Conserving and increasing on-farm genetic diversity: farmer management of varietal bean mixtures in Central Africa. In: Rhoades, R.E. and Moock, J.L. (eds), *Diversity, Farmer Knowledge, and Sustainability.* Cornell University Press, Ithaca, New York, pp. 34–51.

Wahl, I. (1970) Prevalence and geographic distribution of resistance to crown rust in *Avena sterilis. Phytopathology* 60, 746–749.

Watkin, B.R. and Clements, R.J. (1978) The effects of grazing animals on pastures. In: Wilson, J.R. (ed.), *Plant Relations in Pastures.* CSIRO, Australia, pp. 273–289.

Wolfe, M.S. (1985) The current status of prospects of multiline cultivars and variety mixtures for disease resistance. *Annual Review of Phytopathology* 23, 251–273.

Wolfe, M.S. (1989) The use of variety mixtures to control diseases and stabilise yield. In: Simmonds, N.W. and Rajaram, S. (eds), *Breeding Strategies for Resistance to Rusts of Wheat.* CIMMYT, Mexico, DF, pp. 91–118.

Wolfe, M.S. and Barrett, J.A. (1980) Can we lead the pathogen astray? *Plant Disease* 64, 148–155.

Wood, D. and Lenné, J.M. (1993) Dynamic management of domesticated biodiversity by farming communities. In: *Proceedings of the 'Norway/UNEP Expert Conference on Biodiversity'*, May 24–28, 1993, Trondheim, Norway, pp. 84–98.

5 Humidity and Fungal Diseases of Plants – Problems

J.G. Harrison, R. Lowe and N.A. Williams

Scottish Crop Research Institute, Invergowrie, Dundee DD2 5DA, UK.

Importance of Humidity

Atmospheric humidity is an important environmental element in the ecology of microorganisms of aerial plant parts. Indeed, humidity is often the major determinant of the development of pathogenic fungi and therefore also of the diseases they cause. A particular stage in the life cycle of a pathogen may have a specific humidity requirement that has to be met before the disease can be initiated or continue to develop. The diversity of diseases affected by humidity, in terms of both types of pathogen and host species, together with the stage of the life cycle affected, are illustrated in Table 5.1, which represents only a small proportion of the many diseases directly affected by humidity. As is apparent from this table, the predominant effects of humidity are on spore formation, survival, germination and infection. Colonization of host tissues by an established pathogen is usually insensitive to, or only slightly affected by the humidity of air outside a plant, because air in the intercellular spaces of an unwilted plant remains at or very close to saturation, irrespective of atmospheric humidity (Rose, 1966).

The humidity of the air surrounding a plant, together with other factors such as air speed and radiation, determines the rate of drying of wet plant surfaces. The progress of many diseases depends on free surface water, so humidity, as a major factor determining the duration of surface moisture, has an indirect effect on these diseases. There is a large volume of literature describing effects of surface water on disease development (Huber and Gillespie, 1992), possibly reflecting the relative ease with which it can be

Table 5.1. Some diseases directly affected by atmospheric humidity.

Disease	Pathogen	Stage(s) affected	Reference(s)
Flax rust	*Melampsora lini*	Survival of uredospores	Hart (1926)
Powdery mildew of rose	*Sphaerotheca pannosa*	Germination of conidia	Longrée (1939)
Brown spot of rice	*Cochliobolus miyabeanus*	Survival of conidia	Page *et al.* (1947)
Grey mould of lettuce, strawberry etc.	*Botrytis cinerea*	Germination of conidia	Snow (1949)
Blossom blight of plum, peach etc.	*Sclerotinia fructicola*	Germination of conidia	Weaver (1950)
Oak wilt	*Endoconidiophora fagacearum*	Survival of conidia and ascospores	Merek and Fergus (1954)
Rot of peach, strawberry etc.	*Rhizopus stolonifer*	General disease development	Smith and McClure (1960)
Tobacco blue mould	*Peronospora tabacina*	Formation of conidia	Rider *et al.* (1961)
Rust of snap bean	*Uromyces phaseoli*	Survival of uredospores Germination of uredospores	Schein and Rotem (1965) Imhoff *et al.* (1981)
Septoria leaf spot of wheat	*Mycosphaerella graminicola*	Infection	Holmes and Colhoun (1974)
Barley mildew	*Erysiphe graminis*	Formation of conidia Germination of conidia and infection	Ward and Manners (1974)
Late blight of potato	*Phytophthora infestans*	Survival of sporangia Formation of sporangia	Warren and Colhoun (1975) Harrison and Lowe (1989)
Leaf blight of carrots	*Alternaria dauci*	Formation of conidia	Strandberg (1977)
Chocolate spot of field bean	*Botrytis fabae*	Formation and survival of conidia, infection and lesion expansion	Harrison (1980a, 1983a, 1984a,b)
Peach leaf curl	*Taphrina deformans*	Infection	Gautier (1986)
Net blotch of barley	*Pyrenophora teres*	Formation of conidia	Shaw (1986)
Flyspeck of apple fruits	*Zygophiala jamaicensis*	Germination of conidia and ascospores	Ocamb-Basu and Sutton (1988)
Purple blotch of onion	*Alternaria porri*	Formation of conidia	Everts and Lacy (1990)

Table 5.2. Some diseases indirectly affected by atmospheric humidity.

Disease	Pathogen	Stage(s) affected	Reference(s)
Apple scab	*Venturia inaequalis*	Infection	Mills and Laplante (1951)
Wheat glume blotch	*Leptosphaeria nodorum*	Infection	Jeger *et al.* (1985)
Stripe rust of wheat	*Puccinia striiformis*	Infection	Dennis (1987)
Stem canker of oilseed rape	*Leptosphaeria maculans*	Germination of pycnidiospores	Vanniasingham and Gilligan (1988)
		Formation of pycnidia	Vanniasingham and Gilligan (1989)
Shot-hole of almond	*Wilsonomyces carpophilus*	Infection	Shaw *et al.* (1990)
Groundnut rust	*Puccinia arachidis*	Infection	Butler and Jadhav (1991)
Powdery mildew of sweet cherry	*Podosphaera clandestina*	Ascospore release and germination of ascospores	Grove (1991)
Sprinkler rot of pear	*Phytophthora cactorum*	Infection	Grove and Boal (1991)
Vascular-streak dieback of cocoa	*Oncobasidium theobromae*	Basidiospore release	Dennis *et al.* (1992)

controlled and its duration quantified. A few of the many diseases affected by liquid water on aerial plant parts, and therefore indirectly affected by humidity, are shown in Table 5.2. However, this chapter is primarily concerned with effects on diseases of humidity *per se* and surface water will be mentioned later only briefly.

Effects of Humidity on Chocolate Spot of Bean and Potato Foliage Blight

Let us consider two diseases, chocolate spot (*Botrytis fabae*) of field or broad bean and late blight (*Phytophthora infestans*) of potato, to illustrate the importance of humidity as a factor controlling pathogen development and consequently the progress of an epidemic. Although the causal organisms of these diseases are widely different taxonomically, effects of humidity on their life cycles have many similarities and are largely typical of many other diseases. High humidities are needed for the formation of conidia of *B. fabae* on crop debris or on other plants to initiate disease in a crop (Jauch, 1947; Harrison, 1979, 1984b) and for the formation of sporangia of *P. infestans* on shoots from infected tubers or other potato plants (Crosier and Reddick, 1935; Bonde and Schultz, 1943; Harrison and Lowe, 1989). The survival of conidia (Harrison, 1983a) and sporangia (Murphy, 1922; Crosier, 1934; Warren and Colhoun, 1975) is better in humid than in dry air. Infection of bean leaves by conidia requires high humidities, but not liquid water on leaf surfaces (Harrison, 1984a). In contrast, potato leaves do not become infected by *P. infestans* unless they are wet. Infective zoospores are released from the sporangia only in liquid water and they die rapidly without free moisture (Crosier and Reddick, 1935; Rotem *et al.*, 1971; Hartill *et al.*, 1990), so there is an indirect effect of humidity on infection. Unlike most other diseases, increase in size of chocolate spot lesions is highly dependent on humidity. Lesions remain small below about 66% *RH*, but their rate of increase in diameter is directly proportional to relative humidities above about 70% (Harrison, 1980a). The mechanisms of this response to humidity were elucidated by Harrison (1980b, 1983b) and involve differences in the mobility and concentration of phytotoxic compounds originating from infected tissues and the desiccation of infected and surrounding uninfected tissues killed by them, processes that are affected by humidity. Deprived of water essential for growth, *B. fabae* becomes latent until high humidities allow it to continue to invade the leaf. In contrast, colonization of potato leaf tissues by *P. infestans* is not affected by humidity between 80 and 100% *RH*, the range tested by Harrison and Lowe (1989) and has not been studied at lower humidities. Sporulation of both pathogens on lesions of plants within a crop, resulting in inoculum build-up, requires similarly high humidities as those needed to produce spores that initiated the disease.

Diseases Favoured by Low Humidities

Most diseases are encouraged by conditions of high humidity but there are exceptions to this generalization. The authors have observed in Scotland that rust of field bean, caused by *Uromyces fabae*, is most severe during dry summers. Similarly, Tarr (1972) reported that powdery mildew of rose (*Sphaerotheca pannosa*) seems to be favoured by dry weather, although high humidities are needed for germination of conidia (Longrée, 1939). Tarr thought that dry, warm days favoured spore dispersal while sporulation and infection were favoured by dew that formed during the colder nights.

Spores generally require high humidities or even liquid water for germination but the conidia of some powdery mildews have been reported to germinate in air containing very little or even no water vapour (Yarwood, 1936; Delp, 1954). Manners and Hossain (1963) found that, although the germination of conidia of *Erysiphe graminis* was best in saturated air, some spores germinated in completely dry air. The water content of conidia of *E. graminis* depends on the humidity during spore production (Somers and Horsfall, 1966); those formed in moist air have a high water content with conspicuous vacuoles in which the water is stored (Yarwood, 1957), while those formed in dry air have a lower water content. Only conidia with water vacuoles germinate in dry conditions, germination being dependent on water stored within the spore (Carver and Bushnell, 1983). In effect, these spores carry their own water supply for germinating at low humidities, but appear to depend on the germ tube forming an early link with the host as a source of water for continued germling survival.

Physics of Humidity and the Importance of Temperature

An understanding of the physical principles of water vapour in the atmosphere is essential in order to appreciate the difficulties of controlling and measuring humidity, especially near plant surfaces. Much of the following is based on information presented by Lowry (1969), Woodward and Sheehy (1983) and Rundel and Jarrell (1989) to which the reader is referred for more comprehensive accounts of the physics of humidity. Temperature determines the maximum quantity of water vapour that air can

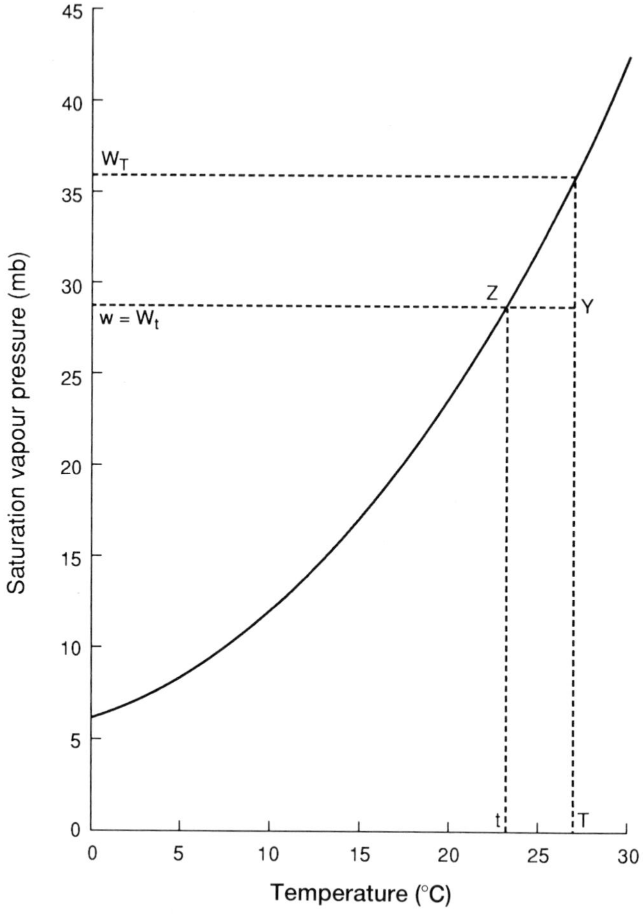

Fig. 5.1. The relation between saturation water vapour pressure and temperature. (Graph drawn from data in hygrometric tables.)

hold (the saturation water vapour pressure), which increases with increasing temperature (Fig. 5.1). At any given temperature, T, relative humidity (RH) is defined as the water vapour content of air, or water vapour pressure, w, expressed as a percentage of the saturation water vapour pressure, W_T, for temperature T:

$$RH = \frac{w}{W_T} \times 100\% \tag{5.1}$$

Air at point Y in Fig. 5.1 is at temperature T and its water vapour pressure is w, 80% of the saturation water vapour pressure, W_T, so the *RH* of this air is 80%. As this same air is cooled and its temperature falls from T towards t, its water vapour pressure, or absolute humidity, remains constant, but its relative humidity increases, reaching 100%, or saturation, at temperature t, i.e. at point Z on the curve, where $w = W_t$. Any further reduction in temperature will result in the removal from the air of excess water vapour as condensation, but in this case as the absolute humidity decreases, the relative humidity remains constant at 100%. Growth and development of pathogenic fungi are determined by relative humidity, not by absolute humidity. Most pathogens are highly sensitive to humidity between 90 and 100% *RH*, but particularly between 98 and 100% *RH*. Figure 5.1 shows the importance of accurate temperature control for the precise control of humidity, especially at humidities approaching saturation. If, for example, in an experiment the target *RH* is 95% at 15.0°C, then the temperature has to fall only 0.8°C for the *RH* to increase to 100%, with the added complication of condensation and the wetting of plant surfaces. The difference between 95 and 100% *RH*, even in the absence of condensation, can have a profound effect on the development of a pathogen (Longrée, 1939; Sirry, 1957; Manners and Hossain, 1963; Harrison, 1984a). The literature abounds with illustrations of a lack of appreciation of the close dependence of relative humidity on temperature and of the importance of the precise control of temperature for good humidity control. Two examples will be considered. In the first, air was saturated with water vapour at 19°C to produce the target of 100% *RH*. However, to avoid condensation the air was then heated to 20°C before it passed through a chamber, also at 20°C, containing test plant material. It was claimed that the air within the test chamber remained at 100% *RH*. Heating the air saturated at 19°C by 1°C had reduced its *RH* to 94%. In the second example, there was an attempt to maintain, among other treatments, a *RH* of 86% at 5°C. It was stated that 'temperature control was better than $T \pm 2$°C'. Temperatures fluctuating from 3 to 7°C would have produced humidities fluctuating between 75 and 99% *RH* – hardly 'controlled relative humidity' as was claimed. The importance of accurate temperature control for good control of humidity cannot be over-emphasized.

Effects of Transpiration and Air Flow on Humidity Close to Foliage

Even with good temperature control, transpiration creates major difficulties in attempting to regulate humidity near aerial parts of plants. Foliage continuously loses water vapour unless the air is saturated, increasing the humidity, particularly of air close to plant surfaces. A pathogen is affected by the humidity of air immediately adjacent to it, not to humidity some distance away. Most pathogens on a leaf, even if sporulating, remain close to the surface, so their development is affected by the humidity within a thin layer of air surrounding a leaf. Water vapour diffuses from saturated, or near-saturated, air in the intercellular spaces of a leaf, through the stomata to the air close to the leaf surface, along a concentration gradient according to Fick's Law of Diffusion:

$$E = -D\frac{dx}{dz} \tag{5.2}$$

where E is the flux of water vapour, D is the molecular diffusion coefficient of water vapour in air and dx/dz is the concentration gradient of water vapour in direction z. Water vapour continues to move away from the leaf by diffusion through the air of the laminar boundary layer, where air flow is parallel to the leaf surface. Further from the leaf, air flow becomes turbulent, with eddying, which mixes the air and moves the water vapour away from the leaf much more rapidly than does diffusion alone. Detailed accounts of the physics of water vapour movement from a leaf and effects of different patterns of air flow are given by Rose (1966), Grace (1977) and Woodward and Sheehy (1983). The net result is that, except in saturated air, there is a gradient of decreasing humidity away from any point on the surface of a leaf. The gradient becomes less steep with increasing distance from the leaf but its form depends on various factors, the two most important being wind speed or air speed, and the humidity of the ambient air, i.e. the air remote from the leaf. Figure 5.2 shows hypothetical effects of different air speeds and ambient

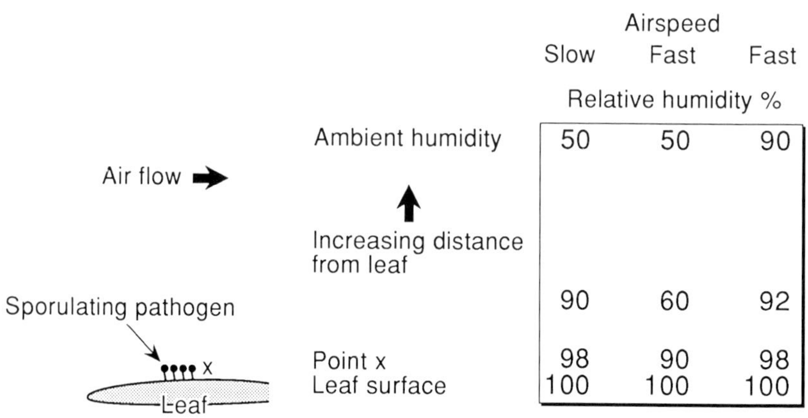

Fig. 5.2. Effects of ambient humidity and air speed on the humidity close to the surface of a leaf (hypothetical relative humidity values).

humidities on humidity gradients away from a leaf and on the humidity at an arbitrary point, x, just above a leaf surface. The humidity between point x, or a similar point close to a leaf, and the leaf surface would be an important factor in determining spore germination, hyphal growth or sporulation of many plant pathogenic fungi. Transpiration ensures that the humidity of air adjacent to the surface of a leaf with open stomata is at, or close to, saturation, irrespective of air speed or ambient humidity. The effect of air speed is based on an increasingly efficient removal of transpired water vapour as the speed increases. In still air, transpired water vapour accumulates around a leaf, humidifying the air, but as the air speed increases, so the water vapour is swept away more rapidly. The relative humidities in the following examples are, of course, completely arbitrary. The first column in Fig. 5.2 shows that, although the ambient air is quite dry (50% *RH*), the slow flow of air over the leaf allows transpired water vapour to accumulate so that the relative humidity at point x is 98%. In the second column, with the same ambient humidity, the faster-moving air removes the transpired vapour more efficiently, producing a steeper gradient of decreasing humidity away from the leaf, so that the relative humidity at point x is 90%. In the third column, the high ambient humidity of 90% *RH* ensures that the humidity gradient from the leaf is less steep, despite the fast air flow, resulting in a relative humidity of 98% at point x.

How does this hypothetical consideration of humidity close to plant surfaces translate into pathogen development in practice? Results from a series of experiments investigating effects of possible interactions between ambient humidity and air speed on sporulation of *P. infestans* on detached potato leaflets are presented in Table 5.3 (Harrison and Lowe, 1989).

Table 5.3. Numbers of sporangia of *P. infestans* produced on detached potato leaflets of cv. Bintje incubated in air at different humidities and speeds.

	Mean number of sporangia per leaflet at air speed of (mm s^{-1})			
Ambient *RH* (%)	0.3	1.4	5.5	13.7
100	66,927	95,804	90,982	139,531
95	83,117	98,221	7,741	4,887
90	76,360	17,147	179	292
85	17,542	4,384	93	417
80	7,857	69	185	1,389

Dotted lines show separation of numbers of sporangia into three distinct groups.

Infection was established using a standardized technique before leaflets were incubated at 15°C in darkness in streams of air at controlled humidities. Leaflets were washed after 10 days and sporangia in suspension were counted. The mean numbers of spores produced per leaflet formed three distinct groups. Slow-moving humid air allowed the pathogen to sporulate profusely, but few sporangia were produced in faster-moving drier air. Intermediate levels of sporulation occurred in relatively slow-moving dry air, or in faster-moving more humid air. Colonization of the tissues was unaffected by these treatments so the results reflect an effect on sporulation *per se*. Further experiments using whole plants, but with fewer treatments, gave an essentially similar outcome. These results support the hypothesis that ambient humidity and air speed interact to determine humidity conditions close to the surface of foliage.

Other Factors Affecting Humidity Close to Foliage

Leaf geometry, including surface topography, as well as effects of aerial hyphae themselves, can all affect air flow close to a leaf and hence also the humidity just above the surface. Transpiration itself may be affected by leaf surface chemistry or by the pathogen blocking stomata. In the field, wind speed fluctuates constantly (Grace, 1977; Woodward and Sheehy, 1983) and leaves sometimes flap (Grace, 1978), leading to problems in defining air flow conditions (Grace, 1985). Furthermore, sunlight falling on a plant heats the tissues, by as much as 20°C above ambient air temperature (Askenasy, 1875; Ansari and Loomis, 1959), which in turn heats the air adjacent to them, lowering the relative humidity. Loss of water by transpiration leads to evaporative cooling of leaves, although the magnitude of this effect is usually relatively minor (Ansari and Loomis, 1959). Thus, a multitude of factors interact to affect humidity close to the surface of a plant.

Humidity Control

How do plant pathologists control humidity? Various systems have been devised based on several different physical principles. They vary widely in their degree of sophistication and accuracy of control and while one system may be suitable for a particular investigation, it may be totally inappropriate for another. For example, spore survival studies are often carried out in the absence of higher plants, perhaps in an attempt to mimic humidity conditions surrounding an airborne spore or one that has alighted on an inert surface. Under these conditions the humidifying effects of transpiration can be ignored and a relatively simple humidity control system used. If, however, pathogen behaviour on or within a plant is being investigated, a humidity control system that removes transpired water vapour should be considered. It should be borne in mind that no system exists, nor ever will, that removes transpired water with an efficiency of 100%. Transpiration always humidifies the air around foliage to a greater or lesser extent. Systems for controlling humidity that are popular, probably because of their simplicity, depend on an equilibration of water potential between an aqueous solution and air within a sealed chamber. Saturated salt solutions are commonly used (e.g. Ward and Manners, 1974; Harrison, 1983a; Shaw, 1986; Vanniasingham and Gilligan, 1989; Everts and Lacy, 1990), the equilibrium humidity being readily obtained from tables (Wexler and Hasegawa, 1954; Winston and Bates, 1960). These solutions have the potential to absorb excess water vapour if an excess of solute is present, but care has to be taken to avoid stratification since a layer of unsaturated solution tends to accumulate at the surface as water vapour is absorbed. The use of an aqueous slurry of the salt helps to eliminate this potential problem. Harrison (1980a) circulated air by bubbling it through saturated salt solutions within sealed systems containing infected leaves to remove most transpired water vapour. Unsaturated solutions of salts, sometimes incorporated into agar, have also been used to control humidity often close to saturation (van den Berg and Lentz, 1968; Harris *et al.*, 1970; Ocamb-Basu and Sutton, 1988). Equilibrium humidities in enclosed systems are calculated from tables of osmotic coefficients (Robinson and Stokes, 1955). Alternatively, sucrose solutions can be used (Clayton, 1942; Delp, 1954). Glycerol/water mixtures offer the potential advantage over salt or sucrose solutions of maintaining very low as well as higher humidities. Air in equilibrium can be kept at any humidity from 0% RH, with glycerol only, to 100%, with water only (Rogers, 1966; Tuite, 1969). Strandberg (1977) used glycerol/water mixtures to control humidities in his studies on sporulation of *Alternaria dauci*. A wide range of equilibrium humidities are also available with different concentrations of H_2SO_4 as the liquid phase (Stevens, 1916), but care has to be taken to avoid contact between acid and organic matter. Paul (1929), Delp (1954) and Merek and Fergus (1954) are among many pathologists who have controlled humidity with H_2SO_4. A serious drawback of the use of unsaturated salt or sucrose solutions, glycerol/water mixtures or H_2SO_4 for humidity control is that any water absorbed from the air or lost from the liquid alters the concentration of the liquid phase itself, which in turn affects the equilibrium humidity. Rogers (1966) recognized this problem and renewed his glycerol/water mixtures every 48 h. Such systems are effectively useless for controlling humidity in the presence of transpiration. Day (1985) advocated the control of humidity with pairs of salt hydrates, which also form an equilibrium with water vapour, claiming that by using solid chemicals, a clean and portable system could be devised, but this suggestion seems to have found little favour with plant pathologists. Carver and Bushnell (1983) and Carver and Adaigbe (1990) dried air with $CaCl_2$ and silica gel respectively to produce the arid conditions they required for their work on germination of conidia of *E. graminis*. Harris and Manners (1983) also used silica gel to dry air which they then mixed in various proportions with air saturated by bubbling it through water, to produce the target humidities of air feeding their plant chamber. Such a system is capable of achieving good control of humidity of air entering the chamber, providing that air flow rates are not too high for efficient drying and humidification. Its accuracy of control also relies on expensive valves and meters to regulate air flow accurately. Humidity within many growth cabinets is controlled by a mist spray activated by a signal from a humidistat, a system used by Weaver (1950) over 40 years ago investigating blossom blight of stone fruits. Even now the accuracy of most humidistats leaves much to be desired and this is reflected in the generally poor humidity control of most cabinets. Delp (1954) controlled humidity in a much more sophisticated way; air was

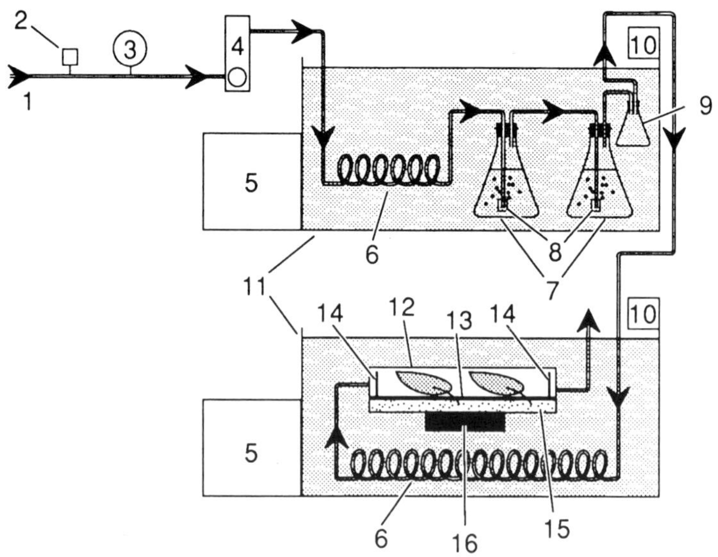

Fig. 5.3. Apparatus used by Harrison and Lowe (1989) to control humidity, air speed and temperature. 1. Direction of air flow. Air from compressed air line; 2. Adjustable pressure relief valve; 3. Pressure gauge; 4. Adjustable flow meter; 5. Refrigerated dip cooler; 6. Copper tubing; 7. Flasks containing distilled water; 8. Aquarium diffuser blocks; 9. Water trap; 10. Heater, stirrer and adjustable thermostat; 11. Water baths; 12. Perspex box with three inlet and three outlet ports. Lid held with bolts and sealed with rubber gasket; 13. Removable Perspex sheet to support leaves. Petioles pass through holes; 14. Six baffle plates, each next to a port; 15. Distilled water; 16. Lead weight.

humidified by steam injection, cooled to the dew point that would give the correct vapour pressure for the target relative humidity, then heated to the target temperature. Efficient humidification, together with both efficient and accurate cooling and re-heating would have given good humidity control before the air was passed over leaves. Harrison and Lowe (1989) used the same principles as Delp to control humidity, but they humidified the air by bubbling it through water. The apparatus they used to control the humidity, flow rate and temperature of air entering chambers containing detached leaflets is shown in Fig. 5.3 and was used to obtain the results presented in Table 5.3. They have also used a similar, but larger, system for conditioning air before passing it over whole potato plants (Harrison and Lowe, 1989). Consideration was given to the fall in pressure as air flowed from just above the water in the flasks to the leaf or plant chambers, and the resulting expansion with a concomitant fall in relative humidity. The temperature of the bath with the flasks was set higher than was indicated from tables of hygrometric data (Anon., 1965) to compensate for this phenomenon that varied in magnitude with the flow rate. The precise control of the temperature of water within the baths largely determines the accuracy of humidity control using this type of apparatus. With the availability of relatively inexpensive modern thermostirrers that control water temperatures within narrow limits, Harrison and Lowe's system offers a good compromise between precision of humidity control and cost. Simpler and less accurate methods that plant pathologists have used for controlling humidity include altering the ventilation within a wet muslin cage containing test plants (Sirry, 1957) and placing trays of water within a plant chamber (Shearer and Zadoks, 1972). Hervey and Horsfall (1931) described a device for regulating humidity using an electric motor to drive wire mesh discs that dipped into water in a tray to throw droplets into the air. A fan blew the air into a plant compartment. A carburettor float chamber controlled the water level and hence the amount splashed, which determined humidity. The authors made the highly dubious claim that humidity control was accurate to 2% *RH*. These unsophisticated systems cannot be used for investigating effects of small humidity differences on pathogen or disease development, but may be useful, for example, in comparing effects of 'dry' and 'moist' air to demonstrate whether or not humidity influences a particular process.

Humidity Measurement

Some of the many different ways of determining humidity are listed in Table 5.4. A detailed account of the physics involved would be inappropriate in this chapter and has been dealt with comprehensively by Day (1985) and Rundel

Table 5.4. Humidity measurement.

Method/instrument	Principle	Advantages	Disadvantages
Dew point meter	Temperature determined at which condensation starts to form as mirror cooled	Accurate	Big
Psychrometry	Rate of evaporative cooling, indicated by temperature difference between wet and dry 'bulb' thermometers, proportional to humidity	Accurate	Big; forced ventilation disturbs air
Infra-red gas analysis	Absorption of specific wavebands of electromagnetic radiation depends on water vapour content of air	Accurate	Big; disturbs air; expensive
Gravimetric or volumetric	Desiccant used to remove water vapour from air	Extremely accurate	Requires large volume of air; expensive
Electrical capacitance	Capacitance depends on humidity	Cheap	Inaccurate, particularly at high humidities; slow response; requires frequent recalibration
Electrical resistance	Equilibrium moisture content of salt depends on humidity and determines resistance	Cheap	Inaccurate; slow response time; suffers from hysteresis
Hair hygrometer	Length of human hair depends on moisture content that is determined by humidity	Cheap and robust	Big; inaccurate, particularly at high humidities; slow response time; requires frequent recalibration; suffers from hysteresis
Ultra-violet spectroscopy	Air irradiated at 122 nm. Intensity of fluorescence at 310 nm from water molecules depends on water vapour content of air	Accurate; remote sensing of humidity possible	Expensive; requires careful calibration
Weight of paper	Equilibrium moisture content, and therefore weight of paper depends on humidity	Cheap and robust	Inaccurate; very slow response time; requires large volume of air
Cobalt salt papers	Colour of salt depends on moisture content in equilibrium with humidity	Cheap and robust	Inaccurate; very slow response time; requires large volume of air

and Jarrell (1989). The best instruments for use in plant pathology are dew point meters, psychrometers and infra-red gas analysers. Modern dew point meters incorporate a small mirror that is alternately heated and cooled electronically to maintain it close to the temperature at which condensation forms, which is detected optically. The relative humidity can be found from tables if the dew point and air temperature are known. Psychrometers incorporate a wet and a dry bulb thermometer. The rate of evaporative cooling, indicated by the difference in readings between the two thermometers, is proportional to the vapour pressure deficit. Again, the relative humidity can be found from tables, or it can be calculated. Infra-red gas analysers, incorporating filters to eliminate wavelengths absorbed by both water and CO_2, measure the water vapour content of air after appropriate calibration. The saturation water vapour content at air temperature is readily obtained from tables. A simple calculation gives the relative humidity. Other methods of measuring humidity all have various drawbacks that can affect their accuracy. However, the main problem with all instruments for measuring humidity is that they are too large. No instrument has a sensor small enough to measure the humidity of air adjacent to the surface of a plant, and, as we have seen, the humidity close to plant surfaces is a major determinant of pathogen development. It is also impossible to estimate with any degree of accuracy the humidity next to a plant because many variables are involved that are difficult to quantify. Infra-red gas analysis or ultra-violet spectroscopy perhaps offer the best possibilities of developing an instrument capable of measuring, with a minimum of disturbance, the moisture content of air within this important zone.

Conclusions

This chapter has highlighted serious problems associated with humidity control and measurement close to plants, and consequently the difficulties of relating pathogen and disease development to humidity. These problems have important implications for the mathematical modelling of epidemics and for disease forecasting, both of which should be treated with caution. Furthermore, most work has attempted to relate disease to constant humidity conditions. Humidity in the field is rarely constant for long. We believe that future research effort should place a greater emphasis on effects of changing humidity to reproduce more closely the dynamic environment experienced by crop plants.

References

Anon. (1965) *Hygrometric Data for Air Conditioning Calculations.* The Institution of Heating and Ventilation Engineers, London.

Ansari, A.Q. and Loomis, W.E. (1959) Leaf temperatures. *American Journal of Botany* 46, 713–717.

Askenasy, E. (1875) Ueber die Temperature welche Pflanzen im Sonnenlicht annehmen. *Botanische Zeitung* 33, 441–444.

Bonde, R. and Schultz, E.S. (1943) Potato refuse piles as a factor in the dissemination of late blight. *Maine Agricultural Experiment Station Bulletin* No. 416, pp. 229–246.

Butler, D.R. and Jadhav, D.R. (1991) Requirements of leaf wetness and temperature for infection of groundnut by rust. *Plant Pathology* 40, 395–400.

Carver, T.L.W. and Adaigbe, M.E. (1990) Effects of oat host genotype, leaf age and position and incubation humidity on germination and germling development by *Erysiphe graminis* f.sp. *avenae*. *Mycological Research* 94, 18–26.

Carver, T.L.W. and Bushnell, W.R. (1983) The probable role of primary germ tubes in water uptake before infection by *Erysiphe graminis*. *Physiological Plant Pathology* 23, 229–240.

Clayton, C.N. (1942) The germination of fungous spores in relation to controlled humidity. *Phytopathology* 32, 921–943.

Crosier, W. (1934) Studies in the biology of *Phytophthora infestans* (Mont.) de Bary. *Cornell University Agricultural Experiment Station Memoir* No. 155, pp. 1–40.

Crosier, W. and Reddick, D. (1935) Some ecological relations of the potato and its chief fungus parasite *Phytophthora infestans*. *American Potato Journal* 12, 205–219.

Day, W. (1985) Water vapour measurement and control. In: Marshall, B. and Woodward, F.I. (eds), *Instrumentation for Environmental Physiology.* Cambridge University Press, Cambridge, pp. 59–78.

Delp, C.J. (1954) Effect of temperature and humidity on the grape powdery mildew fungus. *Phytopathology* 44, 615–626.

Dennis, J.I. (1987) Temperature and wet-period conditions for infection by *Puccinia striiformis* f.sp. *tritici* Race 104E 137A+. *Transactions of the British Mycological Society* 88, 119–121.

Dennis, J.J.C., Holderness, M. and Keane, P.J. (1992) Weather patterns associated with sporulation of *Oncobasidium theobromae* on cocoa. *Mycological Research* 96, 31–37.

Everts, K.L. and Lacy, M.L. (1990) The influence of dew duration, relative humidity, and leaf senescence on conidial formation and infection of onion by *Alternaria porri*. *Phytopathology* 80, 1203–1207.

Gautier, M. (1986) La cloque du pêcher. *Revue Horticole* No. 264, 31–32.

Grace, J. (1977) *Plant Response to Wind.* Academic Press, London, 204 pp.

Grace, J. (1978) The turbulent boundary layer over a flapping *Populus* leaf. *Plant, Cell and Environment* 1, 35–38.

Grace, J. (1985) The measurement of wind speed. In: Marshall, B. and Woodward, F.I. (eds), *Instrumentation for Environmental Physiology.* Cambridge University Press, Cambridge, pp. 103–121.

Grove, G.G. (1991) Powdery mildew of sweet cherry: influence of temperature and wetness duration on release and germination of ascospores of *Podosphaera clandestina*. *Phytopathology* 81, 1271–1275.

Grove, G.G. and Boal, R.J. (1991) Influence of temperature and wetness duration on infection of immature apple and pear fruit by *Phytophthora cactorum*. *Phytopathology* 81, 1465–1471.

Harris, J.G. and Manners, J.G. (1983) Influence of relative humidity on germination and disease development in *Erysiphe graminis*. *Transactions of the British Mycological Society* 81, 605–611.

Harris, R.F., Gardner, W.R., Adebayo, A.A. and Sommers, L.E. (1970) Agar dish isopiestic equilibration method for controlling the water potential of solid substrates. *Applied Microbiology* 19, 536–537.

Harrison, J.G. (1979) Overwintering of *Botrytis fabae*. *Transactions of the British Mycological Society* 72, 389–394.

Harrison, J.G. (1980a) Effects of environmental factors on growth of lesions on field bean leaves infected by *Botrytis fabae*. *Annals of Applied Biology* 95, 53–61.

Harrison, J.G. (1980b) The production of toxins by *Botrytis fabae* in relation to growth of lesions on field bean leaves at different humidities. *Annals of Applied Biology* 95, 63–71.

Harrison, J.G. (1983a) Survival of *Botrytis fabae* conidia in air. *Transactions of the British Mycological Society* 80, 263–269.

Harrison, J.G. (1983b) Growth of lesions caused by *Botrytis fabae* on field bean leaves in relation to foliar bacteria, non-enzymic phytotoxins, pectic enzymes and osmotica. *Annals of Botany* 52, 823–838.

Harrison, J.G. (1984a) Effect of humidity on infection of field bean leaves by *Botrytis fabae* and on germination of conidia. *Transactions of the British Mycological Society* 82, 245–248.

Harrison, J.G. (1984b) Effects of environmental factors on sporulation of *Botrytis fabae*. *Transactions of the British Mycological Society* 83, 295–298.

Harrison, J.G. and Lowe, R. (1989) Effects of humidity and wind speed on sporulation of *Phytophthora infestans* on potato leaves. *Plant Pathology* 38, 585–591.

Hart, H. (1926) Factors affecting the development of flax rust, *Melampsora lini* (Pers.) Lev. *Phytopathology* 16, 185–205.

Hartill, W.F.T., Young, K., Allan, D.J. and Henshall, W.R. (1990) Effects of temperature and leaf wetness on the potato late blight. *New Zealand Journal of Crop and Horticultural Science* 18, 181–184.

Hervey, G.E.R. and Horsfall, J.G. (1931) A simple device for humidity regulation. *Science* 73, 617–618.

Holmes, S.J.I. and Colhoun, J. (1974) Infection of wheat by *Septoria nodorum* and *S. tritici* in relation to plant age, air temperature and relative humidity. *Transactions of the British Mycological Society* 63, 329–338.

Huber, L. and Gillespie, T.J. (1992) Modeling leaf wetness in relation to plant disease epidemiology. *Annual Review of Phytopathology* 30, 553–577.

Imhoff, M.W., Main, C.E. and Leonard, K.J. (1981) Effect of temperature, dew period, and age of leaves, spores, and source pustules on germination of bean rust urediospores. *Phytopathology* 71, 577–583.

Jauch, C. (1947) La 'mancha chocolate' de las habas. *Revista de Investigaciones Agricolas* 1, 65–80.

Jeger, M.J., Griffiths, E. and Jones, D.G. (1985) The effects of post-inoculation wet and dry periods, and inoculum concentration, on lesion numbers of *Septoria nodorum* in spring wheat seedlings. *Annals of Applied Biology* 106, 55–63.

Longrée, K. (1939) The effect of temperature and relative humidity on the powdery mildew of roses. *Cornell University Agricultural Experiment Station Memoir* No. 223, pp. 1–43.

Lowry, P.W. (1969) *Weather and Life. An Introduction to Biometeorology.* Academic Press, New York.

Manners, J.G. and Hossain, S.M.M. (1963) Effects of temperature and humidity on conidial germination in *Erysiphe graminis*. *Transactions of the British Mycological Society* 46, 225–234.

Merek, E.L. and Fergus, C.L. (1954) The effect of temperature and relative humidity on the longevity of spores of the oak wilt fungus. *Phytopathology* 44, 61–64.

Mills, W.D. and Laplante, A.A. (1951) Diseases and insects in the orchard. *Cornell Extension Bulletin* No. 711, pp. 1–100.

Murphy, P.A. (1922) The bionomics of the conidia of *Phytophthora infestans* (Mont.) de Bary. *Scientific Proceedings of the Royal Dublin Society* 16, 442–466.

Ocamb-Basu, C.M. and Sutton, T.B. (1988) Effects of temperature and relative humidity on germination, growth, and sporulation of *Zygophiala jamaicensis*. *Phytopathology* 78, 100–103.

Page, R.M., Sherf, A.F. and Morgan, T.L. (1947) The effect of temperature and relative humidity on the longevity of the conidia of *Helminthosporium oryzae*. *Mycologia* 39, 158–164.

Paul, W.R.C. (1929) A comparative morphological and physiological study of a number of strains of *Botrytis cinerea* Pers. with special reference to their virulence. *Transactions of the British Mycological Society* 14, 118–135.

Rider, N.E., Cruickshank, I.A.M. and Bradley, E.F. (1961) Environment and sporulation in phytopathogenic fungi. III. *Peronospora tabacina* Adam: Field environment, sporulation, and forecasting. *Australian Journal of Agricultural Research* 12, 1119–1125.

Robinson, R.A. and Stokes, R.H. (1955) *Electrolyte Solutions.* Butterworths Scientific Publications, London.

Rogers, W.E. (1966) A microenvironment chamber for critical control of relative humidity. *Phytopathology* 56, 980–982.

Rose, C.W. (1966) *Agricultural Physics.* Pergamon Press, Oxford.

Rotem, J., Cohen, Y. and Putter, J. (1971) Relativity of limiting and optimum inoculum loads, wetting durations, and temperatures for infection by *Phytophthora infestans*. *Phytopathology* 61, 275–278.

Rundel, P.W. and Jarrell, W.M. (1989) Water in the environment. In: Pearcy, R.W., Ehleringer, J.R., Mooney, H.A. and Rundel, P.W. (eds), *Plant Physiological Ecology, Field Methods and Instrumentation*, Chapman and Hall, London, pp. 29–56.

Schein, R.D. and Rotem, J. (1965) Temperature and humidity effects on uredospore viability. *Mycologia* 57, 397–403.

Shaw, D.A., Adaskaveg, J.E. and Ogawa, J.M. (1990) Influence of wetness period and temperature on infection and development of shot-hole disease of almond caused by *Wilsonomyces carpophilus*. *Phytopathology* 80, 749–756.

Shaw, M.W. (1986) Effects of temperature and leaf wetness on *Pyrenophora teres* growing on barley cv. Sonja. *Plant Pathology* 35, 294–309.

Shearer, B.L. and Zadoks, J.C. (1972) The latent period of *Septoria nodorum* in wheat. I. The effect of temperature and moisture treatments under controlled conditions. *Netherlands Journal of Plant Pathology* 78, 231–241.

Sirry, A.R. (1957) The effect of relative humidity on the germination of *Botrytis* spores and on the severity of *Botrytis cinerea* Pers. on lettuce. *Annals of Agricultural Science, University of A'in Shams, Cairo* 2, 247–250.

Smith, W.L. Jr. and McClure, T.T. (1960) *Rhizopus* rot of peaches as affected by post-harvest temperature and moisture. *Phytopathology* 50, 558–562.

Snow, D. (1949) The germination of mould spores at controlled humidities. *Annals of Applied Biology* 36, 1–13.

Somers, E. and Horsfall, J.G. (1966) The water content of powdery mildew conidia. *Phytopathology* 56, 1031–1035.

Stevens, N.E. (1916) A method for studying the humidity relations of fungi in culture. *Phytopathology* 6, 428–432.

Strandberg, J.O. (1977) Spore production and dispersal of *Alternaria dauci*. *Phytopathology* 67, 1262–1266.

Tarr, S.A.J. (1972) The factors which influence the spread of pathogens within crop areas. *Principles of Plant Pathology*, Macmillan, London, pp. 399–415.

Tuite, J. (1969) *Plant Pathological Methods, Fungi and Bacteria*. Burgess Publishing Company, Minneapolis.

van den Berg, L. and Lentz, C.P. (1968) The effect of relative humidity and temperature on survival and growth of *Botrytis cinerea* and *Sclerotinia sclerotiorum*. *Canadian Journal of Botany* 46, 1477–1481.

Vanniasingham, V.M. and Gilligan, C.A. (1988) Effects of biotic and abiotic factors on germination of pycnidiospores of *Leptosphaeria maculans* in vitro. *Transactions of the British Mycological Society* 90, 415–420.

Vanniasingham, V.M. and Gilligan, C.A. (1989) Effects of host, pathogen and environmental factors on latent period and production of pycnidia of *Leptosphaeria maculans* on oilseed rape leaves in controlled environments. *Mycological Research* 93, 167–174.

Ward, S.V. and Manners, J.G. (1974) Environmental effects on the quantity and viability of conidia produced by *Erysiphe graminis*. *Transactions of the British Mycological Society* 62, 119–128.

Warren, R.C. and Colhoun, J. (1975) Viability of sporangia of *Phytophthora infestans* in relation to drying. *Transactions of the British Mycological Society* 64, 73–78.

Weaver, L.O. (1950) Effect of temperature and relative humidity on occurrence of blossom blight of stone fruits. *Phytopathology* 40, 1136–1153.

Wexler, A. and Hasegawa, S. (1954) Relative humidity–temperature relationships of some saturated salt solutions in the temperature range 0°C to 50°C. *Journal of Research of the National Bureau of Standards* 53, 19–26.

Winston, P.W. and Bates, D.H. (1960) Saturated solutions for the control of humidity in biological research. *Ecology* 41, 232–237.

Woodward, F.I. and Sheehy, J.H. (1983) *Principles and Measurements in Environmental Biology*. Butterworth, London.

Yarwood, C.E. (1936) The tolerance of *Erysiphe polygoni* and certain other powdery mildews to low humidity. *Phytopathology* 26, 845–859.

Yarwood, C.E. (1957) Powdery mildews. *The Botanical Review* 23, 235–301.

II METHODOLOGY

6 Interaction of *Bacillus* Species with Phytopathogenic Fungi – Methods of Analysis and Manipulation for Biocontrol Purposes

S.G. Edwards, T. McKay and B. Seddon

Department of Agriculture, University of Aberdeen, School of Agriculture Building, Aberdeen AB9 1UD, UK.

Introduction

The purpose of this chapter is to stimulate and provoke an approach based on the use of well-characterized *Bacillus* spp. for biocontrol and analysis of their interactive ecophysiology with phytopathogenic fungi. Biological control (biocontrol) can be defined as the use of one organism to control another, the pathogen, in the environment. In agriculture, and more specifically in crop protection, a major challenge is the use of biocontrol for fungal diseases (Windels and Lindow, 1985; Andrews, 1992). Chemical fungicides have been used successfully to control these diseases (Jutsum, 1988), but problems have arisen due to the development of fungicide resistance, by lack of specificity and by toxic effects that can be difficult to eliminate, quantify or restrict to the site of application (Dekker and Georgopoulos, 1982; Campbell, 1989). The result is an overuse of fungicides in the environment and some restriction in their use is required (Nelson, 1989). An alternative to use of fungicides is to practise biological control.

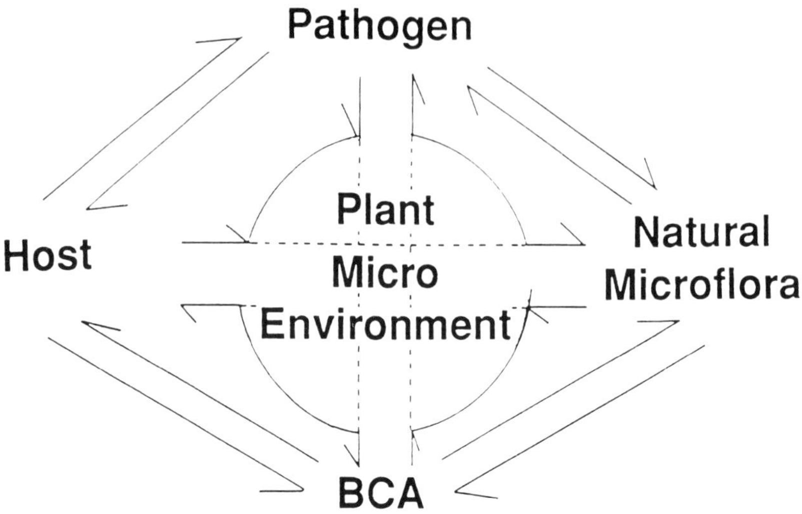

Fig. 6.1. Interactions involved in biocontrol systems *in planta*.

Biological control and interactive ecophysiology of plant pathogen and antagonist

So little is known and understood of the interactive ecophysiology of microorganisms on plant surfaces, irrespective of whether they are soil-borne or foliar, that widespread immediate commercial use of biological control agents (BCAs) is out of the question. The main bottleneck to an understanding and hence commercial application of BCAs for use in the control of phytopathogenic fungi is a knowledge of the ecology and interactive physiology of these systems *in situ* (Blakeman, 1988; Whipps *et al.*, 1988). In the short term, such studies require to be focused on the development of those BCAs showing even a limited degree of success in order that integrated use with fungicides may be possible (Blakeman and Fokkema, 1982). In this way, a reduced use of chemical fungicides with little or no loss in disease control is a realistic goal (Elad and Zimand, 1992). In the long term, development of new methodologies and specific techniques are required for biocontrol to replace the use of chemical fungicides. In many cases, the inconsistent success of biological control systems in the field is due to a lack of knowledge of the pathogen and BCA and how they interact in the microenvironment of the infection court (Fig. 6.1).

Bacillus species as BCAs and antagonism by antibiosis

Bacterial BCAs that have been shown to act against plant pathogenic fungi are mainly from the genera *Pseudomonas* (Schippers, 1983), *Streptomyces* (Tanaka *et al.*, 1987) and *Bacillus* (see Table 6.1). Other genera of bacteria have been shown to act as antagonists, but the majority belong to these groups. Antagonism and the mode of action of the BCA is generally

Table 6.1. Antibiotic-producing *Bacillus* spp. as BCAs against phytopathogenic fungi.

Target pathogen	Crop	Reference
B. subtilis		
Monilinia fructicola	Peach	Pusey and Wilson, 1984
Monilinia fructicola	Cherry	Utkhede and Sholberg, 1986
Alternaria alternata	Cherry	Utkhede and Sholberg, 1986
Rhizoctonia solani	Soil	Olsen and Baker, 1968
Rhizoctonia solani	Cotton	Kloepper, 1991
Uromyces appendiculatus	Bean	Baker *et al.*, 1985
Verticillium dahliae	Maple	Hall *et al.*, 1986
Sclerotium cepivorum	Onion	Utkhede and Rahe, 1983
Puccinia pelargonii-zonalis	Geranium	Rytter *et al.*, 1989
Phytophthora cactorum	Apple	Utkhede and Smith, 1991
Eutypa lata	Grape	Ferreira *et al.*, 1991
Fusarium roseum f.sp. *dianthi*	Carnation	Baker and Aldrich, 1970
B. cereus		
Puccinia allii	Leek	Doherty and Preece, 1978
Alternaria alternata	Tobacco	Fravel and Spurr, 1977
Sclerotium cepivorum	Onion	Wong and Hughes, 1986
Phytophthora megasperma f.sp. *medicaginis*	Alfalfa	Handelsman *et al.*, 1990
B. megaterium		
Fusarium roseum	Rice	Islam and Nandi, 1985
Alternaria alternata	Rice	Islam and Nandi, 1985
Sclerotium cepivorum	Onion	Wong and Hughes, 1986
B. pumilus		
Puccinia spp.	Wheat	Morgan, 1963
Penicillium digitatum	Citrus	Huang *et al.*, 1992
B. licheniformis		
Sclerotium cepivorum	Onion	Wong and Hughes, 1986
Pyrenophora teres	Barley	Scharen and Bryan, 1981
B. brevis		
Botrytis cinerea	Lettuce	Wood, 1951a
Rhizoctonia solani	Lettuce	Wood, 1951b

categorized as antibiosis, competition for nutrients, competition for space, hyperparasitism and induced resistance (Baker, 1987) but these are only superficial statements and in many cases the precise mechanism of action is not fully understood. Antibiosis is generally the mode of antagonism observed with *Bacillus* spp.

Most *Bacillus* spp. produce antibiotics (Katz and Demain, 1977), many of which have antifungal activity (Murray and Seddon, 1986). *Bacillus* spp. also produce spores which are dormant structures resistant to desiccation, heat, UV irradiation and organic solvents (Roberts and Hitchins, 1969), properties that are suitable for formulation and commercialization (Rhodes, 1990). The production of antibiotics and spores by *Bacillus* spp. therefore suggests that they may be attractive BCAs for use against phytopathogenic fungi. The antibiotics from *Bacillus* spp. are nearly always peptide in nature and generally of low molecular weight. In several cases D- rather than L-amino acids are present and the structure can be cyclic or have a cyclic component (Katz and Demain, 1977). Over a hundred antibiotics from the genus *Bacillus* have been identified (Bérdy, 1974) and there is strong potential and scope for use of *Bacillus* spp. in biocontrol systems.

Standard protocol, dogma and *in vitro* studies

The generally accepted protocol in biocontrol studies is to first isolate a large number of potential BCAs from the same ecological niche where disease is to be controlled, e.g. rhizosphere, phylloplane, etc. (Windels and Lindow, 1985), and then to test these *in vitro* for antagonism to the pathogen. The rationale behind this approach is that antagonists obtained in this manner should be well adapted to the environment where they are to be deployed. With a large diversity of microorganisms present in these ecosystems, the chance of obtaining antagonists with potential as BCAs should then be high. The problem with this approach is that any such antagonists are likely to be poorly characterized with no background knowledge of their biology. In such a situation there is little scope for manipulation and improvement in order to exploit the system without considerable effort. In the case of *Bacillus* spp. many such isolates have been obtained and tested as BCAs against phytopathogenic fungi. Table 6.1 identifies a few of those which have shown potential, together with the fungal pathogens and crop plants studied. A success resulting from such studies on a pilot scale has been a patented method of biocontrol of brown rot of stored peach with *B. subtilis* B3 (Pusey and Wilson, 1988). In many cases, however, workers have been unsuccessful in field applications of a BCA, following promising *in vitro* studies, because the environmental interactions have not been fully understood (Fig. 6.1). Studies which omitted (i) testing of antagonism *in planta*, (ii) studies of inoculum dose:disease index and (iii) monitoring of pathogen and BCA population levels during field trials have been referred to as the 'silver bullet' (Spurr and Knudsen, 1985) or 'spray and pray' (Lynch and Ebben, 1986) philosophy. This approach is almost always unsuccessful or unreliable and as a result researchers tend to reject their first candidate microorganisms and screen again for potential antagonists. A more appropriate stratagem would be to determine the population sizes of BCA and pathogen and their activity on the crop in order to bias the system towards dominance of the BCA over the pathogen. The factors influencing this dominance such as temperature, humidity, natural microflora and the plant itself, must also be considered (Andrews, 1992). Development of reliable and effective biocontrol has been restricted by the lack of research into the reasons why the majority of BCAs discovered during screening have poor efficacy in the field. It is possible that these antagonists have poor survival and/or growth characteristics (Knudsen and Spurr, 1987) and are affected by desiccation, UV irradiation, starvation conditions, etc. (Knudsen, 1991). Similarly, if antibiosis is the mode of antagonism *in vitro* there may be little or no antibiotic production *in planta* (Fravel, 1988). More successful biocontrol may be achieved with a detailed *in situ* study of the ecology of the pathogen and its interaction with the antagonist (Pusey, 1990). As stated by Knudsen and Spurr (1988) 'researchers need to accept that biocontrol is a long-term prospect, and abandon the silver bullet approach once and for all'.

Unorthodox Approach – Introduction of a Well-Characterized Foreign Organism

In a review of biocontrol, Andrews (1992) stated that 'Contrary to the dogma that it is the residents of an ecosystem that are the best adapted to it, probably many foreign organisms can colonise the phylloplane under favourable conditions'. Pusey (1990) also suggested that the introduction of non-indigenous antagonists might be successful. Indeed, it is probably not an organism's ability to grow under favourable conditions but its ability to survive unfavourable conditions that strongly determines the success or otherwise of a potential BCA. If a foreign organism is to be introduced, then a well-characterized strain would provide the relevant background information for the manipulation that may be required. For example, it would be desirable to start with a laboratory-based organism where detailed knowledge of the biology, physiology, biochemistry and genetics was known. Unlike an organism directly isolated from the infection court, the viability and activity of this BCA *in situ* would be unknown and this would require examination. The advantage would be that, if any degree of success was achieved, the system would be open to further development and manipulation because much of the necessary background data would be available. In our view this rather unorthodox approach is worth exploring.

The case for *Bacillus brevis*

B. brevis is well-characterized, especially with respect to antibiotic production (Kleinkauf and von Döhren, 1982) where peptide antibiotics are produced by various strains (Table 6.2). Much background work concerning the biosynthesis and physiological parameters regulating production of the

Table 6.2. Peptide antibiotics produced by strains of *Bacillus brevis.*

Strain	Antibiotic	Reference
ATCC 8185	Tyrocidine and linear gramicidin	Okuda *et al.*, 1963
Vm4	Ediene	Kurylo-Borowska, 1967
342-14	Brevistin	Shoji and Kato, 1976
GB	Gramicidin S	Gause and Brazhnikova, 1944
ATCC 9999	Gramicidin S	Winnick and Winnick, 1961
Nagano	Gramicidin S	Saito *et al.*, 1970

(a) (b)

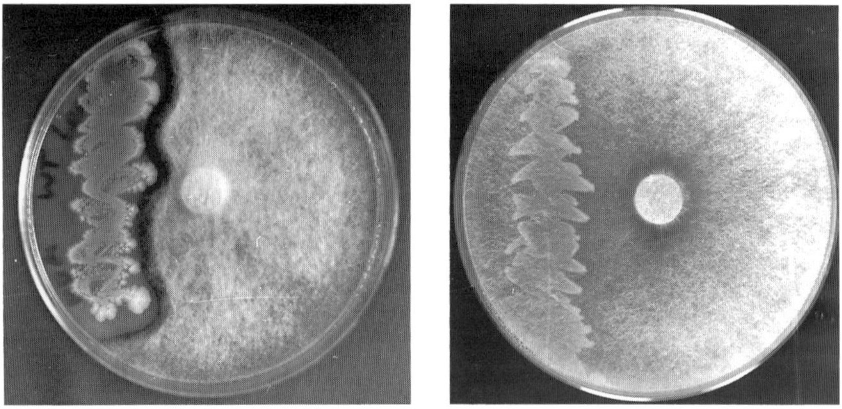

Fig. 6.2. Dual cultures of (a) wild-type and (b) gramicidin S-negative mutant of *Bacillus brevis* against *Pyrenophora graminea.*

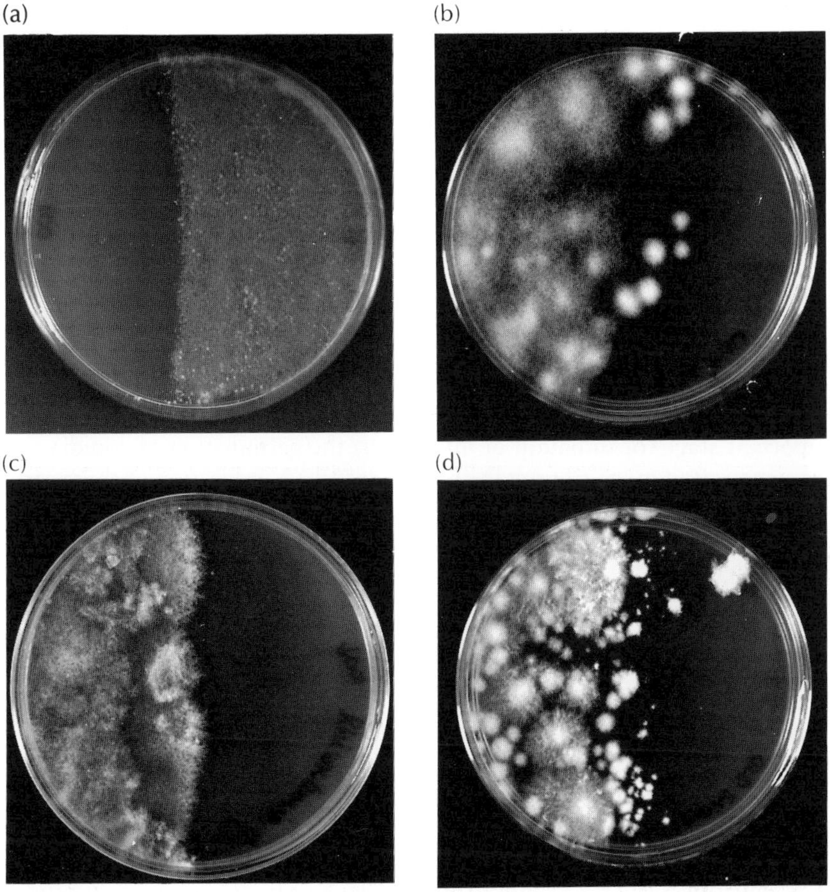

Fig. 6.3. Gradient plates of gramicidin S (0–100 μM) and inhibition of phytopathogenic fungi. (a) *Botrytis cinerea*; (b) *Pyrenophora graminea*; (c) *Rhizoctonia solani*; (d) *Microdochium nivale*. Concentration gradient (0–100 μM gramicidin S) from left to right.

antibiotics has been obtained (Kleinkauf and von Döhren, 1982). There are three main advantages in making *B. brevis* strain Nagano the focus of a study for biocontrol of fungal plant pathogens. Firstly, *B. brevis* strain Nagano produces a single cyclic decapeptide antibiotic, gramicidin S, the structure of which is established (Ovchinnikov and Ivanov, 1982). Well-characterized antibiotic-negative mutants (Iwaki *et al.*, 1972) are also available for comparative study (Fig. 6.2) to determine the significance of antifungal activity as a mode of action. Secondly, gramicidin S is readily available in pure form enabling assessment of antifungal activity (Murray *et al.*, 1986). Thirdly, gramicidin S shows broad spectrum activity against a range of plant pathogens (Fig. 6.3). For target purposes, the fungal pathogen *Botrytis cinerea* was selected to explore biocontrol possibilities.

Botrytis cinerea as target pathogen and manipulation of *B. brevis* for biocontrol

The developmental stages of the pathogen *B. cinerea* involved in disease initiation and spread have been clearly identified (Jarvis, 1977). Conidia of this fungus alight on aerial structures of plants and a variety of surface tissues are attacked when conditions are favourable for germination (high humidity, moderate temperatures) (Blakeman, 1980). Mycelial growth then allows the development of necrotic lesions, and sporulation and conidial dispersion lead finally to spread of inoculum throughout the crop (Jarvis, 1980). The important stage for initiation of infection is the germination of conidia and pre-penetration mycelial growth (Blakeman, 1980). Inhibition of these early stages by a BCA would indicate potential for successful biocontrol. Gramicidin S is more active at inhibiting conidial germination (5 μM) than mycelial growth (50 μM) (Edwards and Seddon, 1992). Antibiotic-negative mutant spores do not carry gramicidin S (Nandi and Seddon, 1978) and evidence strongly indicates that the antibiotic is associated with the surface of the bacterial spore (Lazaridis, 1981) and in this location it can interact with, and antagonize, conidia of *B. cinerea* (Edwards and Seddon, 1992). Spores of *B. brevis* are as active at completely inhibiting conidial germination as the pure antibiotic. Antibiotic-free mutant spores are inactive (Edwards and Seddon, 1992). By varying growth conditions for *B. brevis*, the degree of sporulation and the level of antibiotic produced can be altered (Bentzen and Demain, 1990). In this way, production of *B. brevis* spores with high levels of gramicidin S can be produced for studies *in planta* aimed ultimately at biocontrol (Edwards and Seddon, 1992).

In planta *observations*

Higher levels of gramicidin S, or greater numbers of spores of *B. brevis*, are needed to achieve equivalent inhibition of conidial germination in *B. cinerea in planta* compared to that *in vitro* (Edwards and Seddon, 1992). The use of calcofluor M2R, which does not alter viability and/or pathogenicity in *B. cinerea* (Dolan and McNicol, 1986), allowed the staining of conidia before inoculation of plants and the non-destructive monitoring of conidial germination on leaf surfaces by fluorescence microscopy (Edwards and Seddon, 1992). A feature of conidial germination *in planta* was that extending germ-tubes traversed the surface of the plant cells, often following anticlinal walls at cell boundaries. A similar observation was made for *B. cinerea* on bean leaves (Mansfield and Richardson, 1981). When spores of *B. brevis* were stained with acridine orange and observed on the leaf surface using fluorescence microscopy, the location of these spores was predominantly in the same anticlinal grooves. Blakeman (1988) also observed that saprophytic microorganisms occupied these same locations and concluded that some interaction between pathogen and antagonist was inevitable.

There are several reasons why higher levels of the antagonist were required *in planta* for the inhibition of conidial germination. When *in vitro* experiments are carried out, the antibiotic is readily distributed, whereas on leaf surfaces the *B. brevis* spores carrying the antibiotic and the *B. cinerea* conidia are static and may not be in direct juxtaposition. Additionally, gramicidin S appears to bind to the leaf surface and is not readily available for interaction with *B. cinerea* conidia (Edwards and Seddon, 1992). Thus, experiments carried out solely *in vitro* may be unreliable and detailed studies on the pathogen and antagonist when introduced together in the infection court are needed.

In situ *studies and specific retrieval methods*

The ultimate evaluation of any potential biocontrol system is its performance and reliability in the field. The protected crop, Chinese cabbage (*Brassica campestris* ssp. *pekinensis* cv. Granaat) grown in polyethylene tunnels is susceptible to grey mould infection caused by *B. cinerea* during periods of high humidity particularly in spring and autumn in the UK. The ability of preparations of *B. brevis* spores to act as a BCA and protect this crop was tested under these conditions. Specific methods for the retrieval of the pathogen and the BCA were developed for this. A specific selective medium for *B. brevis* was developed for this purpose (Edwards and Seddon, 1991a). Confirmation of colonies as *B. brevis* was made by extraction and identification of gramicidin S by paper chromatography. Normally, colony morphology was sufficient for identification and enumeration, when compared with other known bacterial isolates naturally present on leaves of Chinese cabbage. Both antibiotic and antibiotic-negative strains of *B. brevis* were easily distinguished from *B. cereus*, *B. megaterium*, *B. licheniformis* and *B. subtilis* by this method. The selective medium was used to determine levels of *B. brevis* spores applied to, and surviving on, leaves of Chinese cabbage in trials under polyethylene. Approximately 10^4 spores cm^{-2} of *B. brevis* were deposited on leaf surfaces by spray treatment (Fig. 6.4); previous studies had shown that such levels were necessary for biocontrol with leaf disks *in vitro*. Numbers of viable spores recovered fell exponentially over a period of 2 weeks and spraying at 14 day intervals was needed to restore initial population levels. These population changes are similar to those observed with *B. thuringiensis* (Knudsen and Spurr, 1987,

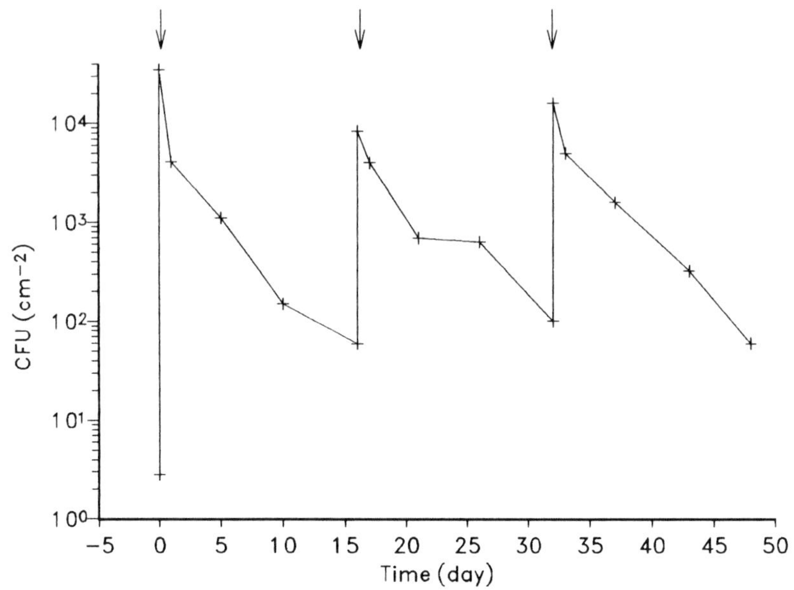

Fig. 6.4. Population levels of viable spores of *B. brevis* on the leaf surface of Chinese cabbage after spray treatments during a polyethylene tunnel trial. Arrows indicate spray treatment with *B. brevis* spore preparation (*c.* 3×10^7 ml^{-1}).

1988); the latter BCA has been used extensively against certain insect pests.

To monitor levels of *B. cinerea* in field trials, a satisfactory retrieval method for the pathogen was also needed; selective media described for *B. cinerea* were tested, but the media of Kritzman and Netzer (1978) and Kerssies (1990) were insufficiently selective and did not allow reliable counts to be obtained. Accordingly, a *Botrytis* selective medium (BSM) was developed and shown to be reproducible and reliable when tested with conidia of *B. cinerea* retrieved from soil and plant surfaces (Edwards and Seddon, 1991b). The BSM was highly selective towards *Botrytis* and radial spread of colonies was sufficiently restricted to allow large numbers of colonies to be counted without difficulty. One of the components of BSM, rose bengal, is photosensitive to daylight and on exposure its toxicity increases (Pady *et al.*, 1960). Therefore, where spore traps had to be placed throughout the tunnel in daylight, rose bengal was omitted from BSM but supplemented with fenarimol to eliminate *Penicillium* spp., which are a major component of the airborne microflora (Kerssies, 1990). This modification, referred to as *Botrytis* spore trap medium (BSTM), was used when sampling from air, while BSM was used for isolating from soil and plant surfaces. Using BSTM daily, spore trap counts could be used to monitor airborne

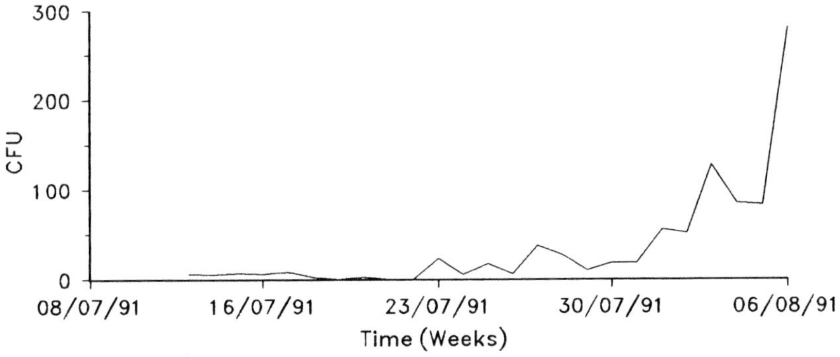

Fig. 6.5. Daily spore trap counts of conidia of *Botrytis cinerea* during a polyethylene tunnel trial. CFU; colony forming units from eight BSTM plates exposed for 24 h at crop level.

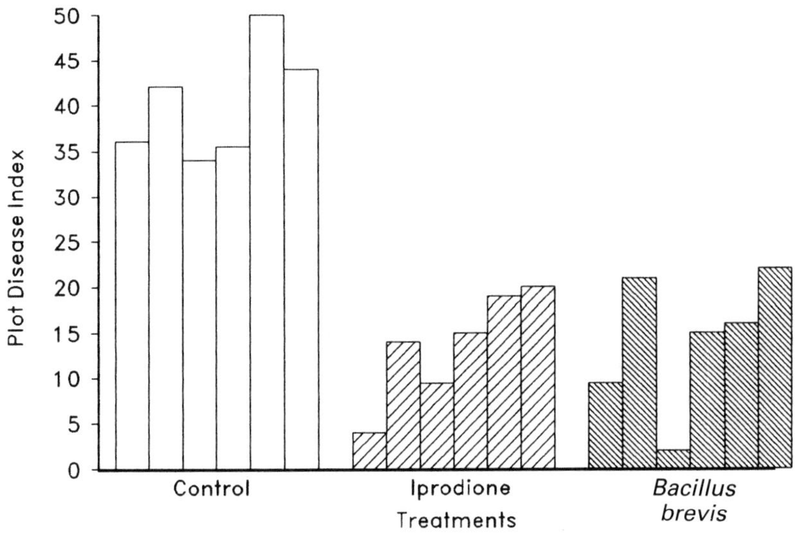

Fig. 6.6. Control of *Botrytis cinerea* infection of Chinese cabbage with *Bacillus brevis* compared to standard fungicide treatment with iprodione. Plot Disease Index; sum of disease indices (0–healthy to 5–dead) of 24 plants within each plot.

levels of conidia of *B. cinerea*. Data from a tunnel trial (Fig. 6.5) showed that less than 10 counts in total were regularly observed under conditions non-conducive to grey mould. However, following a period of cool, damp weather counts then increased after lesions appeared and sporulation resulted in release of high levels of conidia into the air above the crop canopy. Therefore methods can be devised to monitor, retrieve and measure the levels of pathogen and antagonist.

B. brevis was tested as a BCA against *B. cinerea* in four trials in polyethylene tunnels. Grey mould disease failed to develop in two trials due to dry weather conditions, but the remaining two trials were conducive to grey mould infection and in both experiments the BCA was successful in controlling the disease (Edwards and Seddon, 1992). In the second trial, a level of biocontrol with the *B. brevis* spore preparation was as good as the standard fungicide treatment with iprodione (*c.* 68% control) (Fig. 6.6) and in the third trial a similar level of biocontrol (*c.* 73%) was obtained. These observations are promising and show that biocontrol systems can work with protected crops.

Our observations with *B. brevis* and gramicidin S have shown levels of biocontrol in tunnel trials above that expected from laboratory studies *in planta* (Edwards and Seddon, 1992). Only partial inhibition of germination of *B. cinerea* was achieved *in planta*, even at levels of gramicidin S as high as 500 μM (Edwards and Seddon, 1992) but good biocontrol of grey mould in Chinese cabbage was obtained in tunnel trials. Gramicidin S may be more effective under the different conditions prevailing in the tunnel than in the laboratory. Conidia of *B. cinerea* may take longer to germinate in the tunnel because of lower and fluctuating temperatures, humidity and their nutritional status, and this delay might aid inhibition by gramicidin S. Recent observations from our laboratory also indicate that gramicidin S might not be the only mode of antagonism. *B. brevis* spore preparations contain a surfactant which alters leaf wettability. Upon irrigation, water spreads on *B. brevis*-treated leaves to form a thin layer. This layer dries more rapidly than water droplets on untreated plots, resulting in reduced periods of leaf wetness (Edwards and Seddon, 1992). Presence of water is essential to many fungal plant pathogens (see Chapter 5 by Harrison *et al.*) and reducing the period of leaf wetness is a recognized cultural control practice (Jarvis, 1989). Both modes of antagonism, antibiosis and reduced periods of leaf wetness, may be acting simultaneously to give a more effective biocontrol system. In contrast, where incubation under conditions of 100% relative humidity provides continuous wetness, antibiosis alone may not be sufficient to produce a high level of antagonism. It is possible, therefore, that *Bacillus* spp. have a range of mechanisms of antagonism that could be manipulated and exploited for use against fungi in crops. This is supported by the data of Morgan (1963) which indicate that antibiosis is not the sole mode of antagonism.

Conclusion and Future Prospects

Antibiotic-producing *Bacillus* spp. are increasingly being investigated as potential BCAs against fungal plant pathogens. Previous studies met with little success mainly due to a 'spray and pray' approach. Perhaps several of these antagonists should be re-examined in more detail. It would be an advantage to know why the majority of BCAs discovered during screening have poor efficacy in the field. A concerted effort should be put into studying the basic biology of these organisms, in the same way as has been done with well-characterized laboratory isolates.

An alternative approach is to use a well-characterized *Bacillus* sp. with known properties and mechanism of antifungal activity. The major problem here is that even with a bacterium showing strong antagonism *in vitro*, there is no guarantee that it can be successfully introduced into the micro-environment of a crop. Indeed, occupancy of such an ecological niche with an 'intruder' might seem unlikely. In the event that the antagonist can be introduced successfully, then manipulation may lead to modification of its properties to give higher levels of activity. If the organism is not ecologically competent, then only a small amount of effort has been expended in ascertaining this, in comparison with the work needed to enhance a potential, but uncharacterized BCA. Even if a well-characterized BCA is unsuccessful in the target environment, genes responsible for antagonism (e.g. genes for antibiotic synthesis) may be transferred into a bacterium adapted to and competent in that environment. In the case of *B. brevis* this technology is available (Mittenhuber *et al.*, 1990).

The studies described here with *B. brevis* show that such an unorthodox approach can be successful in achieving biological control. These principles may be applicable to other species in the genus. In order to monitor the inoculum of the pathogen and antagonist in the selected environment, specific methods for their numerical estimation will be needed. In most biocontrol studies little is known about the population sizes of a BCA during introduction, establishment and persistence and the interactive microbial ecophysiology of antagonist and pathogen on the plant surface. This area of research offers the greatest potential for perfecting biocontrol, but still requires more work.

B. brevis is only one of many well-characterized *Bacillus* spp. (Ganesan and Hoch, 1986) but workers continue to isolate new strains from the environment in their attempts at biocontrol. It could be that the use of well-characterized, antibiotic-producing strains of *Bacillus* will prove profitable in perfecting workable biocontrol protocols. Scientists would do well to pay heed to the information already available in culture collections and in the literature in order to perfect biocontrol systems. Many BCAs are not amenable to formulation processes but *Bacillus* spp. are well suited for this use by virtue of their resistant spores. The prospects for commercial use of *Bacillus* spp. for biocontrol of fungal plant pathogens looks encouraging.

Acknowledgement

We are grateful to AFRC (Grant PG1/506) for major funding of this work.

References

Andrews, J.H. (1992) Biological control in the phyllosphere. *Annual Review of Phytopathology* 30, 603–635.

Baker, C.J., Staveley, J.R. and Mock, N. (1985) Biocontrol of bean rust by *Bacillus subtilis* under field conditions. *Plant Disease* 69, 770–772.

Baker, K.F. (1987) Evolving concepts of biological control of plant pathogens. *Annual Review of Phytopathology* 25, 67–85.

Baker, R. and Aldrich, J. (1970) Biological control of *Fusarium roseum* f.sp. *dianthi* by *Bacillus subtilis*. *Plant Disease Reporter* 54, 446–448.

Bentzen, G. and Demain, A.L. (1990) Studies on gramicidin S-mediated suicide during germination outgrowth of *Bacillus brevis* spores. *Current Microbiology* 20, 165–169.

Bérdy, J. (1974) Recent developments of antibiotic research and classification of antibiotics according to chemical structure. *Advances in Applied Microbiology* 18, 309–406.

Blakeman, J.P. (1980) Behaviour of conidia on aerial plant surfaces. In: Coley-Smith, J.R., Verhoeff, K. and Jarvis, W.R. (eds), *The Biology of Botrytis*. Academic Press, London, pp. 115–151.

Blakeman, J.P. (1988) Competitive antagonism of air-borne fungal pathogens. In: Burge, M.N. (ed.), *Fungi in Biological Control Systems*. Manchester University Press, Manchester, pp. 141–160.

Blakeman, J.P. and Fokkema, N.J. (1982) Potential for the biological control of plant diseases on the phylloplane. *Annual Review of Phytopathology* 20, 167–192.

Campbell, R. (1989) *Biological Control of Microbial Plant Pathogens*. Cambridge University Press, Cambridge.

Dekker, J. and Georgopoulos, S.G. (1982) *Fungicide Resistance in Crop Protection*. Pudoc Scientific Publishers, Wageningen.

Doherty, M.A. and Preece, T.F. (1978) *Bacillus cereus* prevents germination of uredospores of *Puccinia allii* and the development of rust disease of leek, *Allium porrum* in controlled environments. *Physiological Plant Pathology* 12, 123–132.

Dolan, A. and McNicol, R.J. (1986) Staining conidia of *Botrytis cinerea* with Calcofluor White PMS and its effects on growth and pathogenicity to raspberry. *Transactions of the British Mycological Society* 87, 316–320.

Edwards, S.G. and Seddon, B. (1991a) Development of a selective medium for the specific retrieval of *Bacillus brevis* from the environment. *Journal of Applied Bacteriology* 71, xxx.

Edwards, S.G. and Seddon, B. (1991b) Development of a new selective medium for the specific retrieval of *Botrytis cinerea* from the environment. *Journal of Applied Bacteriology* 71, xxx.

Edwards, S.G. and Seddon, B. (1992) *Bacillus brevis* as a biocontrol agent against *Botrytis cinerea* on protected Chinese cabbage. In: Verhoeff, K., Malathrakis, N.E. and Williamson, B. (eds), *Recent Advances in Botrytis Research.* Pudoc Scientific Publishers, Wageningen, pp. 267–271.

Elad, Y. and Zimand, G. (1992) Integration of biological and chemical control of grey mould. In: Verhoeff, K., Malathrakis, N.E. and Williamson, B. (eds), *Recent Advances in Botrytis Research.* Pudoc Scientific Publishers, Wageningen, pp. 272–276.

Ferreira, J.H.S., Matthee, F.N. and Thomas, A.C. (1991) Biological control of *Eutypa lata* on grapevine by an antagonistic strain of *Bacillus subtilis. Phytopathology* 81, 283–287.

Fravel, D.R. (1988) Role of antibiosis in the biocontrol of plant diseases. *Annual Review of Phytopathology* 26, 75–91.

Fravel, D.R. and Spurr, D.W. (1977) Biocontrol of tobacco brown-spot disease by *Bacillus cereus* subsp. *mycoides* in a controlled environment. *Phytopathology* 67, 930–932.

Ganesan, A.T. and Hoch, J.A. (eds) (1986) *Bacillus Molecular Genetics and Biotechnology Applications.* Academic Press, Orlando.

Gause, G.F. and Brazhnikova, M.G. (1944) Gramicidin S and its use in the treatment of infected wounds. *Nature* 154, 703.

Hall, T.J., Schrieber, L.R. and Leben, C. (1986) Effects of xylem-colonising *Bacillus* spp. on verticillium wilt in maples. *Plant Disease* 70, 521–524.

Handelsman, J., Rafefel, S., Mester, E.H., Wunderlich, L. and Grau, C.R. (1990) Biological control of damping-off of alfalfa seedlings with *Bacillus cereus* UW85. *Applied and Environmental Microbiology* 56, 713–718.

Huang, Y., Wild, B.L. and Morris, S.C. (1992) Postharvest biological control of *Penicillium digitatum* decay on citrus fruit by *Bacillus pumilus. Annals of Applied Biology* 120, 367–372.

Islam, K.Z. and Nandi, B. (1985) Inhibition of some fungal pathogens of host phylloplane by *Bacillus megaterium. Zeitschrift für Pflanzenkrankheiten, Pflanzenpathologie und Pflanzenschutz* 92, 233–240.

Iwaki, M., Shimura, K., Kanda, M., Kaji, E. and Saito, Y. (1972) Some mutants of *Bacillus brevis* deficient in gramicidin S formation. *Biochemical and Biophysical Research Communications* 48, 113–118.

Jarvis, W.R. (1977) Botryotinia *and* Botrytis *Species: Taxonomy, Physiology, and Pathogenicity.* Canadian Department of Agriculture, Ottawa.

Jarvis, W.R. (1980) Taxonomy. In: Coley-Smith, J.R., Verhoeff, K. and Jarvis, W.R. (eds), *The Biology of* Botrytis. Academic Press, London, pp. 1–18.

Jarvis, W.R. (1989) Managing diseases in greenhouse crops. *Plant Disease* 73, 190–194.

Jutsum, A.R. (1988) Commercial application of biological control: status and prospects. *Philosophical Transactions of the Royal Society, London, Series B* 318, 357–376.

Katz, E. and Demain, A.L. (1977) The peptide antibiotics of *Bacillus*: chemistry, biogenesis, and possible functions. *Bacteriological Reviews* 41, 449–474.

Kerssies, A. (1990) A selective medium for *Botrytis cinerea* to be used in a spore-trap. *Netherlands Journal of Plant Pathology* 96, 247–250.

Kleinkauf, H. and von Döhren, H. (eds) (1982) *Peptide Antibiotics Biosynthesis and Functions*. Walter de Gruyter, Berlin.

Kloepper, J.W. (1991) Development of *in vivo* assays for prescreening antagonists of *Rhizoctonia solani* on cotton. *Phytopathology* 81, 1006–1013.

Knudsen, G.R. (1991) Models for the survival of bacteria applied to the foliage of crop plants. In: Hurst, C.J (ed.), *Modelling the Environmental Fate of Microorganisms*. American Society of Microbiology, Washington D.C., pp. 191–216.

Knudsen, G.R. and Spurr, H.W. (1987) Field persistence and efficacy of five bacterial preparations for control of peanut leaf spot. *Plant Disease* 71, 442–445.

Knudsen, G.R. and Spurr, H.W. (1988) Management of bacterial populations for foliar disease biocontrol. In: Mukerji, K.G. and Garg, K.L. (eds), *Biocontrol of Plant Diseases*, Vol.1. CRC Press, Boca Raton, pp. 83–92.

Kritzman, G. and Netzer, D. (1978) A selective medium for isolation and identification of *Botrytis* spp. from soil and onion seed. *Phytoparasitica* 6, 3–7.

Kurylo-Borowska, Z. (1967) Ediene. In: Gottlieb, D. and Shaw, P.D. (eds), *Antibiotics*, Vol. 2. Springer-Verlag, Berlin, pp. 342–352.

Lazaridis, I. (1981) Spore characteristics of wild-type and gramicidin S-negative mutants of *Bacillus brevis* Nagano. Unpublished Ph.D. Thesis, University of Aberdeen.

Lynch, J.M. and Ebben, H. (1986) The use of microorganisms to control plant disease. *Journal of Applied Bacteriology Symposium Supplement 1986* 61, 115S–126S.

Mansfield, J.W. and Richardson, A. (1981) The ultrastructure of interactions between *Botrytis* species and broad bean leaves. *Physiological Plant Pathology* 19, 41–48.

Mittenhuber, G., Krause, M. and Maraheil, M.A. (1990) Peptide antibiotic biosynthesis gene organisation of the gramicidin S and tyrocidine synthetase genes of *Bacillus brevis*. In: Butler, L.O.C., Harwood, C. and Moseley, B.E.B. (eds), *Genetic Transformation and Expression. Ninth European Meeting on Genetic Transformation*. Intercept Ltd, New York, pp. 477–484.

Morgan, F.L. (1963) Infection inhibition and germ tube lysis of three cereal rusts by *Bacillus pumilus*. *Phytopathology* 53, 1346–1348.

Murray, T. and Seddon, B. (1986) Antibiotic-producing bacilli and biocontrol against fungal plant pathogens. *Journal of Applied Bacteriology* 61, xix.

Murray, T., Leighton, F.C. and Seddon, B. (1986) Inhibition of fungal spore germination by gramicidin S and its potential use as a biocontrol agent against fungal plant pathogens. *Letters in Applied Microbiology* 3, 5–7.

Nandi, S. and Seddon, B. (1978) Evidence of gramicidin S functioning as a bacterial hormone specifically regulating outgrowth in *Bacillus brevis* Nagano. *Biochemical Society Transactions* 6, 409–411.

Nelson, M.R. (1989) Biological control: the second century. *Plant Disease* 73, 616.

Okuda, K., Edwards, G.C. and Winnick, T. (1963) Biosynthesis of gramicidin and tyrocidine in the Dubos strain of *Bacillus brevis* 1. Experiments with growing cells. *Journal of Bacteriology* 85, 329–338.

Olsen, C.M. and Baker, K.F. (1968) Selective heat treatment of soil, and its effect on the inhibition of *Rhizoctonia solani* by *Bacillus subtilis*. *Phytopathology* 58, 79–87.

Ovchinnikov, Y.A. and Ivanov, V.T. (1982) The cyclic peptides: structures, conformation and function. In: Neurath, H. and Hill, R.L. (eds), *The Proteins*, Vol. 5. Interscience Publishers, New York, pp. 310–642.

Pady, S.M., Kramer, C.L and Pathak, V.K. (1960) Suppression of fungi by light on media containing rose bengal. *Mycologia* 52, 347–350.

Pusey, P.L. (1990) Control of pathogens on aerial plant surfaces by antagonistic microorganisms. In: Wilcox, W.F. (ed.), *Biological and Cultural Tests for Control of Plant Diseases*, Vol. 5. American Phytopathological Society Press, St. Paul, Minnesota, pp. v–vii.

Pusey, P.L. and Wilson, C.L. (1984) Post-harvest biological control of stone fruit brown rot by *Bacillus subtilis*. *Plant Disease* 68, 753–756.

Pusey, P.L. and Wilson, C.L. (1988) *Postharvest biological control of stone fruit brown rot by* Bacillus subtilis. US Patent No. 4,764,371.

Rhodes, D.J. (1990) Formulation requirements for biological control agents. In: *The Exploitation of Microorganisms in Applied Biology. Aspects of Applied Biology No.24*. Association of Applied Biologists, Warwick, pp. 145–153.

Roberts, T.A. and Hitchins, A.D. (1969) Resistance of spores. In: Gould, G.W. and Hurst, A. (eds), *The Bacterial Spore*. Academic Press, London, pp. 611–670.

Rytter, J.L., Lukezic, F.L., Craig, R. and Moorman, G.L. (1989) Biological control of geranium rust by *Bacillus subtilis*. *Phytopathology* 79, 367–370.

Saito, Y., Otani, S. and Otani, S. (1970) Biosynthesis of gramicidin S. *Advances in Enzymology* 33, 337–380.

Scharen, A.L. and Bryan, M.D. (1981) A possible biological control agent *Bacillus licheniformis* for net blotch *Pyrenophora teres* of barley. *Phytopathology* 71, 902–903.

Schippers, B. (1983) Prospects of biological control of plant pathogens with fluorescent *Pseudomonas* spp. In: *Proceedings of the 10th International Congress of Plant Protection, Vol. 2*. British Crop Protection Council, Swindon, pp. 767–771.

Shoji, J. and Kato, T. (1976) The structure of brevistin. *The Journal of Antibiotics* 9, 380–389.

Spurr, H.W. and Knudsen, G.R. (1985) Biological control of leaf diseases with bacteria. In: Windels, C.E. and Lindow, S.E. (eds), *Biological Control on the Phylloplane*. American Phytopathological Society, St. Paul, Minnesota, pp. 45–62.

Tanaka, Y., Hirata, K., Takahashi, Y., Iwai, Y. and Omura, S. (1987) Globopeptin, a new antifungal peptide antibiotic. *The Journal of Antibiotics* 40, 242–244.

Utkhede, R.S. and Rahe, J.E. (1983) Interactions of antagonist and pathogen in biological control of onion white rot. *Phytopathology* 73, 890–893.

Utkhede, R.S. and Sholberg, P.L. (1986) *In vitro* inhibition of plant pathogens by *Bacillus subtilis* and *Enterobacter aerogenes* and *in vivo* control of two postharvest cherry diseases. *Canadian Journal of Microbiology* 32, 963–967.

Utkhede, R.S. and Smith, E.M. (1991) Biological and chemical treatments for control of phytophthora crown and root rot caused by *Phytophthora cactorum* in high density apple orchard. *Canadian Journal of Plant Pathology* 13, 267–270.

Whipps, J.M., Lewis, K. and Cooke, R.C. (1988) Mycoparasitism and plant disease control. In: Burge, M.N. (ed.), *Fungi in Biological Control Systems*. Manchester University Press, Manchester, pp. 161–187.

Windels, C.E and Lindow, S.E. (1985) *Biological Control on the Phylloplane.* American Phytopathological Society Press, St. Paul, Minnesota.
Winnick, R.E. and Winnick, T. (1961) Biosynthesis of gramicidin S. II. Incorporation experiments with labelled amino acid analogs, and the amino acid activation process. *Biochimica et Biophysica Acta* 53, 461–468.
Wong, W.C. and Hughes, I.K. (1986) *Sclerotium cepivorum* Berk. in onion (*Allium cepa*) crops: isolation and characterisation of bacteria antagonistic to the fungus in Queensland. *Journal of Applied Bacteriology* 60, 57–60.
Wood, R.K.S. (1951a) The control of disease of lettuce by the use of antagonistic organisms. 1. The control of *Botrytis cinerea. Annals of Applied Biology* 38, 203–216.
Wood, R.K.S. (1951b) The control of disease of lettuce by the use of antagonistic organisms. 2. The control of *Rhizoctonia solani. Annals of Applied Biology* 38, 217–230.

7 Methods for Detecting *Armillaria mellea*

R.T.V. Fox, H.M. Manley, A. Culham, K. Hahne and A.I. Tiffin

Schools of Plant and Animal Sciences, University of Reading, 2 Earley Gate, Reading RG6 2AU, UK.

Introduction

Trees and shrubs are attacked by one or more virulent species of the honey fungus, *Armillaria*, in forests, plantations, orchards, gardens (Fig. 7.1) and amenity situations almost anywhere (Fox, 1990b); even remote oceanic islands (Raabe and Trujillo, 1963). Few if any woody plants are immune (Raabe, 1962).

In order to avoid substantial losses, infection must be controlled promptly (Turner and Fox, 1988; Fox *et al.*, 1991). However, though the extent of decay from an established fungal rot can readily be mapped by metal probes, with or without electronic devices, its presence first has to be discovered. Some signs of infection, such as unusually sparse foliage (Fig. 7.1), are recognized by an experienced arboriculturalist and may even be detected from the air (Hunt *et al.*, 1971; Shaw and Kile, 1991). Alternatively, dogs can be trained to bark when they smell rotten wood (Swedjemark, 1989).

Rotten trees are relatively easily detected at the later stages of colonization, but by then they are difficult if not impossible to treat (Swift, 1970; Bray, 1970; Pawsey and Rahman, 1974, 1976a,b; Filip and Roth, 1977). The positive confirmation of the presence of honey fungus at earlier stages generally still depends on being able to trace rhizomorphs that develop around dead hosts. Later these may form a smothering network around the

Fig. 7.1. Basidiocarps of *Armillaria mellea* growing from roots of infected privet in a garden in Reading (from Popoola, 1991).

roots of a prospective host and then grow under the bark (Hood and Sandberg, 1987). Detection of this network may indicate that the host is at risk from infection and closer inspection may reveal effective multiple penetration into the root cambium, as for example during successful colonization by *Armillaria mellea*. However, rhizomorphs may not be in evidence following infection by other species of *Armillaria*, such as *A. ostoyae*, which nonetheless can cause serious rotting, bringing about significant losses of several economically important conifers through windthrow or direct kill. *Armillaria mellea* often infects by root to root contact.

Without the presence of characteristic fruiting bodies (Fig. 7.1) (Pegler and Gibson, 1972; Watling *et al.*, 1982; Roll-Hansen, 1985; Greig *et al.*, 1991) that may be present in some years and then only for a few days in autumn, it is often very difficult to detect the presence of honey fungus in an old dead stump when it no longer has living foliage to exhibit symptoms or the pathogen is deep seated (Shigo and Tippett, 1981; Fox, 1990c).

Although fruiting bodies can be induced *in vivo* and *in vitro* (Fig. 7.2) (Raabe, 1984; Fox and Popoola, 1990), they form too slowly (Intini, 1993) to be used routinely and the timing of their appearance can be unpredictable. Like nearly all root diseases, Armillaria root rots are hard to identify or

Fig. 7.2. *Armillaria ostoyae* basidiocarp fruiting in universal bottle on 3% malt extract agar.

quantify, particularly by the amateur when roots may be deeply buried in soil which frequently obscures any visible symptoms, or even its sporadic luminosity (Guyot, 1927). If a rotten root lacks rhizomorphs that can be used in diagnosis (Morrison, 1989), *Armillaria* will go undetected. In this situation, conventional identification techniques often rely on culturing some of the mycelium present in the rotten wood (Fig. 7.3). Even if successful (Fig. 7.4), this process of isolation on to agar may take days or even weeks to complete and is highly prone to contamination problems as *Armillaria* is relatively slow growing. Cultures of *Armillaria* spp. show considerable intraspecific variation, so interfertility tests with haploid or diploid testers are generally necessary even though these take experience to interpret. As yet, no biochemical methods, such as those investigated by Hutterman *et al.* (1984) or Wargo and Shaw (1985), have proved of much benefit.

Without reliable diagnosis of *Armillaria mellea*, the expense and labour of control measures (Fig. 7.5) will be wasted if the rot is caused by a harmless saprophyte (Fox, 1990b). It is possible to differentiate pathogens from saprophytes readily by baiting with strawberry (Fox and Popoola, 1990) or logs (Mallett and Hiratsuka, 1985); even potato tubers are susceptible

Fig. 7.3. Root rot of privet caused by *Armillaria mellea*. Note rotten tap root with adventitious prop roots.

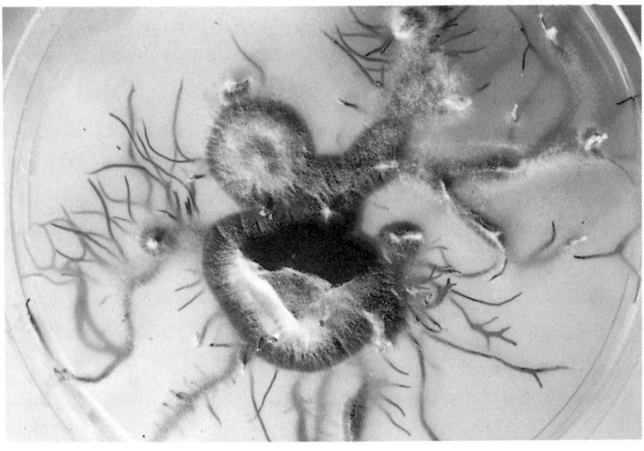

Fig. 7.4. *Armillaria mellea* colony on 3% malt extract agar showing rhizomorphs (photograph courtesy of ICI Agrochemicals).

Fig. 7.5. Treating an infected tree with fungicide (photograph courtesy of ICI Agrochemicals).

(Gregory, 1984). Baiting can be effective in mapping the extent of ground that is infected, but the method cannot give immediate results.

Speed is always crucial when planting bare-rooted trees and is frequently critical when a control agent has to be selected. Few diagnostic tests are as quick as inspecting a specimen visually for symptoms. However, in practice, any time saved in performing the diagnosis will be wasted if sampling is slow or the result is not immediately available. Even though many foliar diseases are often fairly instantly recognized by farmers and growers, providing the symptoms are sufficiently conspicuous, the identification of a pathogen infecting a root is rarely as simple, particularly when soil still obscures the symptoms. Lesions on the roots of a tree therefore take far longer to find than those on its foliage.

The most important criterion for any diagnostic test is reliability (Fox, 1993). Even an apparently straightforward method, the visual inspection of plants for symptoms, is critically dependent on the general condition of the specimen of root that has been sampled, an absence of other pathogens or saprophytes, as well as a minimal level of varietal and phenotypic variation between samples (Fox, 1992). In many routine situations, this potentially simple method has proven inherently unreliable. To a lesser extent, the success of associated microscopical techniques also depends on the quality of the specimens, especially where the pathogen is embedded in relatively tough wood.

Forest soils may often contain several different species of *Armillaria* causing similar symptoms (Morrison, 1989) resulting in further confusion (Guillaumin, 1988). Rotted root material for examination which is contaminated with other microorganisms can greatly frustrate efforts to isolate pure

Table 7.1. Comparison between current and potential (*) methods of diagnosing *Armillaria* spp. in rotten roots without other evidence.

Method	Quality				
	simplicity	speed	clarity	sensitivity	sturdiness
Visual inspection	–	–	–	+	+
Metal probes	+	+	?	?	+
Electronic devices	?	+	?	?	?
Scent	+	+	?	?	+
Baiting	+	–	+	+	+
Light microscopy	–	+	–	–	–
Electron microscopy	–	–	–	–	–
Culture isolation	–	–	+	?	–
Anastomosis and interfertility tests	–	–	+	?	–
Electrophoresis	–	–	?	+	–
Simple serology	+	+	?	–	+
Polyclonal ELISA	+	+	?	?	+
MCA ELISA on plates	+	+	?	?	–
MCA ELISA kit*	+	+	+	?	+
DNA techniques	–	+	+	+	–
PCR + ELISA kit*	+	+	+	+	+

Key: above average merits = +, at or below average = –, variable = ?

cultures (Nobles, 1948; Whitney *et al.*, 1978; Rishbeth, 1986) as well as delaying any subsequent testing with other cultures for anastomosis and interfertility (Korhonen, 1978). Since the outcome of investigations based on such tests is easily affected by often subtle changes in circumstance, their efficacy is less dependable than those methods based on fundamental biochemical differences such as the immunological properties of proteins and the hybridization of nucleic acids. Another disadvantage of these traditional methods is their greater need for information and expertise rather than equipment and reagents which tend to reduce cost-effectiveness.

While the time taken to complete a diagnosis is easily measured, other advantages of one method over another are less clear cut but a comparison (Table 7.1) reveals some critical differences, many of which are common to other soil-borne pathogens (Fox, 1990a).

Several highly sensitive modern methods for the rapid diagnosis of soil-borne pathogens that have been adapted from other branches of biology are reliable and accurate (Fox 1990a, 1993; Dusunceli and Fox, 1992; Duncan and Torrance, 1992), yet require no expertise, even though some such as gel electrophoresis (Morrison *et al.*, 1984; Poon, 1988) are not only slower and

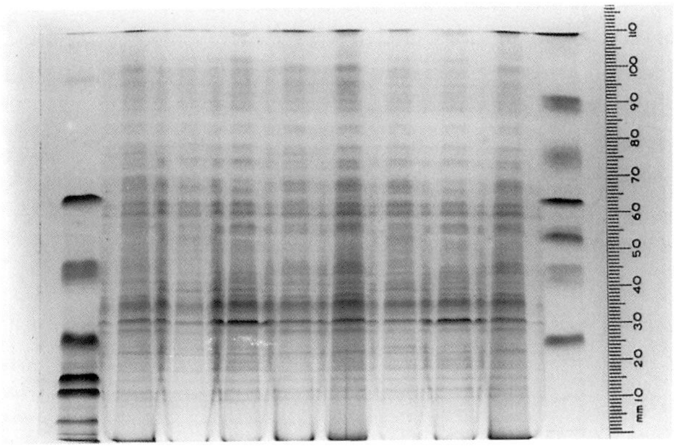

Fig. 7.6. SDS PAGE of *Armillaria* isolates together with protein markers. *Armillaria* isolates from left, AM1* (UK), AM2* (UK), TP4* (UK), 1768 (Fiji), 1881 (India), 61755 (India), 158163 (Kenya), 157774 (Kenya). The identity of those isolates with * has been confirmed as *Armillaria mellea* by interfertility testing (from Poon, 1988).

more laborious than the routine inspection for rhizomorphs by an expert, but also are not very specific as there is little variation between species of *Armillaria* (Fig. 7.6), even though these can readily be distinguished from other fungi that rot wood. Neither contamination nor the presence of soil or wood affects methods based on immunology or nucleic acid hybridization (Fox, 1993). However, apart from immunological techniques, most other laboratory diagnostic procedures have so far proved ill-suited for detecting soil-borne fungi in the field as they are neither sufficiently flexible nor portable (Fox, 1990a).

Techniques

Immunological techniques

The rapid detection and identification of numerous viruses by routine enzyme-linked immunosorbent assays (ELISA) is well established, although until recently few suitable protocols had been adapted for fungi.

Immunological techniques can quickly, clearly and accurately separate those pathogens which cause diseases with variable or latent symptoms on the host plant (Fox, 1993). They have already become widely established for the rapid detection of many pathogens which cannot be easily identified in other routine ways (Martin *et al.*, 1992a). A number of commercial kits are used on a large scale to detect plant pathogens with an indistinct or undistinguished morphology such as viruses and bacteria. Although only a few methods have been developed for soil-borne diseases, some commercial kits for diagnosing soil-borne pathogens are sold for the turf and American agricultural crop markets (Miller *et al.*, 1988).

Reactions between unknown antigens and a repertoire of relatively crude immune sera (antisera) form the basis of many well tried and tested serological techniques which are still widely used even though higher levels of specificity are attained by pure monoclonal antibodies (MCAs). However, antisera have been used successfully by Dewey *et al.* (1984) to detect and identify *Phaeolus schweinitzii* in soil. Preliminary purification of the immunogen is not essential when monoclonal antibodies are raised, since hybridomas producing antibodies specific to any qualitative or quantitative immunogenic differences can be efficiently selected by screening for specificity instead. Hybridomas are cheap to maintain and can readily produce monoclonal antibodies when required yet are prone to contamination. The former characteristics make monoclonal antibody ELISA well suited to the routine serological detection of *Armillaria* (Fox and Hahne, 1989). Hardham *et al.* (1985) used MCAs to detect the surface components of *Phytophthora cinnamomi* and Martin *et al.* (1992a) detected resistance to the fungicide, carbendazim, using MCAs raised to a synthetic immunogen (Robard, 1987).

Several simple kits, which are sold for diagnosing some other soil-borne pathogens on site, are claimed to be cost effective (Miller *et al.*, 1988). These cheap easily-used kits purchase independence from costly advisory services (Miller *et al.*, 1988). As yet, no commercial kit is marketed for the identification of *Armillaria* species, despite their economic importance, though polyclonal (Lung-Escarment *et al.*, 1978, 1985) and monoclonal (Fox and Hahne, 1989) antibodies to *Armillaria* species have been used experimentally in ELISA. However, as the use of SDS-PAGE has already indicated, protein differences between species tend to be quantitative rather than qualitative and so are of limited practical use.

Nucleic acid techniques

Although currently less widely used than immunological techniques, several diagnostic methods based on the detection of similarities between nucleic acids have been successfully adapted for identification of pathogens, particularly to detect viruses which threaten vegetative propagation systems or through seed (Robinson, 1988). Among procedures that have been investigated for *Armillaria* are thermal denaturation and renaturation studies to evaluate overall sequence similarity (Motta *et al.*, 1986), restriction site mapping (Anderson *et al.*, 1987) and DNA Southern hybridizations (Manley, 1992).

Although a well chosen collection of antibodies is capable of differentiating known subspecific differences, serological methods are not able to detect unanticipated strain abnormalities, whereas a single suitable nucleic acid probe will be able to detect a range of unexpected specific and subspecific variants (Smith *et al.*, 1990).

Unfortunately, the routine for carrying out a hybridization is not as simple as an immunological test since it may initially require the prior extraction of nucleic acid from the test sample. Following gel electrophoresis of RNA or restriction fragments of DNA to separate and analyse nucleic acids, molecular hybridization can be used both to detect specific sequences among the bands by Northern and Southern blotting of RNA and DNA respectively (Duchesne and Anderson, 1987), paralleling the use of antibodies in Western blotting to provide additional information.

Most hybridization of fungal nucleic acid has been done with radioactive probes and filter-bound nucleic acids (Wright and Morton, 1989), and hence is time consuming, requiring safety precautions and troublesome to perform, even by experts. Despite some residual practical limitations, these problems have been bypassed in modern commercial DNA-based tests using the sandwich assay (Sylvänen *et al.*, 1986). In the sandwich assay, adjacent sections of the target DNA hybridize to two probes. One is a capture probe which, following hybridization, links the target to the solid support to which the probe is already bound. The other probe, which may be synthesized (Martin *et al.*, 1992b), is labelled with a reporter group, usually a fluorescent molecule, an enzyme or a radioactive atom which hybridizes with an adjacent section of the target DNA. Minute amounts of the reporter group must be readily distinguished. Fluorescent molecules have many advantages, but their use has been restricted since many other molecules in biological samples fluoresce without being labelled, such as certain components of healthy plants. This creates a significant background fluorescence which overpowers any weak signal, making it difficult to obtain adequate sensitivity. Enzyme-labelled probes are used to generate coloured molecules which are then detected. However, thorough washing is necessary to remove impurities derived from the hybridization reaction as they inhibit the enzyme. If the probe is attached to the vitamin biotin this binds to the protein avidin extremely tightly. When extra enzyme bound to more biotin is added, it also binds to the avidin to produce a complex of many enzyme molecules which cluster around each probe molecule. Detection systems using enzymic labels can employ a complex of enzymes or luminescent reactions. Another approach is to fasten more than one reporter group onto the labelled probe. Extra space for several labelling groups on each probe can be accomplished by making the probe either very long or branched. Some non-radioactive methods can attain a sensitivity equivalent to that expected from radioactive label when automated and combined with the polymerase chain reaction (PCR).

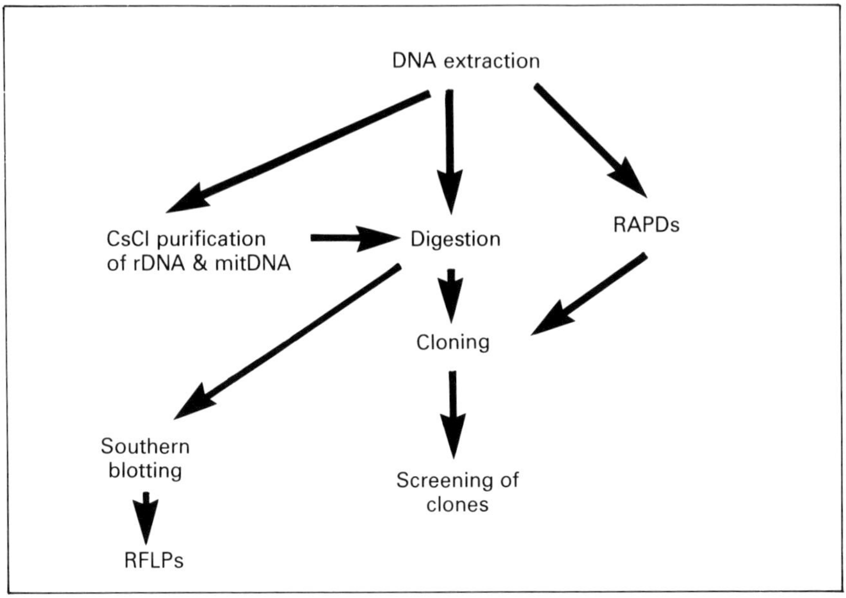

Fig. 7.7. Techniques for species diagnosis.

Nucleic acid hybridization techniques are being evaluated for rapid diagnosis of *Armillaria* in experimental ecological investigations (Manley, 1992), as well as for possible use in routine arboriculture.

DNA polymorphisms are being examined as a method both for distinguishing between the species of *Armillaria* that are found in Britain, and also for distinguishing them from fungi of other genera with which they could be confused (Manley, 1992).

Firstly, the genome of *A. mellea* is being examined for species-specific sequences. A genomic library of *A. mellea* has been constructed in a bacterial host, and selected clones will be examined for their ability to distinguish between DNA from other species of *Armillaria*, other fungal genera, and host plants (Fig. 7.7). This involves labelling cloned inserts and using them separately as probes in Southern hybridization against DNA from a variety of appropriate sources. Probes homologous to sequences in *A. mellea* are expected to hybridize to other *Armillaria* species, as closely related taxa usually show close sequence homology, but different restriction patterns or increasing stringency should allow individual species identification.

Although conventional techniques are currently restricted to the laboratory (Fox, 1992), it is intended to adapt these techniques for use *in situ* using biotin-labelled nucleic acid probes which can then be easily detected by ELISA (Fox, 1993).

By making the recognition of these soil-borne pathogens easier, these innovative techniques should become more widely used to gather information relevant to the biology and control of the diseases they cause.

Conclusions

To be effective a diagnostic test must be simple, accurate, rapid and safe to perform, yet be sensitive enough to avoid 'false positives'. Since DNA, not proteins, forms the basis of its action, DNA hybridization is so sensitive that it should detect a single nucleus or mitochondrion. Nonetheless, although DNA hybridization has been useful for a number of genetic and ecological studies, such as that for identifying the extent of different clones of *Armillaria gallica* (Smith *et al.*, 1992), it has not been adapted for use in the field. Although not yet available for *Armillaria* spp., simple monoclonal antibody ELISA kits have been produced commercially and used widely for other soil-borne pathogens. Some of the latter can detect or diagnose diseases safely and clearly using antigen-coated magnetic beads and chromogens conjugated to specific antibodies. The choice of the most appropriate experimental procedures for detecting root pathogens such as *Armillaria* should widen as innovation continues to solve current problems. Even so, nucleic acid hybridization seems destined to remain for somewhat longer only in the hands of the advisory or consultancy services.

Immunological assays are also entering a period of great change as relatively cheap, easy to use kits are being developed which allow low levels of disease to be monitored on the spot under field conditions. This may enable the initial stages of infection by basidiospores to be observed (Shaw and Kile, 1991). Since these kits are so sensitive, not only could the life cycle (Wargo and Shaw, 1985) and patterns of spread (Rishbeth, 1986) be validated, but it would be possible to treat lower inoculum levels of pathogens than previously. Consequently, if *Armillaria* can be detected earlier this should permit more effective control and the use of fungicides may be avoided where no pathogens are detected.

Formerly difficult to diagnose pathogens, such as *Armillaria* and other soil-borne organisms which are rarely quantified even when recognized, could become a more commercially viable market for fungicides and other control measures.

In addition to benefits for the control of *Armillaria* diseases, it is anticipated that new diagnostic techniques will make an important contribution in the future to our understanding of the ecology and distribution of species in this genus.

References

Anderson, J.B., Petsche, D.M. and Smith, M.L. (1987) Restriction fragment polymorphisms in biological species of *Armillaria mellea* in North America. *Mycologia* 79, 69–76.

Bray, V. (1970) Using creosote for treating honey fungus. *Journal of the Royal Horticultural Society* 95, 27–28.

Dewey, F.M., Barrett, D.K., Vose, I.R. and Lamb, C.J. (1984) Immunofluorescence microscopy for the detection and identification of propagules of *Phaeolus schweinitzii* in infested soil. *Phytopathology* 74, 291–196.

Duchesne, L.C. and Anderson, J.B. (1987) Location and direction of transcription of the 5S rRNA gene in *Armillaria. Mycological Research* 94, 266–269.

Duncan, J.M. and Torrance, L. (1992) *Techniques for the Rapid Detection of Plant Pathogens*. Blackwell Scientific Publications, Oxford, 235 pp.

Dusunceli, F. and Fox, R.T.V. (1992) The accuracy of methods for estimating the size of *Thanatephorus cucumeris* populations in soil. *Soil Use and Management* 8, 21–26.

Filip, G.M. and Roth, L.F. (1977) Stump infections with soil fumigants to eradicate *Armillaria mellea* from young-growth ponderosa pine killed by root rot. *Canadian Journal of Forest Research* 7, 226–231.

Fox, R.T.V. (1990a) Rapid methods for diagnosis of soil borne plant pathogens. In: Hornby, D. (ed.), *Soil-borne Diseases*. Special Issue. *Soil Use and Management* 6, 179–184.

Fox, R.T.V. (1990b) Diagnosis and control of *Armillaria* honey fungus root rot of trees. *Professional Horticulture* 4, 121–127.

Fox, R.T.V. (1990c) Fungal foes in your garden, 10. Honey fungus root rot. *Mycologist* 4, 192.

Fox, R.T.V. (1992) Honey Fungus detected. *Arboricultural Association Journal* 16, 317–326.

Fox, R.T.V. (1993) *Principles of Diagnostic Techniques in Plant Pathology.* CAB International, Wallingford.

Fox, R.T.V. and Hahne, K. (1989) Prospects for the rapid diagnosis of *Armillaria* by monoclonal antibody ELISA. In: Morrison, D.J. (ed.), *Proceedings of the 7th International Conference on Root and Butt Rots; 1988 August 9–16; Vernon and Victoria, BC.* International Union of Forestry Research Organizations, pp. 458–468.

Fox, R.T.V. and Popoola, T.O.S. (1990) Induction of fertile basidiocarps in *Armillaria bulbosa. Mycologist* 4, 70–72.

Fox, R.T.V., McQue, A.M. and Obanya Obore, J. (1991) Prospects for the integrated control of *Armillaria* root rot of trees. In: Beemster, A.B.R. (ed.), *Biotic Interactions and Soil-borne Diseases.* Elsevier Science Publishers, Amsterdam, pp. 154–159.

Gregory, S.C. (1984) The use of potato tubers in pathogenicity studies of *Armillaria* species. In: Kile, G.A. (ed.), *Proceedings of the 6th International Conference on Root and Butt Rots of Forest Trees; 1983 August 25–31; Melbourne, Victoria, and Gympie, Queensland, Australia.* International Union of Forestry Research Organizations, pp. 148–160.

Greig, B.J.W., Gregory, S.C. and Strouts, R.G. (1991) Honey fungus. *Forestry Commission Bulletin 100*. Her Majesty's Stationery Office, Department of the Environment, London, 11 pp.

Guillaumin, J.J. (1988) The *Armillaria mellea* complex. *Armillaria mellea* (Vahl) Kummer *sensu stricto*. In: Smith, I.M., Dunez, J., Phillips, D.H. and Archer, S.A. (eds), *European Handbook of Plant Diseases*. Blackwell Scientific Publications, Oxford, pp. 520–523.

Guyot, R. (1927) Mycelium lumineux de l'Armillaire. *Compte Rendus de la Societé de Biologie* 96, 114–116.

Hardham, A.R., Suzaki, E. and Perkin, J.L. (1985) The detection of monoclonal antibodies specific for surface components on zoospores and cysts of *Phytophthora cinnamomi*. *Experimental Mycology* 9, 254–268.

Hood, I.A. and Sandberg, C.J. (1987) Occurrence of *Armillaria* rhizomorph populations in the soil beneath indigenous forests in the Bay of Plenty, New Zealand. *New Zealand Journal of Forestry Science* 17, 83–99.

Hunt, R.S., Parmeter, J.R., Jr. and Cobb, F.W., Jr. (1971) A stump treatment technique for biological control for forest root pathogens. *Plant Disease Reporter* 55, 659–662.

Huttermann, A., Feig, R. and Trojanowski, J. (1984) Biochemical capabilities and regulation as a basis for species differentiation and ecological behavior of *Armillaria* species. In: Kile, G.A. (ed.), *Proceedings of the 6th International Conference of Root and Butt Rots of Forest Trees; 1983 August 25–31; Melbourne, Victoria, and Gympie, Queensland, Australia*. International Union of Forestry Research Organizations, pp. 57–72.

Intini, M.G. (1993) Development of *Armillaria* carpophores. *Mycologist* 7, 16–24.

Korhonen, K. (1978) Interfertility and clonal size in the *Armillariella mellea* complex. *Karstenia* 18, 31–42.

Lung-Escarment, B., Dunez, J. and Monsion, M. (1978) La differenciation sérologique des formes typique et ostoyae d'Armillaire (*Armillaria mellea*), une preuve supplementaire de la valeur du critere immunologique dans la taxonomie des Champignons. *Compte Rendu Hebdomadaires des Séances de l'Académie des Sciences, Paris, Série D* 287, 475–478.

Lung-Escarment, B., Mohammed, C. and Dunez, J. (1985) Nouvelles méthodes de determination des Armillaires européens: immunologie et electrophorése en gel de polyacrylamide. *European Journal of Forest Pathology* 15, 278–288.

Mallett, K.I. and Hiratsuka, Y. (1985) The 'trap-log' method to survey the distribution of *Armillaria mellea* in forest soils. *Canadian Journal of Forest Research* 15, 1191–1193.

Manley, H. (1992) Detection of Armillaria root rot. Validation Report. University of Reading.

Martin, L-A., Fox, R.T.V. and Baldwin, B.C. (1992a) Rapid methods for the detection of MBC resistance in fungi: I. Immunological approaches. *Proceedings of the 10th International Reinhardsbrunn Symposium*, p. 32.

Martin, L-A., Fox, R.T.V., Baldwin, B.C. and Connerton, I.F. (1992b) Rapid methods for the detection of MBC resistance in fungi: II. Use of the polymerase chain reaction as a diagnostic tool. *Proceedings 1992 Brighton Crop Protection Conference – Pests and Diseases* 1, pp. 207–214.

Miller, S.A., Rittenburg, J.H., Petersen, F.P. and Grothaus, G.D. (1988) Application of rapid, field-usable immunoassays for the diagnosis and monitoring of fungal pathogens in plants. *Proceedings 1988 Brighton Crop Protection Conference – Pests and Diseases* 2, pp. 795–804.

Morrison, D.J. (1989) Pathogenicity of *Armillaria* species is related to rhizomorph growth habit. In: Morrison, D.J. (ed.), *Proceedings of the 7th International Conference on Root and Butt Rots; 1988 August 9–16; Vernon and Victoria, BC.* International Union of Forestry Research Organizations, pp. 584–589.

Morrison, D.J., Thomson, A.J. and Chin, D. (1984) Characterisation of *Armillaria* intersterility groups by isozyme patterns. In: Kile, G.A. (ed.), *Proceedings of the 6th International Conference on Root and Butt Rot of Forest Trees; 1983 August 25–31; Melbourne, Victoria, and Gympie, Australia.* International Union of Forestry Research Organizations, pp. 2–11.

Motta, J.J., Peabody, D.C. and Peabody, R.B. (1986) Quantitative differences in nuclear DNA content between *Armillaria mellea* and *Armillaria bulbosa. Mycologia* 78, 963–965.

Nobles, M.K. (1948) Identification of cultures of wood-rotting fungi. *Canadian Journal of Research, Sect. C* 26, 281–431.

Pawsey, R.G. and Rahman, M.A. (1974) Armillatox field trials. *Gardeners' Chronicle* 175, 29–31.

Pawsey, R.G. and Rahman, M.A. (1976a) Chemical control of infection by honey fungus, *Armillaria mellea*: a review. *Arboricultural Journal* 2, 468–479.

Pawsey, R.G. and Rahman, M.A. (1976b) Armillatox against *Armillariella mellea. Pest Articles and News Summaries* 22, 49–56.

Pegler, D.N. and Gibson, I.A.S. (1972) *Armillariella mellea. Commonwealth Mycological Institute Descriptions of Pathogenic Fungi and Bacteria* No. 321, 2 pp.

Poon, O-S. (1988) Detecting specific proteins for separating and identifying *Armillaria* species. Unpublished MSc Thesis, University of Reading.

Popoola, T.O.S. (1991) The role of stress in *Armillaria* root rot infections. Unpublished Ph.D. Thesis, University of Reading.

Raabe, R.D. (1962) Host list of the root rot fungus, *Armillaria mellea. Hilgardia* 33, 25–88.

Raabe, R.D. (1984) Production of sporophores of *Armillaria mellea* in isolated and pure culture. *Phytopathology* 74, 855.

Raabe, R.D. and Trujillo, E.E. (1963) *Armillaria mellea* in Hawaii. *Plant Disease Reporter* 47, 776.

Rishbeth, J. (1986) Some characteristics of *Armillaria* species in culture. *Transactions of the British Mycological Society* 85, 213–218.

Robard, J. (1987) Synthetic peptides as vaccines. *Nature* 330, 106–107.

Robinson, D.J. (1988) Prospects for the application of nucleic acid probes in plant virus detection. *Proceedings 1988 Brighton Crop Protection Conference – Pests and Diseases* 2, pp. 805–810.

Roll-Hansen, F. (1985) The *Armillaria* species in Europe. *European Journal of Forest Pathology* 15, 22–31.

Shaw, C.G., III and Kile, G.A. (1991) *Armillaria* root disease. *United States Department of Agriculture Forest Service Agriculture Handbook* No. 691.

Shigo, A.L. and Tippett, J.T. (1981) Compartmentalization of decayed wood associated with *Armillaria mellea* in several tree species. *Research Paper NE488.*

US Department of Agriculture, Forest Service. 20 pp.

Smith, M.L., Duchesne, L.C., Bruhn, J.N. and Anderson, J.B. (1990) Mitochondrial genetics in a natural population of the plant pathogen *Armillaria*. *Genetics* 126, 575–582.

Smith, M.L., Bruhn, J.N. and Anderson, J.B. (1992) The fungus *Armillaria bulbosa* is among the largest and oldest living organisms. *Nature* 256, 428–431.

Swedjemark, G. (1989) The use of sniffing dogs in root rot detection. In: Morrison, D.J. (ed.), *Proceedings of the 7th International Conference on Root and Butt Rots; 1988 August 9–16; Vernon and Victoria, BC*. International Union of Forestry Research Organizations, pp. 180–182.

Swift, M.J. (1970) *Armillaria mellea* (Vahl ex Fries) Kummer in central Africa: studies on substrate colonisation relating to the mechanism of biological control by ring-barking. In: Toussoun, T.A., Bega, R.V., Nelson, P.E. (eds), *Root Diseases and Soilborne Pathogens: Proceedings of the Symposium; 1968 July. Imperial College, London*. University of California Press, Berkeley, pp. 150–152.

Sylvänen, A.C., Laaksonen, M. and Söderlund, H. (1986) Fast quantification of nucleic acid hybrids by affinity-based hybrid collection. *Nucleic Acids Research* 14, 5037–5048.

Turner, J.A. and Fox, R.T.V. (1988) Prospects for the successful chemical control of *Armillara* species. *Proceedings 1988 Brighton Crop Protection Conference – Pests and Diseases* 1, pp. 235–240.

Wargo, P.M. and Shaw, C.G. III. (1985) *Armillaria* root rot: the puzzle is being solved. *Plant Disease* 69, 826–832.

Watling, R., Kile, G.A. and Gregory, N.M. (1982) The genus *Armillaria* – nomenclature, typification, and the identity of *Armillaria mellea* and species differentiation. *Transactions of the British Mycological Society* 78, 271–285.

Whitney, R.D., Myren, D.T. and Britnell, W.E. (1978) Comparison of malt agar with malt agar plus orthophenylphenol for isolating *Armillaria mellea* and other fungi from conifer roots. *Canadian Journal of Forest Research* 8, 348–351.

Wright, S.F. and Morton, J.B. (1989) Detection of vesicular-arbuscular mycorrhizal fungus colonization of roots by using a dot-immunoblot assay. *Applied and Environmental Microbiology* 55, 761–763.

8 Identification of Colonizing Forms of *Trichoderma* Species Isolated from Mushroom Compost

P.R. Mills[1] and S. Muthumeenakshi

Biotechnology Centre for Animal and Plant Health, Queen's University of Belfast, Newforge Lane, Belfast BT9 5PX, UK; [1]Present address – Department of Microbial Biotechnology, Horticulture Research International, Wellesbourne, Warwick CV35 9EF, UK.

Introduction

The damaging effects of species of *Trichoderma* weed moulds on commercial mushroom production was first reported over 40 years ago (Kligman, 1950). Changes during the 1980s in the way compost was produced appear to have increased the incidence of weed moulds and losses associated with *Trichoderma* spp. now run into several million pounds (Fletcher, 1990).

Recent articles (Seaby, 1987; Staunton, 1987) describe the problem in Northern Ireland which was originally associated with bag growing and was first recorded in the spring of 1985 when bags were found to turn green 2–5 weeks after the beginning of spawn running. Similar symptoms were reported from the Republic of Ireland and within months of the Irish outbreaks, English growers were experiencing similar problems. It is possible that *Trichoderma* may now also be threatening mushroom production in Canada.

Species of *Trichoderma* are common soil-inhabiting fungi and have been extensively studied by many research workers involved in projects on biological control of plant pathogens. Certain species of *Trichoderma* are either antagonistic to, or parasitic on, some fungi that cause plant diseases. It is therefore not surprising that given the opportunity, *Trichoderma* can affect mushroom (*Agaricus bisporus*) production. The mechanism by which *Trichoderma* affects *Agaricus* is not well understood, but it is apparent that only certain species, or indeed certain isolates of species, can cause problems in commercial production.

Recent work has indicated that the most aggressive strain colonizing compost is *T. harzianum* form 2 (Th2) (Seaby, 1987). Two other forms of *T. harzianum*, Th1 and Th3 are also found commonly in compost, as well as a number of other *Trichoderma* spp.

A system has been developed for identification of *Trichoderma* isolates from compost based primarily on the growth rates of cultures at different temperatures, culture smell, spore shape and the speed with which spores appear in culture (Seaby, 1987). A similar system has been devised by Doyle (1991). Both systems identify isolates of *Trichoderma* spp. with a degree of accuracy within a few days of receipt of samples.

In recent years, taxonomic studies have increasingly used DNA-based methodologies to determine inter- and intra-specific variation. In this chapter we describe experiments using restriction fragment length polymorphism (RFLP) analysis of mitochondrial and ribosomal DNA, randomly amplified polymorphic DNA (RAPD) polymerase chain reaction (PCR) analysis and DNA sequence comparisons of the internal transcribed spacer (ITS) 1 region of *Trichoderma* spp. with specific reference to isolates of *T. harzianum* from mushroom compost.

Restriction fragment length polymorphism

Eighty-one isolates of *T. harzianum* were used in this study. Fifty-nine isolates were obtained from Northern Ireland and provided by Mr D. Seaby, Department of Agriculture for Northern Ireland (DANI), Belfast. Fifteen isolates were obtained from England and provided by Dr J.T. Fletcher, Agricultural Development and Advisory Service (ADAS), Wye. Seven isolates were obtained from the Republic of Ireland and provided by Dr O. Doyle, University College, Dublin. Isolates of *T. viride*, *T. longibrachiatum* and *T. pseudokoningii* were supplied by Mr D. Seaby and Dr J.T. Fletcher.

DNA was extracted from mycelium harvested from liquid shake cultures essentially as described by Raeder and Broda (1985). Total DNA was digested with a variety of restriction endonucleases according to the manufacturers' instructions and restriction fragments were separated electrophoretically in 0.8% agarose gels. Fragments were transferred to nylon membranes (Sambrook *et al.*, 1989) and hybridized to probes which had been labelled with a [α-^{32}P]dATP using a Prime-a-gene Labelling System (Promega).

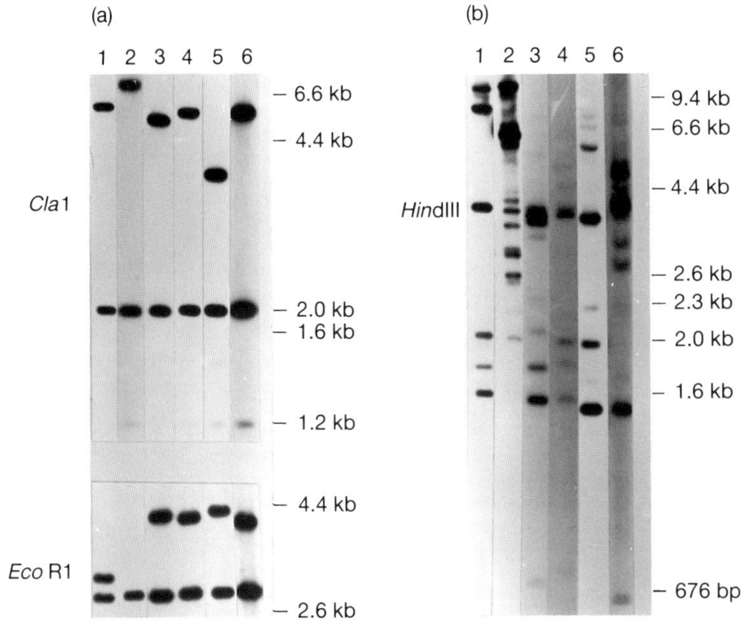

Fig. 8.1. RFLP analysis of *Trichoderma* spp.

a. Restriction fragment patterns of rDNA digested with *Cla*I and *Eco*RI and probed with pMY60.

b. Restriction fragment patterns of mtDNA digested with *Hin*dIII and probed with mtDNA from *T. harzianum* Th2. Lane 1, *T. harzianum* Group 1; Lane 2, *T. harzianum* Group 2; Lane 3, *T. harzianum*, Group 3; Lane 4, *T. viride*; Lane 5, *T. longibrachiatum*; Lane 6, *T. pseudokoningii*. Molecular size markers refer to *Hin*dIII digested lambda DNA.

RFLPs of ribosomal DNA

RFLPs of ribosomal DNA were generated using enzymes *Eco*RI, *Sac*I and *Cla*I. Each *Trichoderma* spp. produced a unique set of restriction enzyme digestion fragments when probed with clone pMY60 containing a complete ribosomal DNA unit from *Saccharomyces carlsbergensis* (Verbeet *et al.*, 1983) with the exception of *T. harzianum* (Fig. 8.1a). The 81 isolates of *T. harzianum* separated into three distinct groups. Within each group all isolates gave a similar pattern with each enzyme tested (data not shown).

Two out of the six enzymes used (*Hin*dIII and *Pst*I) yielded a single digestion product for group 1 and 2 isolates (approximately 9 kb and 12 kb respectively) whereas rDNA from group 3 isolates was not digested. Group 3 isolates could be distinguished by *Bam*HI digestion which resulted in a 9 kb fragment whereas no *Bam*HI site was found in the ribosomal repeat units of either group 1 or group 2 isolates.

By totalling the sizes of fragments from various digestions, the approximate unit length of rDNA was calculated to be 8–9 kb for group 1 and 3 isolates and 11–12 kb for group 2 isolates.

RFLPs of mitochondrial DNA

Restriction enzyme digestion of mitochondrial DNA using *Hin*dIII produced approximately 10–25 discrete fragments for *Trichoderma* spp. When probed with *T. harzianum* group 1, group 2 or group 3 mtDNA labelled with [α-^{32}P]deoxyadenosine 5'-triphosphate, the 81 isolates of *T. harzianum* were separated into three major groups, an example of each shown in Fig. 8.1b. The sum of the fragment sizes gave an estimated molecular size of the mtDNAs of 38–40 kb, 60–62 kb and 28–30 kb for groups 1–3 respectively. Limited polymorphism was apparent within each group but little homology was apparent between groups.

RAPD analysis of *T. harzianum* isolates

Randomly amplified polymorphic DNA (RAPD) analysis can be used to distinguish closely related DNA samples (E. Ward, Chapter 9). Synthetic oligonucleotides (10 mers) supplied by Operon Technologies Inc., California were used for RAPD PCR analysis of *T. harzianum*. Six primers were used to amplify 30 randomly-chosen isolates representing all three major RFLP groups. The amplification products produced by all six primers differentiated, with minor exceptions, the 30 isolates into the same three groups revealed by RFLP analyses. Figure 8.2 shows the RAPD pattern of eight isolates from each group with primer B7. Group 2 isolates showed identical RAPD patterns whereas some variation was apparent within group 1 and 3 isolates. None of the six primers used gave similar banding patterns between groups.

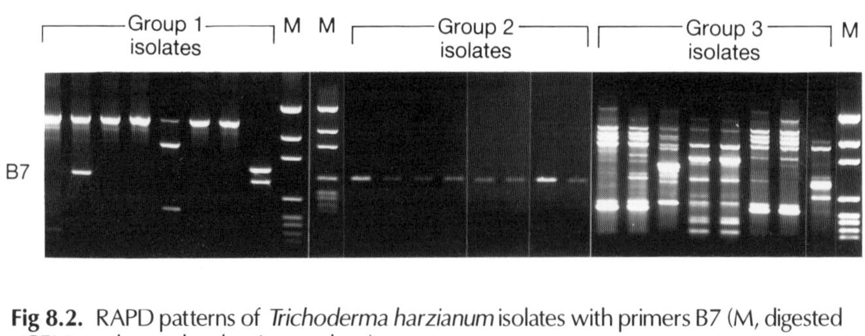

Fig 8.2. RAPD patterns of *Trichoderma harzianum* isolates with primers B7 (M, digested pGEM used as molecular size markers).

```
              10        20        30        40        50        60
Th1  CCGAGTTTACAACTCCCAAACCCAATGTGAACGTTACCAAACTGTTGCCTCGGCGGGATCTCTGCCCC
Th2  ---------------------------------C----------------------------------
Th3  --------------------------------CA-----------------------G--A-*-----

       70        80        90        100       110       120       130
Th1  GGGTGCGTCGCAGCCCCGGACCAAGGCGCCCGCCGGAGGACCAACCTAAAACTCTTATTGTATACCCC
Th2  ---C------------------------------------------C---------------------
Th3  --------------------A-C----------------GA-C-A-C---CTCT--TC----*GT---

         140       150        160       170       180       190      200
Th1  CTCGCGGGTTTTTTT**ATAATCTGAGCCTTTCTCGGCGCCTCTCGTAGGCGTTTCGAAAATGAATCA
Th2  ----------A----T*-CT------------*--------C-----G--------------------
Th3  -------ACG--A--TCT--**-**---***---GA--*********-AAAA--C**-----------
```

Fig. 8.3. Nucleotide sequence of ITS1 region of *Trichoderma harzianum* isolates. Identical bases are indicated by –; * gaps introduced to allow multiple alignment.

Sequence analysis

Functionally and evolutionarily conserved rRNA genes contain both highly- and non-conserved regions which have been used in various studies to determine phylogenetic relationships (Olsen *et al.*, 1986; Förster *et al.*, 1990; Illingworth *et al.*, 1991; Sreenivasaprasad *et al.*, 1992).

PCR amplification of the variable internal transcribed spacer (ITS) 1 region was carried out using biotinylated ITS 1 and ITS 2 primers (White *et al.*, 1990) followed by direct solid phase sequencing of PCR products (Hultman *et al.*, 1989) on Dynabeads (Dynal, Norway) using the chain termination method (Sanger *et al.*, 1977).

Eighteen isolates of *T. harzianum* representing each of the three RFLP groups were sequenced revealing three distinct sequence types (Fig. 8.3). The sequence of all five group 2 isolates (202 bp) was identical. The sequence of all six group 3 isolates (182 bp) was identical. Within the seven group 1 isolates (201–203 bp) insertion/deletion of two thymines was the only variation observed. Group 1 isolates showed 4.7% divergence from group 2 isolates and group 1 and 2 isolates showed 20.0% and 22.9% divergence from group 3 isolates respectively.

Discussion

A number of *Trichoderma* spp. have been isolated from mushroom compost but aggressive colonization appears to be restricted to some isolates of *T. harzianum* (Fletcher, 1990). *T. harzianum* isolates from mushroom compost have been differentiated into three biological forms by Seaby (1987) and Doyle (1991). Delineation of the three forms is not always reliable using biological characteristics, some of which overlap between forms. Unambiguous differentiation of aggressive Th2 isolates from morphologically identical non-(or less) aggressive Th1 and Th3 forms is essential when assessing the risk to mushroom compost from Th2 isolates or when assessing the biological variation within Th2 isolates.

RFLP analysis of the generally conserved ribosomal DNA (rDNA) gene block could identify *Trichoderma* spp. isolated from mushroom compost and separated *T. harzianum* isolates into three distinct groups. RAPD analysis also divided *T. harzianum* isolates into the same three groups. No variation was seen within group 2 isolates but groups 1 and 3 isolates showed some heterogeneity.

The level of variation between *T. harzianum* groups was greatest when comparisons were made using mtDNA. Restriction enzyme fragment patterns again divided *T. harzianum* isolates into three major groups, with each group showing limited variation similar to the levels reported within species of *Phytophthora* (Förster *et al.*, 1988), *Aspergillus* (Moody and Tyler, 1990) and *Colletotrichum* (Sreenivasaprasad *et al.*, 1992). The level of variation between groups was similar to that reported between species of *Phytophthora* (Förster *et al.*, 1988) and *Pythium* spp. (Martin and Kistler, 1990) (data not presented).

Sequence data of the internally transcribed spacer (ITS) 1 region showed no sequence variation amongst group 2 isolates or amongst group 3 isolates and only limited variation amongst group 1 isolates. Divergence values between groups suggest a closer relationship between groups 1 and 2 isolates (4.7%) than the more distant relationship between groups 1 and 2 and group 3 isolates (approximately 20%). These compare with divergence values of only 2% between *Phytophthora* spp. (Lee and Taylor, 1992) and approximately 7% between species of *Colletotrichum* (P.R. Mills, unpublished). However, values of 15% have been reported between two isolates of *Fusarium sambucinum* (O'Donnell, 1992).

The use of molecular diagnostic techniques as discussed in this chapter could make a significant contribution to understanding the ecology of strains of *Trichoderma* weed moulds in mushroom compost. The results have confirmed that the aggressive group 2 strains can be readily distinguished from more benign group 1 or 3 strains and that such differences can be identified by DNA-based as well as by more traditional cultural methods used previously. Use of such rapid and accurate DNA-based methods should facilitate routine monitoring of composts over a range of time intervals from spawning which should help to give a picture of the development of *Trichoderma* strains in mushroom compost as well as providing information on the selective influence of *Agaricus* hyphae on colonization patterns by different *Trichoderma* strains.

Genetic uniformity within the aggressive colonizing group 2 isolates may support the hypothesis that outbreaks of *Trichoderma* within the British Isles could have originated from a single source. Also, these data are not consistent with the view that aggressive forms of *T. harzianum* have been selected from populations of group 1 or 3 isolates, because the extent of the genetic differences are greater than could be explained by a series of recent adaptive mutations within the mushroom compost environment.

It is interesting to note that although cultures received from international culture collections in Europe and New Zealand were similar to group 1 and 3 isolates, none were similar to the aggressive group 2 isolates.

The results reported in this chapter underline the value of molecular techniques in species identification and isolate differentiation. Although the techniques described have been developed for *Trichoderma* weed moulds in mushroom compost, species of *Trichoderma*, especially *T. harzianum* have been widely used experimentally as biocontrol agents against fungal plant pathogens. The DNA-based diagnostic methods discussed here would be equally appropriate for detecting and monitoring the occurrence and persistence of strains used for biocontrol in soil or other habitats.

Acknowledgements

The authors thank Mr D.A. Seaby (DANI), Dr J.T. Fletcher (ADAS) and Dr O. Doyle University College, Dublin for fungal material. This project was partially supported by the Horticultural Development Council.

References

Doyle, O. (1991) *Trichoderma* green mould – Update. *The Irish Mushroom Review* 3, 13–17.

Fletcher, J.T. (1990) *Trichoderma* and *Penicillium* diseases of *Agaricus bisporus*. A literature review for the Horticultural Development Council, ADAS, UK.

Förster, H., Kinscherf, T.G., Leong, S.A. and Maxwell, D.P. (1988) Estimation of relatedness between *Phytophthora* species by analysis of mitochondrial DNA. *Mycologia* 80, 466–478.

Förster, H., Coffey, M.D., Ellwood, H. and Sogin, M.L. (1990) Sequence analysis of the small subunit ribosomal RNAs of three zoosporic fungi and implications for fungal evolution. *Mycologia* 82, 306–312.

Hultman, T., Stahl, S., Hornes, E. and Uhlen, M. (1989) Direct solid phase sequencing of genomic and plasmid DNA using magnetic beads as solid support. *Nucleic Acids Research* 17, 4937–4946.

Illingworth, C.A., Andrews, J.H., Bibeau, C. and Sogin, M.L. (1991) Phylogenetic placement of *Athelia bombacina*, *Aureobasidium pullulans* and *Colletotrichum gloeosporioides* inferred from sequence comparisons of small-subunit ribosomal RNAs. *Experimental Mycology* 15, 65–75.

Kligman, A.M. (1950) *The Handbook of Mushroom Culture*. J.B. Swayne, Kennett Square, Pennsylvania.

Lee, S.B. and Taylor, J.W. (1992) Phylogeny of five fungus-like protoctistan *Phytophthora* species inferred from the internal transcribed spacers of ribosomal DNA. *Molecular Biology Evolution* 9, 636–653.

Martin, F.N. and Kistler, H.C. (1990) Species-specific banding patterns of restriction endonuclease-digested mitochondrial DNA from the genus *Pythium*. *Experimental Mycology* 14, 32–46.

Moody, S.F. and Tyler, B.M. (1990) Restriction enzyme analysis of mitochondrial DNA of the *Aspergillus flavus* group: *A. flavus*, *A. parasiticus* and *A. nomius*. *Applied and Environmental Microbiology* 56, 2441–2452.

O'Donnell, K. (1992) Ribosomal DNA internal transcribed spacers are highly divergent in the phytopathogenic ascomycete *Fusarium sambucinum* (*Gibberella pulicaris*). *Current Genetics* 22, 213–220.

Olsen, G.J., Lane, D.J., Giovannoni, S.J. and Pace, N.R. (1986) Microbial ecology and evolution: a ribosomal RNA approach. *Annual Review of Microbiology* 40, 337–365.

Raeder, U. and Broda, P. (1985) Rapid preparation of DNA from filamentous fungi. *Letters in Applied Microbiology* 1, 17–20.

Sambrook, J., Fritsch, E.F. and Maniatis, T. (1989) *Molecular Cloning. A Laboratory Manual*, 2nd edn, Cold Spring Harbor Laboratory, New York.

Sanger, F.S., Nicklen, S. and Coulsen, A.R. (1977) DNA sequencing with chain-terminating inhibitors. *Proceedings of the National Academy of Sciences of the United States of America* 74, 5463–5467.

Seaby, D.A. (1987) Infection of mushroom compost by *Trichoderma* species. *The Mushroom Journal* 179, 355–361.

Sreenivasaprasad, S., Brown, A.E. and Mills, P.R. (1992) DNA sequence variation and interrelationships among *Colletotrichum* species causing strawberry anthracnose. *Physiological and Molecular Plant Pathology* 41, 265–281.

Staunton, L. (1987) *Trichoderma* green mould in mushroom compost. *The Mushroom Journal* 179, 362–363.

Verbeet, Ph., Klootwijk, J., Heerikhuizen, H., Fontijn, R., Vreugdenhil, E. and Planta, R.J. (1983) Molecular cloning of rDNA of *Saccharomyces rosei* and comparison of its transcription initiation region with that of *Saccharomyces carlsbergensis*. *Gene* 23, 53–63.

White, T.J., Bruns, T., Lee, S. and Taylor, J. (1990) Amplification and direct sequencing of fungal ribosomal RNA genes for phylogenetics. In: Innis, M.A., Gelfand, D.H., Sninsky, J.J. and White, T.J. (eds), *PCR Protocols. A Guide to Methods and Applications*. Academic Press, San Diego, pp. 315–322.

9 Use of the Polymerase Chain Reaction for Identifying Plant Pathogens

E. Ward

Plant Pathology Department, AFRC Institute of Arable Crops Research, Rothamsted Experimental Station, Harpenden, Herts AL5 2JQ, UK.

Introduction

The polymerase chain reaction (PCR) is a very important technique in molecular biology. It is a method for synthesizing millions of copies of a DNA sequence in a few hours. Since its introduction in the mid 1980s (Mullis and Faloona, 1987; Saiki *et al.*, 1988) it has simplified, speeded up and improved molecular tasks previously achieved by recombinant DNA technology. Accounts of the discovery and principles of PCR are found in Mullis (1990) and Cherfas (1990). This chapter firstly describes the principles and method of PCR and then discusses some of the applications of the technique in the identification of organisms. There are many uses and variations of the technique. This chapter describes some of the most common or useful applications of the method in plant pathogen identification; the specific detection of microorganisms by PCR, the use of PCR in analysing ribosomal genes and the technique of RAPDs or PCR with short random primers, which is a simple way of producing 'fingerprints' of organisms.

PCR – Principles and Methods

Figure 9.1 shows a schematic diagram of PCR. Genomic DNA in prokaryotes and eukaryotes consists of two strands held together by hydrogen bonds between pairs of complementary bases. In the first stage of PCR, DNA is heated to 95°C to separate the two strands. Two primers, short pieces of single stranded DNA designed to match the ends of the region to be amplified, are added. The primers bind only to regions of DNA with complementary base sequences. This binding or annealing step is done at between 40°C and 65°C. The second strand of DNA is then extended from the primers by a thermostable DNA polymerase with deoxyribonucleoside triphosphates (dNTPs) as building blocks. This third, elongation step is usually done at 72°C. At the end of this first cycle of the PCR reaction, two copies of the target DNA exist where previously there was one. The two

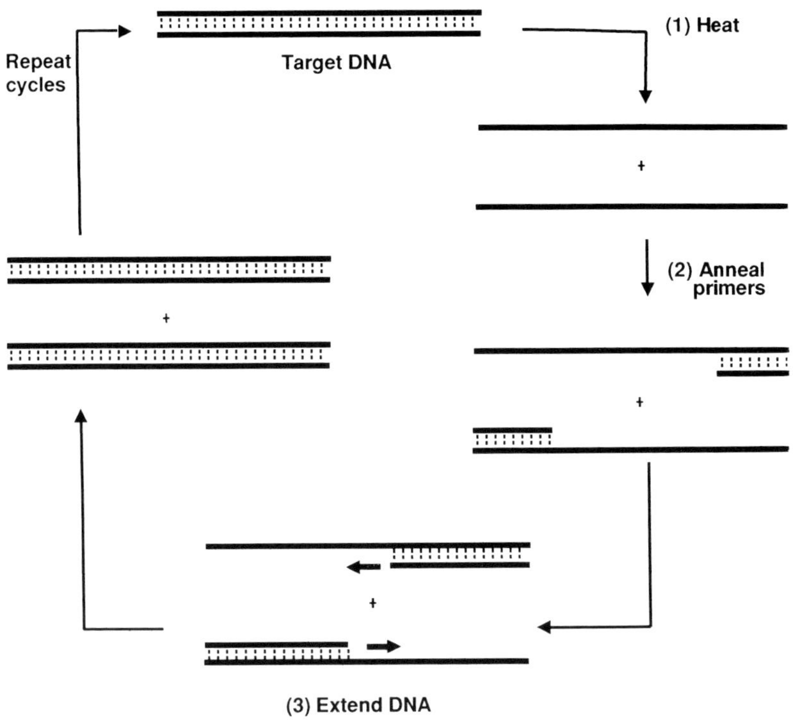

Fig. 9.1. The polymerase chain reaction. The double stranded target DNA is heated to separate the two strands (1). The two primers are then annealed (2) and the second strands of the DNA are synthesized (3). The process is repeated many times, amplifying the DNA exponentially.

copies then enter a second amplification cycle and a third and so on. After 25 such cycles, taking only a few hours, about a million copies of the starting sequence have been produced.

In practice the procedure is very simple. DNA, buffer, dNTPs, primers and DNA polymerase are mixed in a tube and placed in a computer-controlled heating block that can be programmed to switch between the different temperatures required. At the end of the PCR reaction the presence of the amplified DNA is checked by electrophoresing it on an agarose gel and staining with ethidium bromide. The practical details of PCR, optimization of the conditions and variations of the technique are covered in a number of publications, e.g. Erlich (1989); White *et al.* (1989); Innis *et al.* (1990); Bej *et al.* (1991). The crucial component of the system, which determines the specificity of the reaction, is the primers. With one exception (RAPDs, discussed later) it is necessary to know the DNA sequences at each end of the fragment to be amplified so that suitable primers can be chosen. Amplification of DNA occurs only where the primers can anneal.

Applications

The many applications of the technique are covered in various reviews and books (e.g. Erlich, 1989; Innis *et al.*, 1990; Erlich *et al.*, 1991; McPherson *et al.*, 1991; Steffan and Atlas, 1991; Henson and French, 1993). All the applications discussed below are suitable for any organism but they are illustrated with examples of my own work on the soil-borne, root-infecting fungus which causes take-all disease of cereals.

The *Gaeumannomyces–Phialophora* complex

The species *Gaeumannomyces graminis* is subdivided into three varieties. Variety *tritici* (Ggt) is the wheat take-all fungus and the main organism responsible for take-all in cereals in the UK and elsewhere. The disease and its causal organisms are reviewed in Asher and Shipton (1981) and Hornby and Bateman (1991). Variety *avenae* (Gga) is also pathogenic but has different host preferences – in particular it is pathogenic on oats and turf grasses. Variety *graminis* (Ggg) is only very weakly pathogenic on the cereals grown in the UK. These varieties have *Phialophora*-like anamorphs (asexual forms). For Ggt and Gga these are unnamed; the Ggg anamorph is known as *Phialophora* sp. (lobed hyphopodia) and occurs on cereal roots. *Phialophora graminicola*, a closely related species whose teleomorph is *Gaeumannomyces cylindrosporus*, is non-pathogenic, occurs on grasses but is also found on cereal roots. *G. graminis*, *P. graminicola* and other similar non-pathogenic forms of *Phialophora* found on cereal roots constitute the *Gaeumannomyces–Phialophora* complex. A detailed account of the taxonomic relationships between these organisms is found in Walker (1981).

The four organisms mentioned above look similar on infected roots and in culture. To confirm the identity of isolates by conventional methods, i.e. by pathogenicity testing and morphological characteristics (Bateman *et al.* 1992; Mathre, 1992), is slow and often inconclusive. Molecular techniques, including PCR methods, provide a means of overcoming these difficulties.

Specific detection of microorganisms

First, it is necessary to find a DNA sequence, or sequences, unique to the organism and to use these to design primers. One approach is to develop the

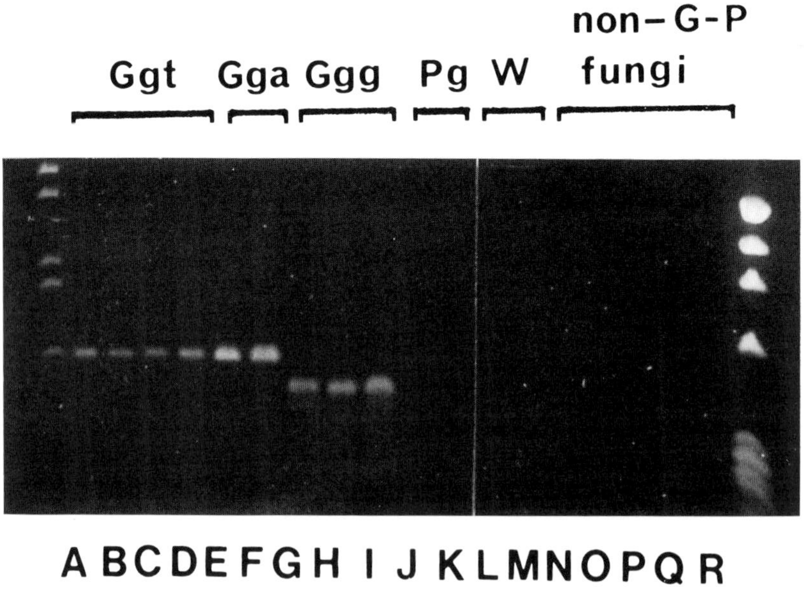

Fig. 9.2. A specific PCR test for *Gaeumannomyces graminis.* Primers KS1F and KS2R (Schesser *et al.*, 1991) were used in PCR assays with DNA from *G. graminis, P. graminicola* and other fungi and plants. Lanes A–D = *G. graminis* var. *tritici* (Ggt) isolates, E and F = *G. graminis* var. *avenae* (Gga) isolates, G–I = *G. graminis* var. *graminis* (Ggg) isolates, J and K = *P. graminicola* (Pg) isolates, L and M = uninfected wheat roots (W), N = *Rhizoctonia solani*, O = *Fusarium culmorum*, P = *Fusarium avenaceum*, Q = *Cochliobolus sativus*, R = unidentified sterile black mycelium. The outside lanes are DNA size markers.

primers from specific probes that already exist to detect the organism (e.g. Bereswill *et al.*, 1992; Seal *et al.*, 1992). For example, a probe was developed (Henson, 1989) that detected only *Gaeumannomyces graminis* and related *Phialophora* species (Bateman *et al.*, 1992; Henson, 1992). The DNA sequence of the probe was determined and used to design PCR primers that delimited a short region in the middle of the probe DNA. These PCR primers were then tested to check that they allowed amplification only of *G. graminis* DNA and not the DNA of unrelated fungi (Schesser *et al.*, 1991; Henson, 1992; Henson *et al.*, 1993).

Figure 9.2 shows an example of a diagnostic test using primers KS1F and KS2R described by Schesser *et al.* (1991) but in modified conditions (E. Ward, unpublished). Using these primers a band is produced with the DNA of isolates of all three varieties of *G. graminis*. Ggg isolates (Fig. 9.2 lanes G–I) produce a band of approximately 0.43kb, whereas that produced for Ggt or Gga isolates (lanes A–F) is slightly larger (approximately 0.56kb). No DNA was amplified from DNA of *P. graminicola* isolates tested (lanes J and K), unrelated fungi (lanes N–R) or uninfected wheat roots (lanes L and M); no sequence complementary to the primers exists in the DNA of these organisms and so amplification cannot occur. The DNA samples that tested negatively in the PCR assay with KS1F and KS2R were tested in a PCR with primers ITS4 and ITS5 (White *et al.*, 1990) that recognize ribosomal sequences in fungi and plants. In every case, bands were amplified, confirming that these DNA samples were sufficiently good for PCR assays provided that suitable primers were present.

PCR using a single pair of primers does not always give a sufficiently specific test, as other, non-related, organisms are also detected. Indeed, the studies of Schesser *et al.* (1991) indicated that the use of KS1F and KS2R in a PCR test was not a specific assay for *G. graminis*; other, non-related, fungi (*Fusarium culmorum* and *Cochliobolus sativus*) were also detected. There are several ways of overcoming this problem and Schesser *et al.* (1991) chose to use a nested primer approach (Mullis and Faloona, 1987). The first round of PCR was done using primers KS1F and KS2R and the product of this was used in a second PCR. This second PCR used primers KS4F and KS5R, which amplify a region of DNA within that amplified by the first pair. The use of nested primers increased the specificity so that only *G. graminis* was detected (Schesser *et al.*, 1991; Henson, 1992; Henson *et al.*, 1993). However, this procedure is more labour-intensive, more costly and more prone to contamination of the assay than the single primer pair method.

We increased the specificity of primers KS1F and KS2R for *G. graminis* by increasing the primer annealing temperature in the PCR. Our experiments have shown that this modification gives the required specificity. Not only did non-related fungi test negatively, including *Fusarium culmorum* and *Cochliobolus sativus* that Schesser *et al.* (1991) found tested positively, but also no band was amplified from isolates of *Phialophora graminicola*, which are close relatives of *G. graminis* (Fig. 9.2 and E. Ward, unpublished).

Analysis of ribosomal DNAs

Ribosomal DNAs (rDNAs) are used widely in identification and taxonomy because they have a number of useful features (Olsen *et al.*, 1986; Woese, 1987; Bruns *et al.*, 1991). They are found in all organisms and they are present at high copy number which improves the sensitivity of detection. In many organisms they have been sequenced (Dams *et al.*, 1988). Comparison of these sequences has shown that the ribosomal genes consist of regions that are identical in many organisms and other regions that are highly variable.

Ribosomal RNAs are encoded in the DNA of both the nucleus and the mitochondrion. The nuclear rDNA region usually consists of three genes, the large subunit gene, the small subunit gene and the 5.8S gene, separated by internal transcribed spacer (ITS) regions, in a unit repeated many times. Some of the rDNAs are well conserved throughout evolution and other parts (e.g. the spacer regions and the mitochondrial rDNAs) are variable even within a species. The conserved regions allow probes from one species to be used to detect the equivalent ribosomal genes from other species, genera, families or even kingdoms (e.g. Walsh *et al.*, 1990; Carder and Barbara, 1991). Regions of the DNA that are always identical in sequence (consensus regions) are used as the primer sites in PCR so that they can amplify DNA from all members of a taxon, e.g. from fungi (White *et al.*, 1990) or bacteria (Barry *et al.*, 1990). However, since they flank variable regions, the DNA fragments amplified have many differences that can be exploited in identification. Three ways of doing this are by: (i) sequencing the amplified DNA; (ii) restriction analysis of the amplified DNA; and (iii) use of the amplified DNA as a probe.

Ribosomal DNA sequencing

Before sequencing, the ribosomal gene regions are amplified from different but related organisms. The sequences from the different organisms can then be aligned with one another so that the regions that are identical and those that are different can be highlighted. This can be done manually, or with the help of computer programs. Analysis of the relatedness between the sequences gives data on which to base a phylogenetic taxonomy for the organisms (e.g. Gunderson *et al.*, 1987; Medlin *et al.*, 1988; White *et al.*, 1990; Sogin, 1990; Bruns *et al.*, 1991; Lee and Taylor, 1992). Although it has been possible for several years to sequence ribosomal DNA or RNA and compare the sequences (Guadet *et al.*, 1989; Peterson, 1991), progress was relatively slow as the techniques involved were laborious and time consuming. The availability of PCR and the use of direct sequencing of PCR products has speeded up the method considerably so that it is now realistic to sequence rDNAs of large numbers of organisms in this way. With the increasing use of automated DNA sequencing machines the advances will be even more swift.

As well as generating data for phylogenetic comparisons, ribosomal DNA sequences can be used in other ways to aid identification. By studying the sequence, short regions of the ribosomal DNA can be found that are unique to particular taxa. Oligonucleotides can be made that are complementary to these regions and these can then be used as specific oligonucleotide probes or as PCR primers in a specific PCR assay for the organism (e.g. Barry *et al.*, 1990; Nazar *et al.*, 1991; Rossau *et al.*, 1991).

This approach has also been used to look at the range of organisms in environments such as soil and water. Ribosomal genes are amplified, by PCR, from the mixture of organisms in the sample. The rDNAs from the individual species present must then be separated from one another and multiplied to produce enough material with which to work. This is done by cloning, where each rDNA is propagated individually by attaching it to a vector DNA molecule that is capable of multiplying itself within a bacterium. Each of the different rDNAs (attached to their vectors) is then extracted from the bacteria, purified and sequenced. The sequences can then be checked with the known sequences of rDNAs in databases to determine what organisms are present (Giovannoni *et al.*, 1990; Ward *et al.*, 1990; Pickup, 1991; Steffan and Atlas, 1991). As more sequences become available this type of approach may be of considerable use in ecological research, especially for non-culturable organisms.

Restriction analysis of PCR-amplified ribosomal DNA

Another way of analysing ribosomal DNAs amplified from various organisms is by restriction analysis of the PCR products to give band patterns (restriction fragment length polymorphisms, RFLPs) that are useful in identification. This involves digestion of the PCR-amplified DNA with one or more restriction enzymes then electrophoresis on a high percentage agarose gel to separate the small DNA fragments produced. It is a very quick, simple strategy for generation of RFLP patterns, and one that has been used successfully for a number of fungi (Cubeta *et al.*, 1991; Chen *et al.*, 1992; Liu and Sinclair, 1992), bacteria (Vilgalys and Hester, 1990) and mycoplasmas (Ahrens and Seemüller, 1992). Figure 9.3 shows results obtained using this approach on isolates of *G. graminis* and *P. graminicola* (Ward and Akrofi, 1993). The internal spacer region (ITS) between the rDNA genes was amplified from isolates of the three varieties of *G. graminis* and *Phialophora*

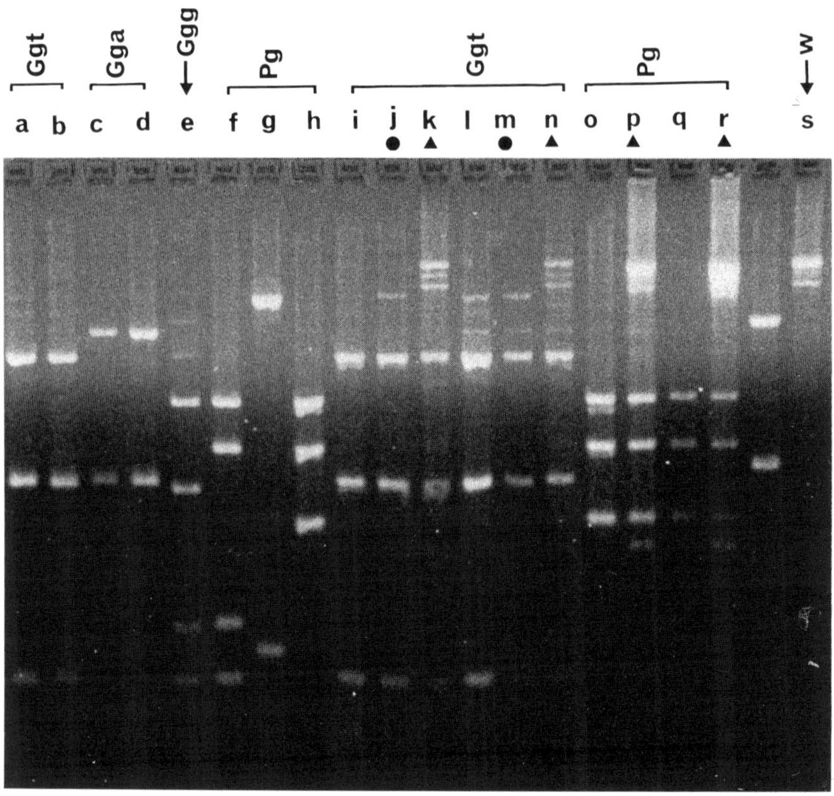

Fig. 9.3. Restriction digestion of PCR-amplified ribosomal DNAs from *G. graminis* and *P. graminicola*. DNA was amplified using primers ITS4 and ITS5 (White *et al.*, 1990) and restricted with *Dde*I. Key: a = Ggt 90/2–4; b = Ggt 90.GLR.17; c = Gga PO86/441; d = Gga PO86/439; e = Ggg 85/23–1; f = Pg 74/1736–2; g = Pg 85/17–3b; h = Pg 92/2–1A; i, j, k = Ggt 92/3–2; l, m, n = Ggt 92/15–4A; o, p = Pg 92/2–1A; q, r = Pg 92/1–1; s = uninfected wheat; ●= DNA obtained by boiling fungal mycelium; ▲ = DNA obtained from infected wheat roots. All isolates not prefixed by 92/ are described in Ward and Gray (1992). Isolates prefixed by 92/ were obtained from R.J. Gutteridge, Rothamsted Experimental Station.

graminicola, using primers ITS4 and ITS5 (White *et al.*, 1990) and cut with the restriction enzyme *Dde*I. Each variety and species had its own characteristic RFLP pattern which could be used to identify unknown isolates (Fig. 9.3).

The quality and quantity of the DNA used for this test, and indeed any other PCR reaction, does not need to be very great. The method even works on DNA released from cells by simply boiling fungal mycelium in tris buffer (pH 8) for 15 min (Fig. 9.3, lanes j and m). It will also work on DNA extracted from infected plant material such as roots (lanes k, n, p and r). Primers ITS4 and ITS5 also amplify DNA from host wheat plant DNA (lane s), but these give very differently sized bands after restriction enzyme digestion that can easily be distinguished from those of the fungus.

Use of amplified ribosomal DNA as a probe

PCR amplified ribosomal DNA, or indeed any other DNA, can also be used as a probe, for example in RFLP analysis. In this technique DNA is prepared from the organism, digested with restriction enzymes and the fragments separated by size on agarose gels. The DNA is then transferred to a membrane and hybridized with a labelled ribosomal probe that has usually been derived from a different organism. In such cases, hybridization conditions are chosen that allow some mismatches of sequence (low stringency conditions), so that many different organisms can be detected by the same probe.

The bands where the probe has bound to the membrane are then detected and the band patterns produced by the different organisms are compared. To make a probe with a higher specificity for a particular

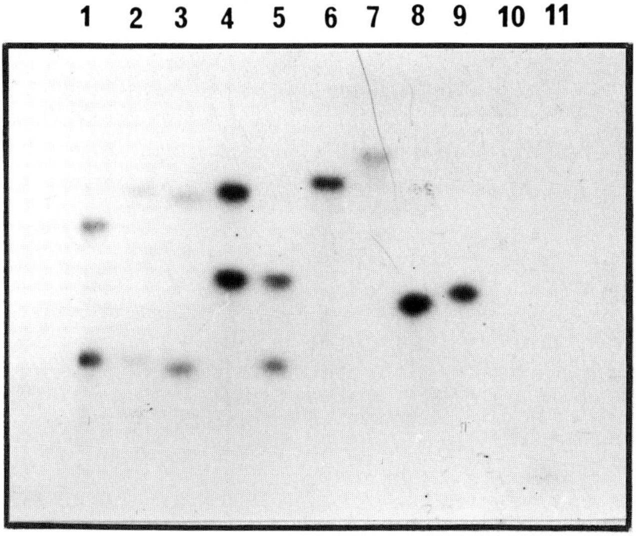

Fig. 9.4. RFLP analysis using a ribosomal probe generated by PCR. DNA from isolates of the *Gaeumannomyces –Phialophora* complex was cut with *Eco*RI, electrophoresed on a 0.8% agarose gel, blotted onto a nylon membrane and hybridized with the PCR-generated mitochondrial ribosomal probe GggMR1 (Ward and Gray, 1992). Detection was by the non-radioactive digoxigenin system (Boehringer-Mannheim) using chemiluminescence. Lane (1) = Ggt 88/10–1, (2) = Ggt Og12N, (3) = Ggt 87/7–4, (4) = Ggt 180, (5) = Gga PO86/439, (6) = Ggg 89/3–1, (7) = Ggg 148, (8) = Ggg 153, (9) = Pg 74/1736–2, (10) = Pg 85/17–3b, (11)= *G. caricis*. These isolates are described in Ward and Gray (1992).

organism, its ribosomal genes can be cloned and used under hybridization conditions that allow only very closely related sequences to be detected (highly stringent conditions). These cloning procedures are, however, relatively slow and a much quicker alternative is to generate the probe by PCR; consensus primers, complementary to highly conserved regions of the rDNA, are used to amplify DNA from the organism being studied.

This method was used to devise a way of differentiating between the three varieties of *G. graminis* and *P. graminicola* (Ward and Gray, 1992). The genomic DNA was cut by the restriction enzyme *Eco*RI and the probe used was complementary to DNA from the small mitochondrial ribosomal subunit of an isolate of *G. graminis* var. *graminis* (Fig. 9.4). As well as producing distinct pattern types for each of the three varieties, the method subdivided the varieties into two or three groups of related isolates. In the case of the Gga isolates this was found to be related to the host plant from which the isolate was derived. All of those in the first group (Fig. 9.4, lane 4) came from wheat or turfgrasses whereas those in the second group (Fig. 9.4, lane 5) came from oats. The method of probe labelling and detection shown in Fig. 9.4 is a non-radioactive one, using chemiluminescence. We have found this to be at least as sensitive, and often more sensitive, than radioactive methods; the results were also obtained more quickly and it was more convenient and safer.

This method can be used for any fungus and it takes only a few hours to produce a probe in this way, making it a much quicker process than by cloning the ribosomal gene. The probe showed a high degree of specificity for *G. graminis*; DNA from several unrelated fungi was not detected.

RAPDs

PCR generally involves amplification of a known target gene or DNA fragment using two highly specific primers. However, there is a form of PCR where, instead, a single short arbitrarily-chosen oligonucleotide is used to produce 'fingerprints', without knowing anything about the sequences of the DNA to be amplified (Welsh and McClelland, 1990; Williams *et al.*, 1990). Various groups have given the technique different acronyms but the best known is RAPDs which stands for Random Amplified Polymorphic DNAs. Other examples of the use of this method are described in Martin *et al.* (1991), Caetano-Anollés *et al.* (1991), Crowhurst *et al.* (1991) and Guthrie *et al.* (1992).

In this technique the primers are much shorter than those generally used, usually about 10 bases rather than around 20 bases long, and one primer is sufficient. Since the primers are so short, it is likely that many sequences complementary to the primer will be present in the target DNA. There is also a high probability that pairs of sequences complementary to the primer will be

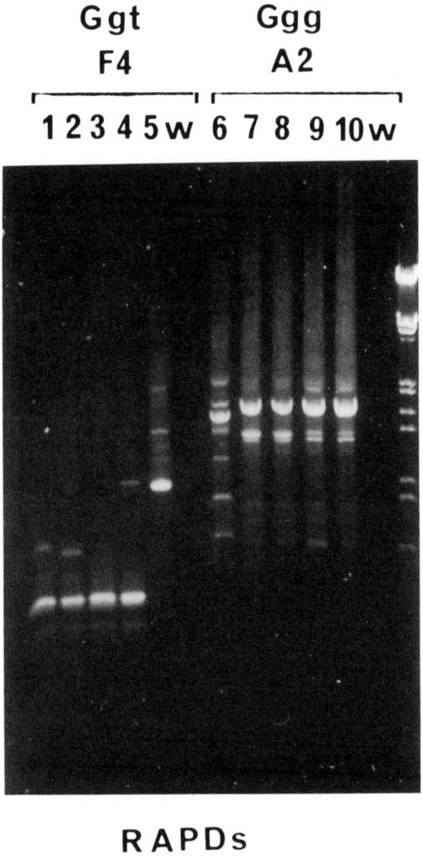

Fig. 9.5. An example of the use of RAPDs. Random primers F4 and A2 (see text) were used on isolates of *G. graminis* var. *tritici* (Ggt, lanes 1–5) and *G. graminis* var. *graminis* (Ggg, lanes 6–10). Key to isolates used: 1 = 184; 2 = 90.GLO(ADAS); 3 = Og12N; 4 = 90.GLR.17; 5 = 180; 6 = 176; 7 = 247; 8 = 89/3–1; 9 = 89/5–3; 10 = 89/5–1. All isolates are described in Ward and Gray (1992). w = negative control with the DNA substituted by water. The track at the far right hand side is the size marker (Lambda DNA cut with *Eco*RI and *Hin*dIII).

arranged close enough together, and in the correct orientation, for PCR amplification of the intervening sequences. By making the conditions of the PCR relatively non-specific, e.g. by reducing the annealing temperature and increasing the number of amplification cycles to 45 (rather than 25), each RAPD PCR reaction is likely to result in amplification of several bands which can be detected by electrophoresing the product on an agarose gel and staining with ethidium bromide. Usually about three to ten bands are seen. The patterns are often highly specific at the isolate level so they can be used as genetic 'fingerprints'.

Figure 9.5 shows results using two separate primers F4 and A2 on isolates of Ggt and Ggg. The sequences of these primers are: F4, GGTGATCAGG; A2, TGCCGAGCTG. The Ggt isolates came originally from a range of host species and locations and all gave different patterns although most had one strong band in common. Earlier tests (Ward and Gray, 1992) had already shown that isolate 180 (Fig. 9.5, lane 5) is an atypical Ggt, being closer to Gga isolates at the DNA level. Figure 9.5 shows that its pattern is very different from the other Ggts. The Ggg isolates in lanes 7–10 (Fig. 9.5) were all cultured from wheat at Rothamsted in the same year and they all gave identical patterns. Isolate 176 (lane 6) is from the USA, isolated from soyabean, and has a completely different pattern from the others. The technique can thus find differences between isolates that it is not possible to detect by conventional methods of identification.

Generally with PCR it is essential to avoid contamination since the technique is so sensitive it can detect minute quantities of foreign DNA. With RAPDs it is even more important because the short primers and low specificity conditions could amplify any DNA that accidentally gets into the sample. Negative controls, where water replaces the DNA, must therefore be included in the tests. In the experiment shown in Fig. 9.5 there was no amplification in the negative control (the track labelled w), but several workers have reported that occasionally bands of DNA are seen in water controls (Reeves and Ball, 1991). If there are contaminating bands in the negative controls, useful results can still be obtained by running them alongside the test samples for comparison, but they need careful interpretation.

Any short 8–12 base sequences might be suitable for RAPDs of a particular organism; useful primers are found by trial and error. Kits are available that each contain 20 different primers (e.g. from Operon Technologies, Alameda, California) and these same primers could be used for fingerprinting any organism. As well as using the technique to obtain 'fingerprints', any bands that appear to be unique to particular taxa can be labelled and tested for use as specific probes. To simplify the cloning of useful DNA fragments, primers can be used that already have restriction enzyme sites as part of their sequence.

Concluding Remarks

PCR is a very important and useful technique as well as being very quick and easy to do. The method is very sensitive, and needs only minute quantities of DNA, e.g. from single fungal spores (Lee and Taylor, 1990). The quality of the starting material for PCR assays does not need to be very great. Thus sample preparation is very quick and easy. Indeed, boiling bacterial cells or fungal mycelia in water or buffer can release sufficient DNA for PCR tests. DNA for PCR analysis can also be readily extracted from environmental samples (Steffan and Atlas, 1991) such as water (Atlas and Bej, 1990) and soil (Pillai *et al.*, 1991; Henson *et al.*, 1993) although some modification of the standard procedures is generally needed to overcome the effect of inhibitory substances. The possibility of direct extraction of DNA from environmental samples makes it a particularly useful technique for organisms that cannot be cultured on synthetic media.

However, care is needed when using PCR. Its great sensitivity can be a disadvantage; careless sample preparation or PCR technique can result in contamination of the assay leading to erroneous results. It is particularly important to take great care when the primers used will amplify DNA from many organisms (e.g. those based on consensus sequences and RAPD primers). Since PCR is an *in vitro* method, it is not subject to the cell's normal safeguards against errors of base incorporation (mutations) during DNA synthesis. Thus some errors can occur resulting in the amplified DNA not being identical to the starting material (Eckert and Kunkel, 1991). These error rates, however, are very low and only likely to be important if the amplified DNA is to be sequenced or expressed as a protein.

PCR has also been used to quantify DNA, mRNA and therefore numbers of organisms in samples (Gilliland *et al.*, 1990; Wang and Mark, 1990; Dallman and Porter, 1991; Simon *et al.*, 1992) but the methods are not simple. This is because the PCR proceeds through an exponential phase, so minute differences in the efficiency of the reaction between samples can give rise to dramatically different amounts of final product. Also, it is important to determine the quantity of product before the reaction reaches its eventual plateau phase, since then the amount of product is no longer proportional to the original amount of template. There is some controversy about the best approaches to use and indeed as to whether PCR can be truly a quantitative technique. Certainly, when using PCR for quantification, great care is needed in the optimization of the reaction conditions, the choice of appropriate controls and the validation of the results. However, several publications have demonstrated that the results achieved using quantitative PCR correlate closely with those obtained using other methods for quantification such as hybridization studies (Ferre, 1992).

However, despite these minor problems, PCR is a very effective technique and is being increasingly used for many molecular biological tasks. Its main uses in ecological work are likely to be to produce sensitive and simple methods of detecting and identifying organisms in the environment. This includes discrimination at many taxonomic levels and the monitoring of natural and introduced populations of organisms. It will allow the study of complex population dynamics in media such as soil where approaches using biological methods have been difficult and laborious. However, PCR is still a relatively new technique and there are likely to be many further advances and additions to the methods available in the next few years.

Acknowledgements

I thank Geoff Bateman, John Antoniw and David Hornby for advice and helpful comments on the manuscript and acknowledge the technical assistance of my research students Rachel Gray, Virginia Raleigh and Andrews Akrofi.

References

Ahrens, U. and Seemüller, E. (1992) Detection of DNA of plant pathogenic mycoplasma-like organisms by a polymerase chain reaction that amplifies a sequence of the 16S rRNA gene. *Phytopathology* 82, 828–832.

Asher, M.J.C. and Shipton, P.J. (ed.) (1981) *Biology and Control of Take-all.* Academic Press, London, 538 pp.

Atlas, R.M. and Bej, A.K. (1990) Detecting bacterial pathogens in environmental water samples by using PCR and gene probes. In: Innis, M.A., Gelfand, D.H., Sninsky, J.J. and White, T.J. (eds), *PCR Protocols. A Guide to Methods and Applications.* Academic Press, San Diego, pp. 399–406.

Barry, T., Powell, R. and Gannon, F. (1990) A general method to generate DNA probes for microorganisms. *Biotechnology* 8, 233–236.

Bateman, G.L., Ward, E. and Antoniw, J.F. (1992) Identification of *Gaeumannomyces graminis* var. *tritici* and *G. graminis* var. *avenae* using a DNA probe and non-molecular methods. *Mycological Research* 96, 737–742.

Bej, A.K., Mahbubani, M.H. and Atlas, R.M. (1991) Amplification of nucleic acids by polymerase chain reaction and other methods and their applications. *Critical Reviews in Biochemistry and Molecular Biology* 26, 301–334.

Bereswill, S., Pahl, A., Bellemann, P., Zeller, W. and Geider, K. (1992) Sensitive and species-specific detection of *Erwinia amylovora* by polymerase chain reaction analysis. *Applied and Environmental Microbiology* 58, 3522–3526.

Bruns, T.D., White, T.J. and Taylor, J.W. (1991) Fungal molecular systematics. *Annual Review of Ecology and Systematics* 22, 525–564.

Caetano-Anollés, G., Bassam, B.J. and Gresshoff, P.M. (1991) DNA amplification fingerprinting using very short arbitrary oligonucleotide primers. *Biotechnology* 9, 553–557.

Carder, J.H. and Barbara, D.J. (1991) Molecular variation and restriction fragment length polymorphisms (RFLPs) within and between six species of *Verticillium. Mycological Research* 95, 935–942.

Chen, W., Hoy J.W. and Schneider, R.W. (1992) Species-specific polymorphisms in transcribed ribosomal DNA of five *Pythium* species. *Experimental Mycology* 16, 22–34.

Cherfas, J. (1990) Genes unlimited. *New Scientist* 126, No. 1712, pp. 29–33.

Crowhurst, R.N., Hawthorne, B.T., Rikkerink, E.H.A. and Templeton, M.D. (1991) Differentiation of *Fusarium solani* f.sp. *cucurbitae* races 1 and 2 by random amplification of polymorphic DNA. *Current Genetics* 20, 391–396.

Cubeta, M.A., Echandi, E., Abernethy, T. and Vilgalys, R. (1991) Characterization of anastomosis groups of binucleate *Rhizoctonia* species using restriction analysis of an amplified ribosomal RNA gene. *Phytopathology* 81, 1395–1400.

Dallman, M.J. and Porter, A.C.G. (1991) Semi-quantitative PCR for the analysis of gene expression. In: McPherson, M.J., Quirke, P. and Taylor, G.R (eds), *PCR. A Practical Approach*. Oxford University Press, Oxford, pp. 215–224.

Dams, E., Hendriks, L., Van de Peer, Y., Neefs, J.-M., Smits, G., Vandenbempt, I. and De Wachter, R. (1988) Compilation of small ribosomal subunit RNA sequences. *Nucleic Acids Research* 16 (Supplement), r87–r173.

Eckert K.A. and Kunkel T.A. (1991) The fidelity of DNA polymerases used in the polymerase chain reactions. In: McPherson, M.J., Quirke, P. and Taylor, G.R. (eds), *PCR. A Practical Approach*. Oxford University Press, Oxford, pp. 225–244.

Erlich, H.A. (ed.) (1989) *PCR Technology. Principles and Applications for DNA Amplification*. Macmillan Publishers, Basingstoke, UK, 246 pp.

Erlich, H.A., Gelfand, D. and Sninsky, J.J. (1991) Recent advances in the polymerase chain reaction. *Science* 252, 1643–1651.

Ferre, F. (1992) Quantitative or semi-quantitative PCR: reality versus myth. *PCR Methods and Applications* 2, 1–9.

Gilliland, G., Perrin, S. and Bunn, H.F. (1990) Competitive PCR for quantitation of mRNA. In: Innis, M.A., Gelfand, D.H., Sninsky, J.J. and White, T.J. (eds), *PCR Protocols. A Guide to Methods and Applications*. Academic Press, San Diego, pp. 60–69.

Giovannoni, S.J., Britschgi, T.B., Moyer, C.L. and Field, K.G. (1990) Genetic diversity in Sargasso Sea bacterioplankton. *Nature* 345, 60–63.

Guadet, J., Julien, J., Lafay, J.F. and Brygoo, Y. (1989) Phylogeny of some *Fusarium* species as determined by large-subunit rRNA sequence comparison. *Molecular Biology and Evolution* 6, 227–242.

Gunderson, J.H., Elwood, H., Ingold, A., Kindle, K. and Sogin, M.L. (1987) Phylogenetic relationships between chlorophytes, chrysophytes and oomycetes. *Proceedings of the National Academy of Sciences of the United States of America* 84, 5823–5827.

Guthrie, P.A.I., Magill, C.W., Frederiksen, R.A. and Odvody, G.N. (1992) Random amplified polymorphic DNA markers: a system for identifying and differentiating isolates of *Colletotrichum graminicola*. *Phytopathology* 82, 832–835.

Henson, J.M. (1989) DNA probe for the identification of the take-all fungus *Gaeumannomyces graminis*. *Applied and Environmental Microbiology* 55, 284–288.

Henson, J.M. (1992) DNA hybridization and polymerase chain reaction (PCR) tests for the identification of *Gaeumannomyces*, *Phialophora* and *Magnaporthe* isolates. *Mycological Research* 96, 629–636.

Henson, J.M. and French, R. (1993) The polymerase chain reaction and plant disease diagnosis. *Annual Review of Phytopathology* 31, 81–109.

Henson, J.M., Goins, T., Grey, W., Mathre, D.E. and Elliott, M.L. (1993) Use of polymerase chain reaction to detect *Gaeumannomyces graminis* DNA in plants grown in artificially and naturally-infested field soil. *Phytopathology* 83, 283–287.

Hornby, D. and Bateman, G. (eds) (1991) *Take-all Disease of Cereals. Home-Grown Cereals Authority Research Review No. 2.* Home-Grown Cereals Authority, London, 147 pp.

Innis, M.A., Gelfand, D.H., Sninsky, J.J. and White, T.J. (eds) (1990) *PCR Protocols. A Guide to Methods and Applications.* Academic Press, San Diego.

Lee, S.B. and Taylor, J.W. (1990) Isolation of DNA from fungal mycelia and single spores. In: Innis, M.A., Gelfand, D.H., Sninsky, J.J. and White, T.J. (eds), *PCR Protocols. A Guide to Methods and Applications.* Academic Press, San Diego, pp. 282–287.

Lee, S.B. and Taylor, J.W. (1992) Phylogeny of five fungus-like protoctistan *Phytophthora* species, inferred from the internal transcribed spacers of ribosomal DNA. *Molecular Biology and Evolution* 9, 636–653.

Liu, Z.L. and Sinclair, J.B. (1992) Genetic diversity of *Rhizoctonia solani* Anastomosis Group 2. *Phytopathology* 82, 778–787.

McPherson, M.J., Quirke, P. and Taylor, G.R. (eds) (1991) *PCR. A Practical Approach.* Oxford University Press, Oxford, 253 pp.

Martin, G.B., Williams, J.G.K. and Tanksley, S.D. (1991) Rapid identification of markers linked to a *Pseudomonas* resistance gene in tomato by using random primers and near-isogenic lines. *Proceedings of the National Academy of Sciences of the United States of America* 88, 2336–2340.

Mathre, D.E. (1992) *Gaeumannomyces.* In: Singleton, L.L., Mihail, J.D. and Rush, C.M. (eds), *Methods for Research on Soilborne Phytopathogenic Fungi.* American Phytopathological Society Press, St. Paul, Minnesota, pp. 60–63.

Medlin, L., Elwood, H.J., Stickel, S. and Sogin, M.L. (1988) The characterization of enzymatically amplified eukaryotic 16S-like rRNA-coding regions. *Gene* 71, 491–499.

Mullis, K.B. (1990) The unusual origin of the polymerase chain reaction. *Scientific American* 262, No. 4, 36–43.

Mullis, K.B. and Faloona, F.A. (1987) Specific synthesis of DNA *in vitro* via a polymerase-catalysed chain reaction. *Methods in Enzymology* 155, 335–350.

Nazar, R.N., Hu, X., Schmidt, J., Culham, D. and Robb, J. (1991) Potential use of PCR-amplified ribosomal intergenic sequences in the detection and differentiation of verticillium wilt pathogens. *Physiological and Molecular Plant Pathology* 39, 1–11.

Olsen, G.J., Lane, D.J., Giovannoni, S.J. and Pace, N.R. (1986) Microbial ecology and evolution: a ribosomal RNA approach. *Annual Review of Microbiology* 40, 337–365.

Peterson, S.W. (1991) Phylogenetic analysis of *Fusarium* species using ribosomal RNA sequence comparisons. *Phytopathology* 81, 1051–1054.

Pickup, R.W. (1991) Development of molecular methods for the detection of specific bacteria in the environment. *Journal of General Microbiology* 137, 1009–1019.

Pillai, S.D., Josephson, K.L., Bailey, R.L., Gerba, C.P. and Pepper, I.L. (1991) Rapid method for processing soil samples for polymerase chain reaction amplification of specific gene sequences. *Applied and Environmental Microbiology* 57, 2283–2286.

Reeves, J.C. and Ball, S.F.L. (1991) Preliminary results on the identification of *Pyrenophora* species using DNA polymorphisms amplified from arbitrary primers. *Plant Varieties and Seeds* 4, 185–189.

Rossau, R., Duhamel, M., Jannes, G., Decourt, J.L. and Van Heuverswyn, H. (1991) The development of specific rRNA-derived oligonucleotide probes for *Haemophilus ducreyi*, the causative agent of chancroid. *Journal of General Microbiology* 137, 277–285.

Saiki, R.K., Gelfand, D.H., Stoffel, S., Scharf, S.J., Higuchi, R., Horn, G.T., Mullis, K.B. and Erlich, H.A. (1988) Primer-directed enzymatic amplification of DNA with a thermostable DNA polymerase. *Science* 239, 487–491.

Schesser, K., Luder, A. and Henson, J.M. (1991) Use of polymerase chain reaction to detect the take-all fungus *Gaeumannomyces graminis*, in infected wheat plants. *Applied and Environmental Microbiology* 57, 553–556.

Seal, S.E., Jackson, L.A. and Daniels, M.J. (1992) Isolation of a *Pseudomonas solanacearum*-specific DNA probe by subtraction hybridization and construction of species-specific oligonucleotide primers for sensitive detection by the polymerase chain reaction. *Applied and Environmental Microbiology* 58, 3751–3758.

Simon, L., Levesque, R.C. and Lalonde, M. (1992) Rapid quantitation by PCR of endomycorrhizal fungi colonizing roots. *PCR Methods and Applications* 2, 76–80.

Sogin, M.L. (1990) Amplification of ribosomal RNA genes for molecular evolution studies. In: Innis, M.A., Gelfand, D.H., Sninsky, J.J. and White, T.J. (eds), *PCR Protocols. A Guide to Methods and Applications.* Academic Press, San Diego, pp. 307–314.

Steffan, R.J. and Atlas, R.M. (1991) Polymerase chain reaction: applications in environmental microbiology. *Annual Review of Microbiology* 45, 137–161.

Vilgalys, R. and Hester, M. (1990) Rapid genetic identification and mapping of enzymatically amplified DNA from several *Cryptococcus* species. *Journal of Bacteriology* 172, 4238–4246.

Walker, J. (1981) Taxonomy of take-all fungi and related genera and species. In: Asher, M.J.C. and Shipton, P.J. (eds), *Biology and Control of Take-all.* Academic Press, London, pp. 15–74.

Walsh, S.R.A., Tyrrell, D., Humber, R.A. and Silver, J.C. (1990) DNA restriction fragment length polymorphisms in the rDNA repeat unit of *Entomophaga. Experimental Mycology* 14, 381–392.

Wang, A.M. and Mark, D.F. (1990) Quantitative PCR. In: Innis, M.A., Gelfand, D.H., Sninsky, J.J. and White, T.J. (eds), *PCR Protocols. A Guide to Methods and Applications.* Academic Press, San Diego, pp. 70–75.

Ward, D.M., Weller, R. and Bateson, M.M. (1990) 16S rRNA sequences reveal numerous uncultured microorganisms in a natural community. *Nature* 345, 63–65.

Ward, E. and Akrofi, A.Y. (1993) Identification of fungi in the *Gaeumannomyces–Phialophora* complex by RFLPs of PCR-amplified ribosomal DNAs. *Mycological Research* (in press).

Ward, E. and Gray, R. (1992) Generation of a ribosomal DNA probe by PCR and its use in identification of fungi within the *Gaeumannomyces–Phialophora* complex. *Plant Pathology* 41, 730–736.

Welsh, J. and McClelland, M. (1990) Fingerprinting genomes using PCR with arbitrary primers. *Nucleic Acids Research* 18, 7213–7218.

White, T.J., Arnheim, N. and Erlich, H.A. (1989) The polymerase chain reaction. *Trends in Genetics* 5, 185–189.

White, T.J., Bruns, T., Lee, S. and Taylor, J. (1990) Amplification and direct sequencing of fungal ribosomal RNA genes for phylogenetics. In: Innis, M.A., Gelfand, D.H., Sninsky, J.J. and White, T.J. (eds), *PCR Protocols. A Guide to Methods and Applications*. Academic Press, San Diego, pp. 315–322.

Williams, J.G.K., Kubelik, A.R., Livak, K.J., Rafalski, J.A. and Tingey, S.V. (1990) DNA polymorphisms amplified by arbitrary primers are useful as genetic markers. *Nucleic Acids Research* 18, 6531–6535.

Woese, C.R. (1987) Bacterial evolution. *Microbiological Reviews* 51, 221–271.

10 Quantifying *Polymyxa betae* and BNYVV in Soil

G. Tuitert

Sugar Beet Research Institute, Bergen op Zoom, The Netherlands
Present address: Department of Phytopathology, Wageningen Agricultural University, P.O. Box 8025, 6700 EE Wageningen, The Netherlands.

Introduction

Ecological and epidemiological studies of plant pathogens and the diseases they cause depend on adequate detection methods. For detection of obligate parasites in soil, bioassays using bait plants are often applied.

Polymyxa betae (Keskin, 1964), belonging to the Plasmodiophoraceae, is an obligate root parasite. It is the vector of beet necrotic yellow vein virus (BNYVV), which causes rhizomania or 'root madness' of sugarbeet, the common name of the disease being derived from the specific root bearding symptom on the beet. The disease reduces sugar content and root weight, depending on the level of infestation in the soil, the susceptibility of the beet cultivar and environmental conditions.

The resting spores of the vector are very persistent (Abe and Tamada, 1986). In the presence of beet roots, virus-containing resting spores release viruliferous zoospores that infect epidermal root cells. BNYVV multiplies in the plant cell and virus particles are acquired by the developing plasmodia, which may either form zoosporangia, releasing secondary zoospores in a rapid (40–80 h) infection cycle, or differentiate into clusters of resting spores.

There are still no reliable methods for directly detecting either *P. betae* or BNYVV in soil. The general method of detection is by bioassay. Field levels of infestation of soil are often characterized by using crop data obtained from the particular field, or bioassay parameters like the number of infected bait plants or the virus content in their rootlets. In order to study the effects of different factors on ecology and epidemiology of the disease and the population dynamics of the pathogen, an adequate means of quantitative assessment of BNYVV in soil is essential.

In this chapter I report on the assessment of the inoculum potential of both virus and vector in soil (Tuitert, 1990) by applying a bioassay to serial dilutions of infested soil, thus estimating most probable numbers (MPN) of infective units (Cochran, 1950). The use of the MPN method in epidemiological research on rhizomania is illustrated by data on the inoculum potential of BNYVV in soil after growing one or two susceptible beet crops (Tuitert and Hofmeester, 1992).

Quantification of Inoculum

Most probable number or dilution method

Quantification of soil-borne inoculum by means of a bioassay can be achieved by using serial dilutions of the soil. The presence (positive) or absence of a microorganism can be determined in aliquots taken from the different dilutions of soil. Using specific tables or computer programs, the number of infective units of the organism in the original sample can be estimated from the series of numbers of positives per dilution (Cochran, 1950). This method has been used for bioassays to estimate the population density of several fungal plant pathogens (Maloy and Alexander, 1958; Pfender *et al.*, 1981) and mycorrhizal fungi (Porter, 1979). Studies were performed on the applicability of the MPN method to estimate *P. betae* populations, but these did not include BNYVV (Asher and Blunt, 1987; Goffart *et al.*, 1987; Ciafardini and Marotta, 1989). In this study, since the virus is the actual pathogen, the assessment of BNYVV-infection of individual bait plants was included in the assay, to estimate the inoculum potential of the virus in soil (Tuitert, 1990).

Bioassay on rhizomania

Pots were filled with 200 ml of the diluted soil and in each pot one sugarbeet seedling was planted as a bait plant. After growth in the greenhouse for 6 weeks, the roots were washed free of soil and examined for the presence of *P. betae* with a light microscope. Sap was extracted from the tap root with lateral roots remaining attached using a handpress. The presence of BNYVV in the sap was tested by double antibody sandwich ELISA. The numbers of plants infected by *P. betae* and BNYVV were then used to estimate the MPN of both fungus and virus.

Comparison of dilution series with different dilution ratios and numbers of bait plants

From one naturally-infested soil sample four dilution series were prepared with coarse sand (v/v), using a dilution ratio of 5 or 10 and two different numbers of bait plants per dilution. Corresponding estimates for the MPNs of *P. betae* and BNYVV (Table 10.1) were obtained from the four dilution series (Tuitert, 1990).

For practical application of the method, it was preferable to use a small number of replicates per dilution and a high dilution ratio, which reduces the required number of dilutions within the relevant dilution range. However, increasing the practicality of the method in this way results in an increased standard error and, therefore, a decrease in the likelihood of obtaining significant differences between treatments. Throughout this study a dilution ratio of 10 with 6 replicates per dilution was used. This involved analysing 30 plants per soil sample.

Table 10.1. Number of BNYVV-infected bait plants (positives) at different dilutions of infested soil and most probable numbers (MPN) of infective units of BNYVV estimated with dilution ratios of 5 and 10, N = replicates per dilution (Tuitert, 1990).

Dilution ratio = 5			Dilution ratio = 10		
	Number of positives			Number of positives	
Dilution	N = 7	N = 4	Dilution	N = 10	N = 6
5^{-1}	7	4	10^{-1}	10	6
5^{-2}	7	4	10^{-2}	10	6
5^{-3}	7	4	10^{-3}	5	5
5^{-4}	7	3	10^{-4}	3	1
5^{-5}	3	2	10^{-5}	0	0
5^{-6}	0	0			
5^{-7}	0	0			
MPN ml^{-1} soil[a]	9.2	5.5		5.0	8.8

[a]MPNs are not significantly different at $P = 0.05$ (Cochran, 1950).

Storage of soil samples

The inoculum potential of *P. betae* and BNYVV was not affected by conditions during storage of soil samples for 28 months. A duplicate dilution series was made of samples stored in moist/cool (5°C) or dry/warm (20°C) conditions. Differences between replicates and between the differently stored soil samples were not statistically significant (Tuitert, 1990).

Viruliferous *P. betae*

MPNs of *P. betae* were always higher than those of BNYVV, indicating that not all resting spores contained the virus. The quotient of the MPN of BNYVV and that of *P. betae* can be used to characterize the proportion of viruliferous *P. betae* propagules. This is based on the assumptions that: (i) the rate of germination and infection is the same for non-viruliferous and viruliferous *P. betae*; (ii) the same number of non-viruliferous or viruliferous *P. betae* propagules is required for detectable infection of the bait plant by *P. betae* and BNYVV, respectively. In the soil samples used in the experiments just described, 10–15% of the infective population of *P. betae* in the soil was viruliferous. In a field experiment, the viruliferous proportion ranged from approximately 1 to 12% (Tuitert and Hofmeester, 1992). In comparison, for zoospores of *P. graminis*, transmitting barley mild mosaic virus, Jianping *et al.* (1991) found that 1–2% contained virus particles.

Applications in Epidemiological Research

The epidemiology of BNYVV is largely determined by the behaviour of its vector. Ecological factors, including the influence of soil moisture and temperature on fungal development, have been studied (Abe, 1987; Asher and Blunt, 1987; Blunt *et al.*, 1991; De Heij, 1991). Quantitative epidemiology of the disease received little attention.

The setting up of a field experiment

In 1988, a field experiment was set up to examine disease development at different initial inoculum levels of BNYVV. The influence of soil moisture conditions was investigated by applying drip irrigation. The build-up of inoculum in soil during 2 successive years was quantified (Tuitert and Hofmeester, 1992). The effects of different levels of inoculum on disease incidence and yield were examined (Tuitert and Hofmeester, 1994).

In field plots free of rhizomania, five inoculum levels of BNYVV were created by application of infested soil. For the highest inoculum level approximately 50 g m^{-2} of infested soil was added, the lowest inoculum level, except for the uninfested control, received 0.05 g m^{-2}. The infested soil was spread evenly over the surface and incorporated by harrowing. A susceptible sugarbeet cultivar (cv. Regina) was grown for 3 years consecutively.

Directly after application of the infested soil to the plots, soil samples were taken with a frequency of 1 core per 1.4 m^2. A 6-week bioassay was performed on the 50% diluted soil samples. BNYVV was traced in only a few plots at the two highest inoculum levels.

Increase of inoculum of BNYVV in soil

After growing sugarbeet for 1 and 2 years, soil samples were taken and analysed by the MPN method. Mean MPNs of BNYVV after 1 year of sugarbeet growing ranged from 0.6 and 7 infective units per 100 g soil for the plots which received the smallest amount of inoculum, non-irrigated and irrigated, respectively, to 1100 infective units per 100 g soil for the irrigated plots at the highest inoculum level. Analysis of variance of the $\log_{10}$-transformed data indicated that, after 1 and after 2 years, the MPNs were

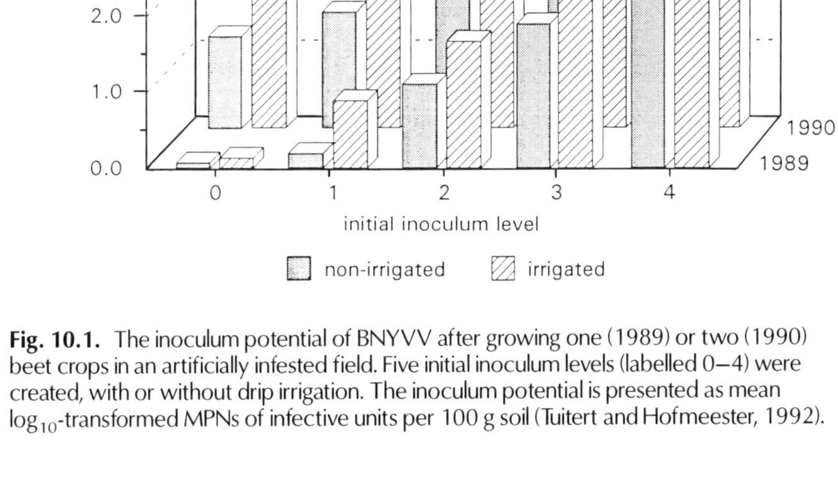

Fig. 10.1. The inoculum potential of BNYVV after growing one (1989) or two (1990) beet crops in an artificially infested field. Five initial inoculum levels (labelled 0–4) were created, with or without drip irrigation. The inoculum potential is presented as mean $\log_{10}$-transformed MPNs of infective units per 100 g soil (Tuitert and Hofmeester, 1992).

significantly determined by the initial inoculum level applied (Fig. 10.1). Irrigation for 1 or 2 years resulted in approximately threefold higher MPNs than when irrigation was not applied (Tuitert and Hofmeester, 1992).

Relationship between soil inoculum and yield

MPNs of BNYVV of all plots in spring 1989 were used to investigate the relationship with yield of beet obtained in this year. Sugar yield decreased with increasing $\log_{10}$ MPN of BNYVV, the relationship being described by a Gompertz or logistic curve (Tuitert and Hofmeester, 1994).

Conclusions

Application of the dilution or MPN method on the bioassay for detection of rhizomania in soil proved to be an adequate method for measuring the inoculum potential in the soil. With different dilution series of one soil sample, corresponding MPNs of infective units were obtained for BNYVV, as well as for its vector *P. betae*.

The inoculum potential, as measured by a bioassay method, was not affected by conditions during storage of soil samples.

MPNs of BNYVV were lower than those of *P. betae* in all soil samples tested. The ratio of the two MPNs can be used as an indication of the proportion of the infective population of *P. betae* that is viruliferous, when certain assumptions are made.

The MPN method allowed a study of the dynamics of inoculum of BNYVV in soil, as affected by the growth of a susceptible beet crop and application of drip irrigation. The method also enabled the establishment of a relationship between soil inoculum and disease incidence and yield.

References

Abe, H. (1987) Studies on the ecology and control of *Polymyxa betae* Keskin, as a fungal vector of the causal virus (beet necrotic yellow vein virus) of rhizomania disease of sugar beet. *Report of the Hokkaido Prefectural Agricultural Experiment Station* No. 60, 90 pp.

Abe, H. and Tamada, T. (1986) Association of beet necrotic yellow vein virus with isolates of *Polymyxa betae* Keskin. *Annals of the Phytopathological Society of Japan* 52, 235–247.

Asher, M.J.C. and Blunt, S.J. (1987) The ecological requirements of *Polymyxa betae*. *Proceedings 50th Winter Congress IIRB, Brussels*, pp. 45–55.

Blunt, S.J., Asher, M.J.C. and Gilligan, C.A. (1991) Infection of sugar beet by *Polymyxa betae* in relation to soil temperature. *Plant Pathology* 40, 257–267.

Ciafardini, G. and Marotta, B. (1989) Use of the most-probable-number technique to detect *Polymyxa betae* (Plasmodiophoromycetes) in soil. *Applied and Environmental Microbiology* 55, 1273–1278.

Cochran, W.G. (1950) Estimation of bacterial densities by means of 'the most probable number'. *Biometrics* 6, 105–116.

De Heij, A. (1991) The influence of water and temperature on the multiplication of *Polymyxa betae*, vector of beet necrotic yellow vein virus. In: Beemster, A.B.R., Bollen, G.J., Gerlagh, M., Ruissen, M.A., Schippers, B. and Tempel, A. (eds), *Biotic Interactions and Soil-Borne Diseases*. Elsevier, Amsterdam, pp. 83–90.

Goffart, J.P., van Bol, V. and Maraite, H. (1987) Quantification du potentiel d'inoculum de *Polymyxa betae* Keskin dans les sols. *Proceedings 50th Winter Congress IIRB, Brussels*, pp. 295–306.

Jianping, C., Swaby, A.G., Adams, M.J. and Yili, R. (1991) Barley mild mosaic virus inside its fungal vector, *Polymyxa graminis*. *Annals of Applied Biology* 118, 615–621.

Keskin, B. (1964) *Polymyxa betae* n.sp., ein Parasit in den Wurzeln von *Beta vulgaris* Tournefort, besonders während der Jugendentwicklung der Zuckerrübe. *Archiv für Mikrobiology* 49, 348–374.

Maloy, O.C. and Alexander, M. (1958) The 'most probable number' method for estimating populations of plant pathogenic organisms in the soil. *Phytopathology* 48, 126–128.

Pfender, W.F., Rouse, D.I. and Hagedorn, D.J. (1981) A 'most probable number' method for estimating inoculum density of *Aphanomyces euteiches* in naturally infested soil. *Phytopathology* 71, 1169–1172.

Porter, W.M. (1979) The 'most probable number' method for enumerating infective propagules of vesicular arbuscular mycorrhizal fungi in soil. *Australian Journal of Soil Research* 17, 515–519.

Tuitert, G. (1990) Assessment of the inoculum potential of *Polymyxa betae* and beet necrotic yellow vein virus (BNYVV) in soil using the most probable number method. *Netherlands Journal of Plant Pathology* 96, 331–341.

Tuitert, G. and Hofmeester, Y. (1992) Epidemiology of beet necrotic yellow vein virus in sugar beet at different initial inoculum levels in the presence or absence of irrigation: dynamics of inoculum. *Netherlands Journal of Plant Pathology* 98, 343–360.

Tuitert, G. and Hofmeester, Y. (1994) Epidemiology of beet necrotic yellow vein virus in sugar beet at different initial inoculum levels in the presence or absence of irrigation: disease incidence, yield and quality. *European Journal of Plant Pathology* (in press).

III ASPECTS OF AUTECOLOGY

11 Spore Dispersal: Environmental and Biological Factors

H.A. McCartney

Department of Plant Pathology, AFRC Institute of Arable Crops Research, Rothamsted Experimental Station, Harpenden, Herts AL5 2JQ, UK.

Introduction

A knowledge of the dispersal of fungal spores is important in understanding the ecology of fungi, especially those which cause plant, human or animal diseases. This chapter briefly considers how biological and physical processes can influence the dispersal of fungal spores by using examples mainly drawn from studies of plant fungal pathogens, but many of the points illustrated will have applications to the dispersal of other microorganisms.

Dispersal can be divided into three interrelated phases: spore removal, dispersal and deposition. Biological processes influence dispersal by determining the size and shape of the spore; where the spores originate; the main mode of dispersal (wind or rain); how the spores are released; the timing of release and what happens after deposition. Other biological factors such as the local vegetation type and morphology of the host can also significantly affect the fate of spores. Physical factors such as wind, rain, temperature, humidity and light can affect the manner of spore release, the mode of dispersal and the method of deposition.

Spore Removal

Of the three dispersal processes, spore removal is perhaps the most affected by the biology of the fungus and the host. For example, the morphology of the fungal fruiting body largely determines the method of release of the spore, while environmental factors, particularly wind speed and turbulence, humidity and water availability, largely control spore removal. Spores can be removed from their host by simply being blown or shaken off, by an active release mechanism or by being washed or splashed off.

Active release mechanisms in fungi are many and diverse and have been extensively described by Ingold (1971). Active release ejects spores into the air above the fruiting body. In many cases, the mechanism by which spores are ejected is not fully understood, although it is frequently due to a response to a change in the environment, such as humidity or temperature. All mechanisms require an adequate supply of water (Gregory, 1973). Thus, rainfall and dew may play an important role in spore release. Active release mechanisms, especially in the ascomycetes, can be very effective in releasing spores, even in still air. For example, ascospores of *Sclerotinia sclerotiorum*, an important pathogen of many broadleaf crops, can be ejected 2–3 cm from apothecia (Harthill and Underhill, 1976). Spore release in this species and other ascomycetes, such as *Pyrenopeziza brassicae*, appears to be a response to drying because spore release in both cases can occur in response to decreasing relative humidity (McCartney and Lacey, 1990, 1992); as a result, *S. sclerotiorum* ascospores are usually released during the morning or after rain (McCartney and Lacey, 1991a). Rain also largely determines the release of *P. brassicae* ascospores because water is required to allow apothecia to develop and spore release occurs as the apothecia dry (McCartney and Lacey, 1990). Drying is also responsible for the release of conidia of *Peronospora tabacina*. The spores are held above the leaf surface on stalk-like structures which twist violently as they dry, throwing spores away from the leaf (Ingold, 1971).

Many spores are simply blown or shaken from their hosts. To be removed in this way, the aerodynamic or mechanical forces acting on the spore must overcome the forces holding the spore to the surface (Aylor and Parlange, 1975). This may include breaking a mechanical link which joins the spore to the fruiting body. Spores of fungi which are adapted to be removed in this way often grow in chains, e.g. conidia of *Erysiphe graminis* (Hammett and Manners, 1973), or on stalks, e.g. conidia of *Helminthosporium maydis* (Aylor, 1975) so that they are held above the surface. The strength of spore attachment or the wind speeds needed to remove spores is not known for many fungi. However, it seems likely that the wind speeds needed to remove spores can be relatively large (Grace, 1977). Conidia of *H. maydis* need wind speeds in the order of 5 m s^{-1} to remove them (Aylor, 1975), while conidia of *E. graminis* are released in wind exceeding about 0.5 m s^{-1} (Hammett and Manners, 1974). Mean wind speeds within plant canopies are generally smaller than those needed to release spores (Monteith, 1973; Grace, 1977), but spore dispersal by wind is common.

Airflow within canopies is highly turbulent and gusts greatly exceeding mean values occur frequently (Shaw *et al.*, 1979; Shaw and McCartney, 1985). It is likely that gusts are responsible for removing spores, and other biological particles (Aylor, 1978; Aylor *et al.*, 1981). The enhanced deposition of conidia of *E. graminis* observed in a barley crop by McCartney and Bainbridge (1987) was attributed to spore release in gusts. Mechanical shaking has also been shown to be able to dislodge spores (Bainbridge and Legg, 1976). Wind, especially gusts, is also responsible for the movement of foliage, thus a combination of wind gusts and plant movement may cause the passive removal of many fungal spores.

In addition to providing water for spore development, rain can also disperse spores and other biological organisms such as bacteria (Fitt *et al.*, 1989). Spores can either be washed from surfaces by run-off water or carried in droplets splashed by the impact of rain drops. Rain can quickly deplete bacterial populations on leaf surfaces (Hirano and Upper, 1983; Butterworth and McCartney, 1991, 1992) and it is possible that fungal spores may also be removed by run-off. Thousands of spores can be removed by a single splash (Gregory *et al.*, 1959; Fitt and Lysandrou, 1984) and splash can also disperse bacteria from leaf surfaces (Butterworth and McCartney, 1991, 1992; McCartney and Butterworth, 1992).

Many fungi have evolved spores which can only be dispersed by water. The spores are typically produced in mucilage which prevents dispersal by wind alone (Gregory, 1973; Fitt *et al.*, 1989). On wetting, the mucilage dissolves releasing the spores into a thin film on the host surface (Fitt *et al.*, 1989) and when a water drop strikes the film, spores are removed in the splash droplets produced. Splash-dispersed spores are characteristically smooth, have thin hyaline walls and are frequently elongate (Fitt and McCartney, 1986a). They also may be able to stick to surfaces in the presence of water, increasing the chances of deposition after dispersal. Spores which are usually wind-dispersed may also be removed by the impact of raindrops (Hirst and Stedman, 1963; Ramalingam and Rati, 1979).

The effectiveness of rain splash in removing spores depends on the drop size and its velocity and on the orientation and mechanical properties of the surface, but the physical mechanisms are not well understood. Drop size plays an important role in removing spores. Gregory *et al.* (1959) found that 5 mm diameter drops removed between four and six times as many spores of *Fusarium solani* as did 3 mm drops. In experiments on the dispersal of conidia of *Rhynchosporium secalis* from barley leaves, 1 mm drops splashed less than 3% of the number of conidia splashed by 3 mm drops (Fitt *et al.*, 1988). Reynolds *et al.* (1989) found that the number of sporangia of *Phytophthora cactorum* splashed from strawberry fruit increased with the size of the impacting drop. The removal of bacteria by splash also depends on rain drop size. Graham and Harrison (1975) detected about seven to eight times as many cells of *Erwinia carotovora* in the air downwind of infected potato tissue when splashed by 4 or 5 mm drops than by 2 mm drops;

Butterworth and McCartney (1992) showed that rain drops of 4.7 mm removed populations of *Bacillus subtilis* from leaves more rapidly than 2.4 mm drops. The force of impact of rain drops may determine the effectiveness of spore removal (Ghadiri and Payne, 1979; Walklate, 1989). Thus, large slow moving droplets dripping from leaves, may remove organisms as effectively as smaller raindrops falling at their terminal velocity. Rainfall intensity probably determines the rate of removal of organisms (Fitt *et al.*, 1986), but the size distribution of the raindrops is also important. Rain, therefore, with a large proportion of large drops (>3 mm) will remove and disperse organisms from surfaces more effectively than rain with mostly small drops.

Spore Dispersal

Once removed from their source the fate of spores is determined by purely physical processes. The potential for dispersal by wind depends on the wind speed and on its turbulent intensity. The turbulent nature of wind causes individual spores to follow different paths and travel different distances, even if released from the same source under the same wind conditions (Fig. 11.1). Therefore, the concentration of spores in the air, and the numbers deposited, decrease downwind of the release point as the spore plume expands and as spores are deposited on foliage and other surfaces. This decrease in

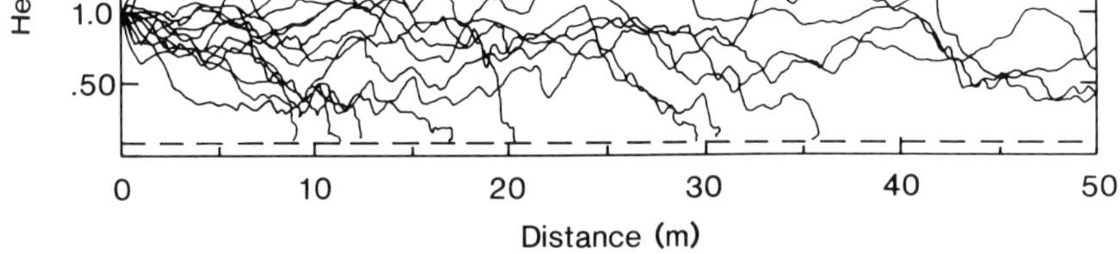

Fig. 11.1. Simulation of spores dispersing from a point 1 m above the ground calculated using a random walk model (Walklate, 1987). Each line represents the trajectory of individual spores. Spore fall speed was taken as 5 cm s^{-1}.

Table 11.1. Measured values of dispersal half distance, $d_{1/2}$, for wind-dispersed spores and pollen and for splash-dispersed spores and bacteria.

Spores (wind-borne)	$d_{1/2}$ (m)	Pollen	$d_{1/2}$ (m)	Splash	$d_{1/2}$ (m)
Bovista plumbea[1]	5.6–8.2	*Ambrosia artemisiifolia*[1]	9.4	*Pseudocercosporella herpotrichoides*[1]	0.07–0.14
Podaxis pistillaris[1]	6.0–7.0	*Beta vulgaris*[1]	64.0	*Pyrenopeziza brassicae* (conidia)[1]	0.10
Tilletia tritici[1]	7.5	*Betula pubescens*[1]	433.0	*Septoria nodorum*[1]	0.11–0.15
Tilletia caries[1]	5.5	*Corylus avellana*[1]	27.0	*Erwinia carotovora* (bacteria)[1]	0.06
20 μm droplets (barley)[1]	0.6–1.4	*Picea abies*[1]	130.0	*Bacillus subtilis* (bacteria)[3]	0.05–0.14
Pyrenopeziza brassicae (ascospores)[2]	9.0	*Zea mays*[1]	47.0		
		Brassica napus[2]	5.0		

[1]Fitt *et al.* (1987); [2]McCartney (1990a); [3]Butterworth and McCartney (1992).

concentration is frequently referred to as a 'concentration (or deposition) gradient' (Gregory, 1973) and has been described by a number of different equations (McCartney and Fitt, 1985; Fitt *et al.*, 1987). Gradients are often found to follow a negative exponential relationship:

$$C = C_0 \exp(-\alpha x) \tag{11.1}$$

where C is the concentration or deposition rate, x is the distance down wind and C_0 and α are constants. The coefficient α determines the rate of decrease in spore concentration (or deposition) with distance. Although this relationship may not strictly be valid in some situations (Aylor, 1987), α gives a convenient method of visualizing gradients as it can be expressed as a half distance, $d_{1/2}$ (= 0.693/α), that is the distance in which C decreases by half.

Gradients of wind-dispersed particles can range from a few centimetres to hundreds of metres depending on wind, particle size, position and size of the source (Fitt *et al.*, 1987; Table 11.1). Spores released within close stands of vegetation, such as cereal crops, tend to travel shorter distances than those released in more open locations, for example, isolated stands of trees. Concentration or deposition gradients for plant pathogens measured within crops are usually of the order of a few metres, but, measured from the edges of the crops tend to be larger. For example, $d_{1/2}$ values for 20 µm particles released within a barley crop were in the range 0.3 to 1.8 m (McCartney and Bainbridge, 1984), whereas, values were about 9 m for *P. brassicae* ascospores and 5 m for oilseed rape (*Brassica napus*) pollen dispersing from an oilseed rape crop (McCartney, 1990a). Anemophilous pollens, particularly those released from trees, can travel considerable distances and $d_{1/2}$ values of hundreds of metres have been recorded (Fitt *et al.*, 1987; Table 11.1).

Wind disperses spores downwind from the source and also transports them up into the atmosphere where they have the potential to disperse over large distances. Indeed, pollen and fungal spores have been found at heights of between 500 and 1000 m in the atmosphere (Gregory and Monteith, 1967; Hirst *et al.*, 1967). Spore trapping during a morning flight from Yorkshire to Denmark (Gregory and Monteith, 1967) showed that concentrations of *Cladosporium* spores, typical of daytime air spora, decreased with distance away from the coast. After about 60 km they were replaced by *Sporobolomyces* and ascospores characteristic of night spora, probably released the previous night. Clouds of *Cladosporium* spores were again found after about 250 and 600 km and *Sporobolomyces* after 500 km. This pattern was interpreted as wind-borne remnants of alternating day and night spore clouds released from the English land surface. The spores were found at heights up to 1000 m. The effect of vertical mixing is illustrated in Fig. 11.2 which compares vertical profiles of concentration of *P. brassicae* ascospores and *B. napus* pollen measured over an oilseed rape crop. It has been estimated, using an atmospheric dispersal model, that more than 90% of the ascospores and about 70% of the pollen grains would still have been airborne

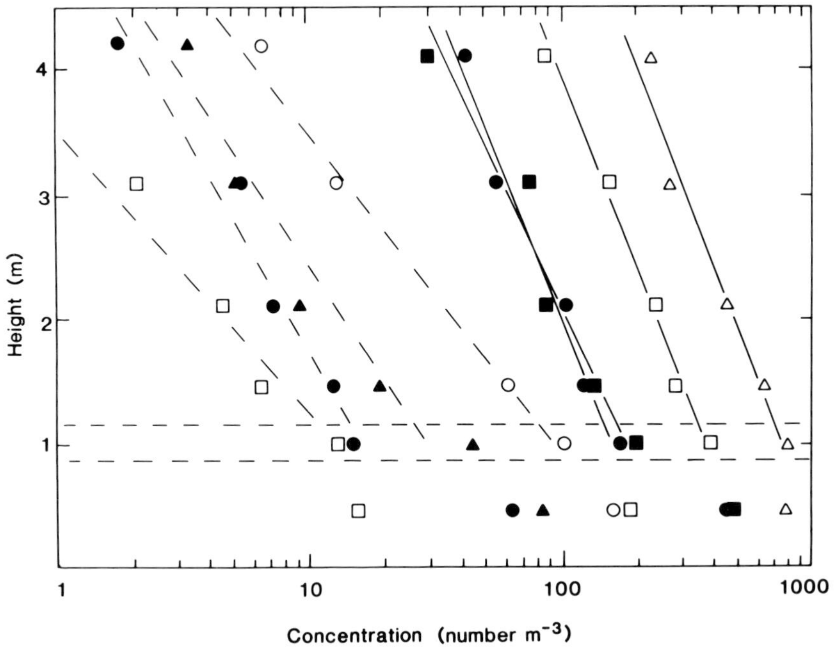

Fig. 11.2. Vertical spore and pollen concentration profiles measured above an oilseed rape crop on different days in 1987. The solid lines are for *Pyrenopeziza brassicae* ascospores and the broken lines are for oilseed rape pollen. The horizontal broken lines indicate crop height (from McCartney, 1990a).

100 m downwind of the source (McCartney, 1990a).

Figure 11.2 shows that the potential for dispersal also depends on the size of the particle. *P. brassicae* spores are ellipsoid, about 12 µm long and 2.3 µm in diameter (Lacey *et al.*, 1987), and are much smaller than pollen grains of oilseed rape which are spherical and 25 µm in diameter (McCartney and Lacey, 1991b) and are consequently more easily transported upwards by turbulent eddies. The settling speed, V_s, of the ascospores was estimated to be about 0.1 cm s^{-1} and that of the pollen to be about 1.7 cm s^{-1} (McCartney, 1990a). V_s for most fungal spores generally lies in the range 0.1 to 3 cm s^{-1} (Gregory, 1973). For small spherical particles (< 100 µm), V_s is proportional to the square of the diameter d, (Chamberlain, 1975):

$$V_s = (d^2 g\rho)/(18\nu\rho_a) \tag{11.2}$$

where ρ and ρ_a are the densities of the particle and air; g is gravitational acceleration (9.01 m s^{-2}) and ν the kinematic viscosity of air. Equation 11.2 can be used to estimate V_s for spherical spores (Gregory, 1973). The fall speed of compact particles with simple shapes such as ellipses, cylinders and spheres can also be estimated from their dimensions (Green and Lane, 1957; Fuchs, 1964; Chamberlain, 1975).

Many spores and pollens have complex non-spherical shapes (Gregory, 1973). Such particles usually fall more slowly than spherical ones of the same volume and density. Conidia of *Alternaria* species are typically club shaped and can also be elongated. For example, the length/diameter ratio of *A. linicola* conidia is about 10 compared with about 3 for *A. alternata*. The fall speed of *Alternaria* spores can be approximated as that of a cylinder of the same length and mean diameter of the spore (McCartney *et al*., 1993). The spores of some fungi can also be dispersed in clusters (Bainbridge and Stedman, 1979; Ferrandino and Aylor, 1984; McCartney and Bainbridge, 1987). Clusters also form complex shapes, but the values of V_s for compact clusters appears to be proportional to the square root of the number of spores in the cluster (Ferrandino and Aylor, 1984; McCartney and Bainbridge, 1987).

Dispersal by wind has been extensively studied, especially in relation to gaseous and particulate atmospheric pollutants (Pasquill, 1974; Pasquill and Smith, 1983). Several approaches have been used to model these processes: some purely descriptive and some based on physical principles of dispersal. These approaches have also been applied to spore dispersal (McCartney and Fitt, 1985; Fitt and McCartney, 1986b). The details of the methods used are too complex to describe here, but, briefly, three approaches have been used: the statistical approach, based on the 'Gaussian Plume' model of gaseous dispersal (Gregory, 1945; Reddi, 1975; Aylor, 1978); the diffusion approach, based on gradient diffusion theory (Legg and Powell, 1979; Aylor, 1982; Aylor and Taylor, 1982; McCartney, 1987); and the stochastic approach, based on random walk models (Wilson *et al*., 1981; Ley, 1982; Legg, 1983; Walklate, 1986, 1987). 'Gaussian Plume' models are the most straightforward to use, but have limited value for predicting dispersal in vegetative canopies or close to sources because it is difficult to account adequately for particle deposition but can be useful for long distance predictions. Models based on diffusion theory are more versatile, but there are theoretical doubts of their validity when applied to dispersal very close to sources and within vegetation. These models can easily account for deposition processes but not when the particles are predominantly released in gusts of wind. Recently, random walk models based on Markov chain theory (Wilson *et al*., 1981; Legg, 1983; Walklate, 1987) have been applied to particle dispersal by wind. The paths of individual particles are simulated as a pseudo-random walk, using knowledge of the turbulence statistics of the air flow. Although these models are still being developed, they have the potential to describe dispersal in the highly turbulent flows often found in nature, they are valid at all distances from the source and can allow for different modes of spore release.

The potential of rain splash as a factor in the spread of disease in crops has long been recognized (Faulwetter, 1917a,b) and many fungal plant pathogens rely on rain splash to disperse spores. As most of the splash droplets which carry spores are between 200 and 800 μm in diameter (Fitt and McCartney, 1986a; Fitt *et al.*, 1989) most of the spores dispersed by splash travel relatively short distances. Dispersal half distances are usually in the range 10–30 cm (Table 11.1), although distances may be larger when wind is associated with rain (Brennan *et al.*, 1985).

Raindrop size may play an important role in defining splash dispersal patterns. For example, Gregory *et al.* (1959) found that 50% of the droplets splashed by 3 mm drops hitting a 1 mm thick water film travelled about 14 cm compared to 21 cm for 5 mm drops. In experiments using strawberry fruit as targets, Yang *et al.* (1991) found that the average distance travelled by splash droplets was about 6 cm for 2 mm incident drops and about 8 cm for 4 mm drops. Butterworth and McCartney (1992) found that 50% of *B. subtilis* cells splashed from leaves by 4.7 mm drops travelled about 13 cm compared to 7 cm for 2.4 mm incident drops. Walklate *et al.* (1989) have shown that the maximum height reached by splashed droplets may be related to the product of the incident drop diameter and the impact velocity. The further a splash droplet travels the greater chance it has of being entrained by the wind. Therefore, it seems likely that a larger proportion of spores splashed by large raindrops have the potential to be further dispersed by wind.

Dispersal patterns appear to be little affected by rainfall intensity by itself (Fitt *et al.*, 1988). However, in some situations, such as convective showers, the size distribution of raindrops is related to rainfall intensity, with high intensity rain containing a larger proportion of large drops, and in these circumstances dispersal patterns and intensity may be related. For example *Septoria nodorum* and *S. tritici* have been shown to be spread from lower to upper leaves in winter wheat by short intense summer showers (Royle *et al.*, 1986; Shaw, 1987), which are likely to contain many large drops.

Many splash droplets are large enough for their dispersal to be only slightly affected by wind turbulence, especially within plant canopies. Therefore, their paths tend to follow ballistic trajectories which can be modelled using conventional Newtonian dynamics (Macdonald and McCartney, 1987). In such models, droplet trajectories are calculated from the initial velocities of the droplets by considering the gravitational and aerodynamic forces acting on the drops. There have been few studies of splash droplet velocities (Ghadiri-Khorzoojhi, 1978; Reynolds *et al.*, 1987; Macdonald, 1988; Macdonald and McCartney 1988; Yang *et al.*, 1991), but, it appears that droplets can be ejected by splash at speeds in excess of 10 m s^{-1}.

Spore Deposition

The removal of spores from the air by deposition on to surfaces can be thought of as being a continuation of two mechanisms; gravitational settling and impaction (Legg and Powell, 1979; Bache, 1979; Aylor, 1982; McCartney and Aylor, 1987). The rate at which spores settle on to surfaces, S, is proportional to the spore fall speed, V_s, and the spore concentration above the surface, C, ($S = CV_s$). Because V_s for spores is largely determined by their shape, size and density, spore biology plays an important role in controlling deposition. Surface characteristics of spores are also important as they will determine whether the spores will adhere to the surface after the initial deposition. Little appears to be known about the retention of spores on surfaces but it is generally felt that many fungal spores readily stick to leaf surfaces after deposition.

As the mass of spores, or water droplets carrying spores, is much larger than the molecules of the air in which they are travelling, they cannot follow the fluctuations of the air exactly. Therefore, some spores will strike an object when air flows round it because they cannot follow the same path as the air molecules. This process is known as impaction. The rate of deposition by impaction, I, depends on spore concentration, C, and wind speed, u:

$$I = CuE \tag{11.3}$$

where E is the efficiency of impaction and is defined as the rate at which spores impact on a surface, divided by the rate at which the spores pass through the same area as the object projected along the mean wind direction.

Both biological and environmental factors affect the value of impaction efficiency. E increases with spore size (V_s) and wind speed (u) but decreases with increasing width of the surface (w). In low-turbulence flows E is a function of a non-dimensional parameter ($S = V_s u/gw$) called the Stokes' number. The relationship between S and E for laminar flow is well known (May and Clifford, 1967; Chamberlain, 1975), but this relationship may not hold in the field, especially within crops, where efficiencies several times those expected from estimated values of S have been found (Bainbridge and Stedman, 1979; McCartney and Aylor, 1987; McCartney and Bainbridge, 1987). Aylor (1978) pointed out that if spores are only removed in gusts of wind (see above) then for the first few metres of travel they will only be carried in winds exceeding the threshold needed to remove them. Thus, as E is dependent on wind speed, spores will be impacted more efficiently than would be expected from the mean wind speeds (Aylor *et al.*, 1981; McCartney *et al.*, 1983). In crops, where mean wind speeds are small, gust release may enhance impaction rates by up to an order of magnitude (McCartney and Bainbridge, 1987).

Spore dispersal patterns may also be affected when spores are released in gusts. Enhanced deposition close to the source will remove spores faster than if they were released at all wind speeds, thus steepening deposition gradients (Legg, 1983; McCartney, 1987). Above crops, where deposition is less important, spores will travel initially at speeds exceeding their removal threshold and therefore may be expected to go further than those released at slower wind speeds. In this case, deposition or concentration gradients would be expected to be enhanced by gust release. Calculations from random walk models suggest that gradients may be enhanced by an amount roughly proportional to the threshold wind speed (McCartney, 1990b).

Conclusions

Spore dispersal is determined by a complex interaction of biological and environmental factors. Dispersal patterns are varied and appear to be different for different species and for the same species at different locations and times. The effects of biological factors differ between species and location, but the physical mechanisms of dispersal are essentially the same for all spores. Our understanding of both biological and environmental dispersal processes is far from complete. Much still has to be done to understand spore release mechanisms and little is known of spore retention processes. Methods have still to be developed to understand dispersal in complex flows such as near field boundaries, hedges or obstructions or within natural vegetation.

In such a short chapter it is impossible to cover all aspects of spore dispersal in great depth, but perhaps this review will serve as a useful introduction to a complex biological and environmental phenomenon.

References

Aylor, D.E. (1975) Force required to detach conidia of *Helminthosporium maydis*. *Plant Physiology* 55, 99–101.

Aylor, D.E. (1978) Dispersal in time and space: aerial pathogens. In: Horsfall, J.G and Cowling, E.B. (eds), *Plant Disease: an Advanced Treatise*. Academic Press, New York, pp. 159–180.

Aylor, D.E. (1982) Modelling spore dispersal in a barley crop. *Agricultural Meteorology* 26, 215–219.

Aylor, D.E. (1987) Deposition gradients of *Puccinia recondita* near a source. *Phytopathology* 77, 1442–1448.

Aylor, D.E. and Parlange, J-Y. (1975) Ventilation required to entrain small particles from leaves. *Plant Physiology* 56, 97–99.

Aylor, D.E. and Taylor, G.S. (1982) Aerial dispersal and drying of *Peronospora tabacina* conidia in tobacco shade tents. *Proceedings of the National Academy of Sciences of the United States of America* 79, 697–700.

Aylor, D.E., McCartney, H.A. and Bainbridge, A. (1981) Deposition of particles liberated in gusts of wind. *Journal of Applied Meteorology* 10, 1212–1221.

Bache, D.H. (1979) Particle transport within plant canopies. I: A framework for analysis. *Atmospheric Environment* 13, 1257–1262.

Bainbridge, A. and Legg, B.J. (1976) Release of barley-mildew conidia from shaken leaves. *Transactions of the British Mycological Society* 66, 495–498.

Bainbridge, A. and Stedman, O.J. (1979) Dispersal of *Erysiphe graminis* and *Lycopodium clavatum* spores near to the source in a barley crop. *Annals of Applied Biology* 91, 187–198.

Brennan, R.M., Fitt, B.D.L., Taylor, G.S. and Colhoun, J. (1985) Dispersal of *Septoria nodorum* pycnidiospores by simulated rain and wind. *Phytopathologische Zeitschrift* 112, 291–297.

Butterworth, J. and McCartney, H.A. (1991) The dispersal of bacteria from leaf surfaces by water splash. *Journal of Applied Bacteriology* 71, 484–496.

Butterworth, J. and McCartney, H.A. (1992) Effect of drop size on the removal of *Bacillus subtilis* from foliar surfaces by water splash. *Microbial Releases* 1, 177–185.

Chamberlain, A.C. (1975) The movement of particles in plant communities. In: Monteith, J.L. (ed.), *Vegetation and the Atmosphere*, Vol. 1. Academic Press, London, pp. 155–203.

Faulwetter, R.C. (1917a) Dissemination of the angular leafspot of cotton. *Journal of Agricultural Research* 8, 457–475.

Faulwetter, R.C. (1917b) Wind-blown rain, a factor in disease dissemination. *Journal of Agricultural Research* 10, 639–648.

Ferrandino, F.J. and Aylor, D.E. (1984) Settling speed of clusters of spores. *Phytopathology* 74, 969–972.

Fitt, B.D.L. and Lysandrou, M. (1984) Studies on mechanisms of splash dispersal of spores, using *Pseudocercosporella herpotrichoides* spores. *Phytopathologische Zeitschrift* 111, 323–331.

Fitt, B.D.L. and McCartney, H.A. (1986a) Spore dispersal in splash droplets. In: Ayres, P.G. and Boddy, L. (eds), *Water, Fungi and Plants*. Cambridge University Press, Cambridge, pp. 87–104.

Fitt, B.D.L. and McCartney, H.A. (1986b) Spore dispersal in relation to epidemic models. In: Leonard, K.J. and Fry, W.E. (eds), *Plant Disease Epidemiology, Vol.1: Population Dynamics and Management*. Macmillan, New York, pp. 311–345.

Fitt, B.D.L., Creighton, N.F., Lacey, M.E. and McCartney, H.A. (1986) Effects of rainfall intensity and duration on dispersal of *Rhynchosporium secalis* conidia from infected barley leaves. *Transactions of the British Mycological Society* 86, 611–618.

Fitt, B.D.L., Gregory, P.H., Todd, A.D., McCartney, H.A. and Macdonald, O.C. (1987) Spore dispersal and plant disease gradients: a comparison between two empirical models. *Journal of Phytopathology* 118, 227–242.

Fitt, B.D.L., McCartney, H.A., Creighton, N.F., Lacey, M.E. and Walklate, P.J. (1988) Dispersal of *Rhynchosporium secalis* conidia from barley leaves or straw by simulated rain. *Annals of Applied Biology* 112, 49–59.

Fitt, B.D.L., McCartney, H.A. and Walklate, P.J. (1989) Role of rain in the dispersal of pathogen inoculum. *Annual Review of Phytopathology* 27, 241–270.

Fuchs, N.A. (1964) *The Mechanics of Aerosols*. Pergamon, Oxford, (Translation C.N. Davis).

Ghadiri-Khorzoojhi, H. (1978) Raindrop impact and soil splash. Unpublished PhD Thesis, University of Reading.

Ghadiri, H. and Payne, D. (1979) Raindrop impact and soil splash. In: Lal, R. and Greenland, D.J. (eds), *Role of Soil Physical Properties in Maintaining Productivity of Tropical Soils*. Wiley, Chichester, pp. 94–104.

Grace, J. (1977) *Plant Response to Wind*. Academic Press, London.

Graham, D.C. and Harrison, M.D. (1975) Potential spread of *Erwinia* spp. in aerosol. *Phytopathology* 65, 739–741.

Green, H.L. and Lane, W.R. (1957) *Particulate Clouds, Dusts, Smokes and Mist*. E. and F.N. Spon, London.

Gregory, P.H. (1945) The dispersion of air-borne spores. *Transactions of the British Mycological Society* 28, 26–72.

Gregory, P.H. (1973) *Microbiology of the Atmosphere*, Leonard Hill, Aylesbury.

Gregory, P.H. and Monteith, J.L. (1967) *Airborne Microbes*. Cambridge University Press, London.

Gregory, P.H., Guthrie, E.J. and Bunce, M.E. (1959) Experiments on splash dispersal of fungus spores. *Journal of General Microbiology* 20, 328–354.

Hammett, K.R.W. and Manners, J.G. (1973) Conidium liberation in *Erysiphe graminis* II: conidial chain and pustule structure. *Transactions of the British Mycological Society* 61, 121–133.

Hammett, K.R.W. and Manners, J.G. (1974) Conidium liberation in *Erysiphe graminis*. III: Wind tunnel studies. *Transactions of the British Mycological Society* 62, 267–282.

Harthill, W.F.T. and Underhill, A.P. (1976) 'Puffing' in *Sclerotinia sclerotiorum* and *S. minor*. *New Zealand Journal of Botany* 14, 355–358.

Hirano, S.S. and Upper, C.D. (1983) Ecology and epidemiology of foliar bacterial plant pathogens. *Annual Review of Phytopathology* 21, 243–269.

Hirst, J.M. and Stedman, O.J. (1963) Dry liberation of fungus spores by raindrops. *Journal of General Microbiology* 33, 375–393.

Hirst, J.M., Stedman, O.J. and Hurst, G.W. (1967) Long distance spore transport: vertical sections of spore clouds over the sea. *Journal of General Microbiology* 48, 357–377.

Ingold, C.T. (1971) *Fungal Spores, their Liberation and Dispersal*. Clarendon Press, Oxford.

Lacey M.E., Rawlinson, C.J. and McCartney, H.A. (1987) First record of the natural occurrence in England of the teleomorph of *Pyrenopeziza brassicae* on oilseed rape. *Transactions of the British Mycological Society* 89, 135–140.

Legg, B.J. (1983) Movement of plant pathogens in the crop canopy. *Philosophical Transactions of the Royal Society, London* B 302, 559–574.

Legg, B.J. and Powell, F.A. (1979) Spore dispersal in a barley crop: a mathematical model. *Agricultural Meteorology* 20, 47–67.

Ley, A.J. (1982) A random walk simulation of two dimensional turbulence in the neutral surface layer. *Atmospheric Environment* 16, 2799–2808.

Macdonald, O.C. (1988) Splash on leaves and the dispersal of spore carrying droplets.

Unpublished PhD Thesis, University of London.

Macdonald, O.C. and McCartney, H.A. (1987) Calculation of splash droplet trajectories. *Agricultural and Forest Meteorology* 39, 95–110.

Macdonald, O.C. and McCartney, H.A. (1988) A photographic technique for investigating splashing of water drops on leaves. *Annals of Applied Biology* 113, 627–638.

May, K.R. and Clifford, R. (1967) The impaction of aerosol particles on cylinders, spheres, ribbons and discs. *Annals of Occupational Hygiene* 10, 83–95.

McCartney, H.A. (1987) Deposition of *Erysiphe graminis* conidia on a barley crop, II: consequences for spore dispersal. *Journal of Phytopathology* 118, 258–264.

McCartney, H.A. (1990a) The dispersal of plant pathogen spores and pollen from oilseed rape crops. *Aerobiologia* 6, 147–152.

McCartney, H.A. (1990b) Dispersal mechanisms through the air. In: Bunce, R.G.H. and Howard, D.C. (eds), *Dispersal in Agricultural Habitats.* Belhaven Press, London, pp. 133–158.

McCartney, H.A. and Aylor, D.E. (1987) Relative contribution of sedimentation and impaction to deposition of particles in a crop canopy. *Agriculture and Forest Meteorology* 40, 343–358.

McCartney, H.A. and Bainbridge, A. (1984) Deposition gradients close to a point source. *Phytopathologische Zeitschrift* 109, 219–236.

McCartney, H.A. and Bainbridge, A. (1987) Deposition of *Erysiphe graminis* conidia on a barley crop. I: Sedimentation and impaction. *Journal of Phytopathology* 118, 243–257.

McCartney, H.A. and Butterworth, J. (1992) Effect of humidity on the dispersal of *Pseudomonas syringae* from leaves by water splash. *Microbial Releases* 1, 187–190.

McCartney, H.A. and Fitt, B.D.L. (1985) Construction of dispersal models. In: Gilligan, C.A. (ed.), *Advances in Plant Pathology, Vol. 3: Mathematical Modelling of Crop Disease*. Academic Press, London, pp. 107–143.

McCartney, H.A. and Lacey, M.E. (1990) The production and release of ascospores of *Pyrenopeziza brassicae* Sutton et Rawlinson on oilseed rape. *Plant Pathology* 39, 17–32.

McCartney, H.A. and Lacey, M.E. (1991a) The relationship between the release of ascospores of *Sclerotinia sclerotiorum*, infection and disease in sunflower plot in the United Kingdom. *Grana* 30, 486–492.

McCartney, H.A. and Lacey, M.E. (1991b) Wind dispersal of pollen from crops of oilseed rape (*Brassica napus* L.). *Journal of Aerosol Science* 22, 467–477.

McCartney, H.A. and Lacey, M.E. (1992) Release and dispersal of *Sclerotinia* ascospores in relation to infection. In: *Brighton Crop Protection Conference: Pests and Diseases, 1992*, British Crop Protection Council, pp. 109–116.

McCartney, H.A., Bainbridge, A. and Aylor, D.E. (1983) The importance of gusts in distributing fungal spores among crops foliage. *EPPO Bulletin* 13, 133–137.

McCartney, H.A., Schmechel, D. and Lacey, M.E. (1993) Aerodynamic diameter of conidia of *Alternaria* species. *Plant Pathology* 42, 280–286.

Monteith, J.L. (1973) *Principles of Environmental Physics.* Edward Arnold, London.

Pasquill, F. (1974) *Atmospheric Diffusion*, 2nd edn. Ellis Horwood, Chichester.

Pasquill, F. and Smith, F.B. (1983) *Atmospheric Diffusion*, 3rd edn. Ellis Horwood, Chichester.

Ramalingam, A. and Rati, E. (1979) Role of water in dispersal of nonwettable spores. *Indian Journal of Botany* 2, 8–11.

Reddi, C.S. (1975) Lateral and vertical dispersion of *Lycopodium* and *Podaxis* spores released from an artificial point source. *Journal of Palynology* 11, 111–120.

Reynolds, K.M., Madden, L.V., Reichard, D.L. and Ellis, M.A. (1987) Method for study of raindrop impaction on plant surfaces with application for predicting inoculum dispersal by rain. *Phytopathology* 77, 226–232.

Reynolds, K.M., Madden, L.V., Reichard, D.L. and Ellis, M.A. (1989) Splash dispersal of *Phytophthora cactorum* from infected strawberry fruit by simulated canopy drip. *Phytopathology* 79, 425–432.

Royle, D.J., Shaw, M.W. and Cook, R.J. (1986) The natural development of *Septoria nodorum* and *Septoria tritici* in some winter wheat crops in Western Europe, 1981–1983. *Plant Pathology* 35, 201–213.

Shaw, M.W. (1987) Assessment of upward movement of rainsplash using a fluorescent tracer method and its application to the epidemiology of cereal pathogens. *Plant Pathology* 36, 210–213.

Shaw, R.H. and McCartney, H.A. (1985) Gust penetration into plant canopies. *Atmospheric Environment* 19, 827–830.

Shaw, R.H., Ward, D.P. and Aylor, D.E. (1979) Frequency of occurrence of fast gusts of wind inside a corn canopy. *Journal of Applied Meteorology* 18, 168–177.

Walklate, P.J. (1986) A Markov-chain particle dispersion model based on airflow data: extension to large water droplets. *Boundary-Layer Meteorology* 37, 313–318.

Walklate, P.J. (1987) A random-walk model for dispersion of heavy particles in turbulent air flow. *Boundary-Layer Meteorology* 39, 175–190.

Walklate, P.J. (1989) Vertical dispersal of plant pathogens by splashing: I. the theoretical relationship between rainfall and upward rain-splash. *Plant Pathology* 38, 56–63.

Walklate, P.J., McCartney, H.A. and Fitt, B.D.L. (1989) Vertical dispersal of plant pathogens by splashing: II. experimental study of the relationship between rain drop size and maximum splash height. *Plant Pathology* 38, 64–70.

Wilson, J.D., Thurtell, G.W. and Kidd, G.E. (1981) Numerical simulation of particle trajectories in inhomogeneous turbulence. *Boundary-Layer Meteorology* 21, 423–441.

Yang, X., Madden, L.V., Reichard, D.L., Fox, R.D. and Ellis, M.A. (1991) Motion analysis of drop impaction on a strawberry surface. *Agriculture and Forest Meteorology* 56, 67–92.

12 Latency and Quiescence in Survival and Success of Fungal Plant Pathogens

B. Williamson

Scottish Crop Research Institute, Invergowrie, Dundee DD2 5DA, UK.

Introduction

All fungi which derive nutrients from live plants establish a particular relationship with the host which can usefully be discussed from an ecological viewpoint. The compelling and often specific requirement of the fungus for exogenous nutrients at successive stages of its growth and differentiation must be satisfied by release of extracellular enzymes which degrade the live substrate, but this process takes place in constant competition with the host and other microorganisms in the vicinity. In fact, the infection court of a pathogen is a microcosm on the surface of the plant organ; the chemistry, topography and competing microflora of the plant surface are integral parts of the infection process. In contrast to saprophytes, plant pathogens as primary colonizers occupy a niche which constantly changes in chemical composition with age and morphological complexity and, more significantly, can change chemically in response to the presence of the pathogen as 'signalling' triggers an array of host resistance responses.

The position on the host surface, the microhabitat, and the host's location within a plant community has a bearing on the extent to which environmental conditions affect the outcome of the pathogenic relationship. For a pathogen, resistance mechanisms operating in the host, within certain environmental conditions or constraints, are therefore probably as important as the genetic control of pathogenicity determinants in the fungus.

All the events from the time that the fungal propagule makes contact with the plant surface may affect the success or otherwise of the interaction. There has been a tendency to think of the infection process as a constant-rate progression from germination of the propagule, penetration of the epidermis, growth of hyphae through tissues, to production and release of further propagules. From the pathologists' perspective it is an advantage to be able to measure the time it takes for each of these growth stages to be achieved and record any changes in the appearance, physiology or growth rate – the symptoms of disease.

The progress of many pathogen–plant relationships is characteristically interrupted when the fungus stops growing, before resuming active growth. This tendency of pathogens to enter a latent or quiescent phase in their life cycles will be examined to highlight the substantial ecological advantages this life style confers and the bearing this has on the problems encountered in breeding for disease resistance or in the design of control strategies.

Latent Period and Incubation Period

This entire discussion inevitably focuses on the ecological imperatives of timeliness. With a dynamic process such as infection, the timing of certain defined events during colonization of the host (e.g. penetration of the cuticle, release of propagules) must be established. Unfortunately there is no general consensus about the definition of many of these important events and processes (see reviews by Verhoeff, 1974; Swinburne, 1983 and Butt and Royle, 1980). Therefore, for the purpose of this discussion I must define some terms unambiguously.

The latent period was defined by van der Plank (1963) as 'the time newly infected tissue takes to become infectious', i.e. the time needed for a generation of the pathogen. This simple definition is also the one used in medical epidemiology and, despite its deficiencies as an 'operational term' (Butt and Royle, 1980), I will use van der Plank's definition as a starting point for this ecological discussion. It is still used widely in studies of epidemics caused by biotrophs such as rusts, powdery mildews and downy mildews. The term provides a useful variate for measuring incomplete resistance of a host genotype (Parlevliet, 1979; Leonard and Mundt, 1984). For example, the rust *Phragmidium rubi-idaei* at the uredinial stage has a precise latent period of 12.0 days on red raspberry cv. Glen Clova after inoculation under glasshouse conditions (Anthony *et al*., 1986). In this case the propagule is a urediniospore, but the same definition could be applied to a wind-blown petal ramified by mycelium of *Botrytis cinerea* arriving on the surface of a strawberry fruit and leading to grey mould and release of conidia.

The incubation period was defined by van der Plank (1963) as 'the time needed for symptoms to develop'. This term implies that processes in the early stages of infection are macroscopically invisible or symptomless for a period, and that subsequently a disease will become detectable. All fungal diseases of plants can be said to have a symptomless phase. It may be extremely brief, particularly if the pathogen elicits a hypersensitive response, but the incubation period and latent period are often markedly prolonged. This is a typical feature of many fungal diseases of woody perennials and can be seen to have survival value for the pathogen in phasing spore dispersal in time and space to the appearance of susceptible host tissues, particularly in cold regions where the host enters a dormancy period or in arid regions between rainy seasons.

Quiescent Infection and Latent Infection

In the context of post-harvest diseases Swinburne (1983) precisely defined quiescent infection as 'the inhibition of development of the pathogen through physiological conditions imposed by the host until some stage of maturation has been accomplished'. Therefore, this definition implies that the pathogen must stop growing to be described as quiescent. Hirst and Schein (1965) suggested that the term 'infection' should include the processes of spore germination, germ tube elongation, appressorium formation, penetration and subsequent colonization of tissues. Swinburne (1983) pointed out that a period of quiescence can be enforced in a pathogen at any stage during infection and endorsed the definition of infection given by Hirst and Schein. It is a broader view of infection than I would initially accept, but since 'signal transfer' between pathogen and host is now widely considered to begin immediately after germination of the spore (Kolattukudy, 1992), I now readily endorse Swinburne's view of infection and quiescence, with the reservation that it is usually difficult, with present knowledge of host–pathogen interactions, to be sure whether the quiescence is entirely the outcome of control imposed by the host until a critical stage in host maturation has occurred.

There has been no general agreement about the use of the term latent infection. Symptomless growth of bacterial pathogens or saprophytes was described as latent infection by Hayward (1974). Verhoeff (1974) defined latency as 'a quiescent or dormant parasitic relationship which after a time changes to an active one'. After Swinburne (1983) gave his restricted definition of quiescent infection, he suggested that the term latent infection could still be left for wider uses. However, he did not specify these. In discussing infection of soyabean plants and seeds Sinclair (1991) said that latent infection 'is the association or colonization of tissues by the pathogens for a prolonged period without visible symptoms'. This seems to be an acceptable definition of the condition of the infection during the incubation period. It has the added advantage that it does not specify whether the pathogen stops growing during the symptomless phase, or how its growth may be regulated by the host.

Therefore, from this discussion we can exclude auto-inhibition and endogenous dormancy, as found in many fungal propagules. The physiological dormancy of rust teliospores which require a period of chilling and irradiation by specific wavelengths before germination can occur are good examples of the latter. Such mechanisms have a clear ecological advantage to the rust *Phragmidium rubi-idaei* in that its sole overwintering survival structure is primed and timed to germinate only when tender young leaves are available for ephemeral basidiospores to infect (Anthony *et al.*, 1985). I do not think, however, we should exclude the possibility that some pathogens may have rest periods, other than a tendency to form chlamydospores or sclerotia, which may be equivalent to diapause in insects and allows them to spread their activities over 2 or more years to improve their chances of survival in rapidly changing habitats or environments.

Long latent periods: a survival tactic

Infections of leaves

The leaves of conifers contain large quantities of polyphenols and tannins and consequently are decomposed slowly in the litter phase (Millar, 1974; Mitchell *et al.*, 1978a). The surface of pine needles could be regarded as a hostile habitat for deposition of propagules and penetration because the surface of these xerophytes is protected from desiccation and microbial attack by a thick layer of epicuticular wax and a thick cuticle which almost completely surrounds each epidermal cell. For these reasons many fungal pathogens seem to have evolved to attack rapidly growing needles at the base immediately above the fascicular sheath (they have a basal meristem) in early summer.

Lophodermella sulcigena releases ascospores from hysterothecia on 1-year-old needles of Corsican pine (*Pinus nigra* var. *maritima*) which are clustered within a few centimetres of the new flushing needles (Millar, 1970; Williamson *et al.*, 1976). The ascospores have thick gelatinous sheaths to facilitate attachment to the needle surface and appressoria are produced rapidly if the needles are kept moist for 24 h. Symptoms of infection appear 1 month later (incubation period) as spots or transverse resinous bands c. 5 mm above the sheath. Apart from the proximal region of the needle, all the tissues are killed and in the following season the teleomorph differentiates in time to discharge ascospores when young needles are flushing. It seems to be important for the survival and success of this pathogen that needles remain attached to the tree for one season until spore dispersal is complete; there is a clear advantage in accurately targeting the infection court.

This pathogen is vulnerable to competition for the substrate it acquired by parasitism; *Hendersonia acicola* preferentially attacks the lesions and systematically utilizes the cellulosic materials of the dead needle tissues in a manner which prevents the sporulation of the primary pathogen (Mitchell *et al.*, 1976).

Other needlecast pathogens have extended latent periods and are well adapted to this microhabitat. *L. conjuncta* also infects first-year needles of *P. nigra*, but no symptoms appear for a year and sporulation and needlecast occur after 2 years (Mitchell *et al.*, 1978b). These authors could not trace the earliest infection stages but hyphae were later located in the endodermis of the needle and from there invaded the intercellular spaces of the mesophyll as lesions developed. In this example of latent infection, the fungus may remain totally inactive, or hyphae may develop extremely slowly without producing macrosymptoms.

There are numerous examples of diseases of conifers with prolonged incubation periods and the seasonal physiological status of the tree must be a factor. The pathogens that cause these diseases could be said to show latency, but it is not clear whether growth of the pathogen ceases at any stage.

Rhabdocline pseudotsugae which causes a needlecast of Douglas fir (*Pseudotsuga menziesii*) has a symptomless phase of development persisting c. 4 months; yellow blotches develop in autumn, these turn purple-brown in winter and sporulation occurs in summer. This fungus is clearly pathogenic. Severely infected needles fall very shortly after spore discharge, moderately infected needles later in the summer, while lightly infected needles may stay on the tree and produce fruit bodies in the second, or exceptionally in the third spring (Peace, 1962). In general a severe attack involves the loss of most of the previous year's needles and with annual needle loss the tree may survive merely 5 years.

In contrast, the fungus *Rhabdocline parkeri*, described as an endophyte, infects needles of the same host at any time during their life on the tree, but macroscopically the needles lack symptoms (Stone, 1987). Elegant ultrastructural studies (Stone, 1988) showed that the infections are confined to a single epidermal cell; this fungus has an unusually long incubation period. Active growth only resumes at the onset of needle senescence 2–7 years after penetration with the production of small hyphae ('haustoria') which penetrate the wall of adjacent healthy cells. Further colonization and rapid sporulation occurs after needle fall.

I believe this is an example of quiescent infection. It can be shown clearly that the fungus is inactive for a long period and resumes growth, perhaps in response to a metabolic trigger from adjacent senescent cells. However, it is not clear whether the period of apparent inactivity is primarily under the genetic control of the fungus or of the host, or a balance between them. Only a small proportion of the total epidermal cells of the needle are infected – at worst this could be regarded as a weak parasite which has almost a commensal relationship with the host for its presence deters a leaf-mining fly (*Contarinia* spp.). Therefore, *Rhabdocline* species show, in the same habitat, the full range of possible life styles from aggressive parasitism to endophytism.

Endophytes are beyond the scope of this chapter, but there have been several recent reviews dealing with this subject (Petrini, 1986; Clay, 1986; Carroll, 1986). Nevertheless, it is instructive to appreciate the possible ecological significance of these organisms because closely related pathogens may have similar effects on their hosts and benefit in similar ways. Although endophytic fungi cause no disease symptoms, their metabolites function as antifeedants or affect the viability or fecundity of insect pests and therefore confer valuable pest resistance to the host. Unfortunately, grasses infected by species of *Balansia*, *Epichloë* and *Myriogenospora* cause poisoning of domestic animals, possibly because they contain alkaloids (see Clay, 1986). Carroll (1986) proposed that the relationship with a short-cycle endophyte represents a response of long-lived hosts (e.g. Douglas fir which may live 500–800 years) to the challenge presented by rapidly evolving pests.

In the case of pine needle pathogens it appears that many fungi have evolved to adopt a 'waiting game' to take best advantage of the nutrients that become available at senescence. They could be regarded as specialized primary colonizers of a substrate which is particularly resistant to microbial attack. Early establishment and survival may give them a competitive advantage against saprophytic species in the phyllosphere and in the litter phase of needle decomposition.

Leaves of a wide range of angiosperms are infected by facultative parasites only when fully expanded or mature. These infections may not cause great yield loss, but often represent an important link in the life cycle allowing the pathogen to survive cold or dry periods before susceptible host tissues are again available. *Botrytis cinerea* is well-adapted for this lifestyle and survives in dead foliage or grows from infected leaves to other plant organs to overwinter. For example, it infects mature raspberry (*Rubus idaeus*) leaves, grows through the petiole and then establishes a lesion in the primary cortex of the stem before sclerotia are formed (Williamson and Hargreaves, 1981; Williamson and Jennings, 1986); the sclerotia are the major sources of inoculum in spring for infection of flowers (see later).

On leaves of strawberry (*Fragaria* × *ananassa*) in Ontario, sclerotia of *B. cinerea* are unimportant (Sutton, 1991). However, in this work it was shown that infection cycles of this pathogen occur on leaves which are produced in three main flushes in July, September and April to May. The leaves of these flushes die mainly in September to April, May and June. Mycelium is present only in a few epidermal cells, but causes no lesion or any acceleration of leaf senescence (Sutton, 1991). The mycelium which survives in dead or senescent leaves produces conidia which are important for infection of flowers, fruits and other leaves in spring. This is a form of quiescence, probably under genetic control by the host, but an aggressive parasitic phase does not seem to develop before the host tissue ages and dies. Nevertheless, on resumption of growth from a 'saprophytic base' in the leaves, the pathogen can sporulate, infect styles, petals and anthers in spring and attack ripening fruits in summer (Bristow *et al.*, 1986). Symptomless and quiescent infection therefore plays a vital part in the life cycle. *B. cinerea* is vulnerable to the activity of microbial antagonists during colonization and survival in leaves (Sutton, 1991) and this is perhaps one of the penalties of subsisting at a low metabolic rate for a long period. Infection of fruits will be discussed in a later section.

Transport of plants during the incubation period of pathogens: the surprise element

Cut flowers

Export of cut flowers is important to the economy of some countries, particularly the trade in roses, gerberas and orchids. The substantial losses due to infection of rose petals by *Botrytis cinerea* have stimulated work on the biology and control of post-harvest grey mould in Israel (Elad, 1988) and California (Hammer and Marois, 1989). The conidia of *B. cinerea* germinate on the petals in droplets of condensation, or at high humidity, and penetrate the epidermis without appressoria. The resultant infection is invisible for several days after harvest, but symptoms develop during storage at low temperature before sale or during transport, as grey mould rot or as a hypersensitive flecking reaction of the host (Elad, 1988). In gerberas, the problem is mainly that of hypersensitivity and flecking of the ray florets which renders them unsaleable (Salinas *et al.*, 1989). This type of infection has been described as latent (e.g. Elad, 1988). In most cases the conidia which alighted on petals probably lacked the environmental conditions necessary for infection, but the flowers were harvested and then transported at high humidity during the incubation period of the fungus.

Quarantine procedures were introduced primarily to control the inadvertent spread of disease with plants and their products. One of the most likely ways in which this can occur is when symptomlessly infected plants are moved during the incubation period of a pathogen.

Symptomless infection of flowers: the prelude to fruit infection

It has been known for 30 years that *B. cinerea* effectively attacks strawberry and raspberry fruits by first invading the flowers symptomlessly. Powelson (1960) and Jarvis (1962) provided indirect, though convincing, evidence that infections of these fruits could be reduced or controlled by removal of infected floral organs after fertilization, or by use of fungicides. Their work was the basis of present day control measures for post-harvest grey mould in which fungicides are sprayed repeatedly during the blossom period in a wide range of crops in which it was subsequently shown to be an important early stage in the establishment of the pathogen, e.g. grapes (McClellan and Hewitt, 1973; Nair and Parker, 1985), apples (Tronsmo *et al.*, 1977), raspberry (Dashwood and Fox, 1988). With the increasing and widespread problem of fungicide cross-resistance in this pathogen (Katan, 1982; Pak *et al.*, 1990; Gullino, 1992) the emphasis on breeding for host resistance and biological control (Dubos, 1992) has increased. Unfortunately, in soft fruit crops such as raspberry, there has been no general agreement between breeders as to which plant genotypes constitute reliable sources of resistance to use as parental material in a crossing programme and there is a need for a more precise understanding of the infection process and the long periods during which the pathogen remains symptomless and apparently inactive in the flowers and fruits (Williamson and McNicol, 1986).

The use of a fluorescence microscopy technique with aniline blue has facilitated the examination of floral organs during their infection by *B. cinerea* (McNicol *et al.*, 1985). In strawberry, the infection of petals is a common starting point for invasion of the receptacle (the edible portion). The petals are relatively large and after infection often adhere to the surface of the developing fruit, giving rise to grey mould as the fruit ripens. The fungus also colonizes the stamens and spreads through the filaments to the proximal end of the fruit to establish a stem-end rot at fruit maturation (Bristow *et al.*, 1986).

Stylar infection occurs in all the soft fruits examined. Even under dry environmental conditions, conidia of *B. cinerea* are able to germinate in the fluid secreted by stigmas and its hyphae grow slowly through the transmitting tissues of the styles, in the presence or absence of pollen tubes, to establish a symptomless infection (McNicol *et al.*, 1985; Bristow *et al.*, 1986; Williamson et al., 1987). Except in blackcurrant styles (McNicol and Williamson, 1989), the hyphae rarely enter the cortex of the style and, by their short side branches and stunted growth, seem to be either inhibited by the host or find insufficient nutrients for the infection to become aggressive.

Stylar infection by *B. cinerea* occurred in 23 genotypes of strawberry dry-inoculated with conidia, but hyphae never reached the ovary wall or ovules (Bristow *et al.*, 1986) and strong host resistance to this pathway of infection seems to be operating. In raspberry and related species, there was wide variation in the extent to which transmitting tissues of styles were infected and in many, the fungus entered the ovary wall and sometimes the ovules (Williamson *et al.*, 1987). Styles of these flowers are ephemeral organs and as soon as fertilization is achieved they die but remain attached to the developing drupelets which are strongly resistant to grey mould until fruit maturation. From a 'saprophytic base' in dead stylar tissues, mycelium of the pathogen can spread rapidly at high humidity to invade susceptible fruit tissue and generate a new population of conidia on old stigmas (Williamson *et al.*, 1987).

The infection process leading to grey mould in raspberries described above should perhaps be regarded as quiescent infection. However, in these examples, unlike those described in many tropical fruits by Swinburne (1983), there is a two-stage infection in which the pathogen is apparently strongly inhibited by stylar tissues until that tissue dies: immature fruits are also highly resistant and although the pathogen survives in dead attached styles, rotting only develops when fruits ripen or in the post-harvest period. The possible biochemical basis of this fruit resistance is discussed later.

In blackcurrants, symptomless infection of styles by *B. cinerea* can occur as soon as flowers become 'receptive to pollen' and, in a high proportion of flowers, hyphae spread to the ovary walls and ovules. Premature flower drop in blackcurrants, known as 'running-off', has now been linked with this type of infection; ethylene production by infected flowers causes the abscission of developing fruitlets and there is a good correlation between ethylene proneness of blackcurrant genotypes and their susceptibility to the running-off disorder (McNicol *et al.*, 1989). If inoculation of stigmas in emasculated flowers was delayed in the field for 6 days, until pollination and fertilization had occurred, growth of hyphae as far as the ovary was strongly inhibited (McNicol and Williamson, 1989). Premature abscission is a well-known plant response to stress or infection and in this case probably prevents the spread of the infection to many other fruitlets on the long raceme. From the perspective of the pathogen, the shed fruitlet is a dominated substrate with little prospect of microbial competition.

The flowers of Rosaceae of the 'wet stigma' type clearly have developed effective antimicrobial systems to ensure that pollen tube growth and the fertilization process are not jeopardized, but the nature of this defensive system in styles is unknown. *B. cinerea* as a primary colonizer of this specialized substrate has the ability to establish a limited infection and subsist until further spread is possible.

Potent host resistance mechanisms involving the production of preformed antimicrobial substances also operate against *Botrytis* spp. in styles of other hosts. Tuliposides, which represent up to 32.3% of the dry weight of tulip pistils, are present in the cell vacuoles; when *B. cinerea* increases the permeability of cell membranes, the molecule is released and converted rapidly into lactones with high biological activity towards this pathogen

(Schönbeck and Schroeder, 1972). This mechanism probably accounts for the strong resistance of tulips to *B. cinerea*. In contrast, *B. tulipae* has evolved as an efficient specialized pathogen of tulips. It does not increase the permeability of membranes to the same extent as *B. cinerea*; it is much less sensitive to the lactones, therefore it can grow intercellularly and, if tuliposides are released, it cleaves them into corresponding acids with low biological activity.

Fungal quiescence in fruits

In seeking explanations for the extended period of quiescence of many fungal pathogens that might be under genetic control of the host, it is necessary to look at the evidence for changes in the concentration or activity of antimicrobial substances or changes in the structure of potential enzyme substrates during the maturation process that signal the switch to the aggressive phase in pathogenesis.

The synthesis, secretion and activity of extracellular enzymes may be influenced strongly by the precise chemical structure of the substrate and this may change during ripening. For example, as apples ripen, the pectic fraction is replaced by newly synthesized pectins of different structure; these new polysaccharides are highly esterified, of high molecular weight, and have a low proportion of neutral residues, particularly galactose (Knee, 1978; Knee and Bartley, 1981). Neutral sugar side chains of rhamnogalacturonans can inhibit the ability of a pectin methyl esterase to de-esterify galactosyluronic acid residues in the backbone of the molecule (Matsuura, 1987). Modification of polysaccharide structures can affect the rate and extent to which enzymes can attack cell wall polysaccharides, and the degree to which fungal extracellular enzymes are able to attack substrates undoubtedly affects the nutrition of the pathogen and determines the release of oligosaccharide elicitors of plant defence systems (see Hahn *et al.*, 1989).

A large number of ripening fruits soften mainly by degradation of pectin, with renewed synthesis limited, absent, or confined to early stages of ripening (see Jarvis, 1984). The loosening and separation of live parenchyma cells by plant pectinases will facilitate the growth of intercellular hyphae and the access of fungal enzymes to their substrates. Polygalacturonases (PGs) which degrade the backbone of the pectin polymer are thought to play a key role in fungal pathogenesis because they are the first wall-degrading enzymes to be secreted by several fungi when grown on isolated cell walls (English *et al.*, 1971; Cooper and Wood, 1975).

Raspberry grey mould: inhibitor of polygalacturonase

On pectin as sole carbon source, *B. cinerea* produces four PGs, two endo-PGs which are constitutively expressed and two exo-PGs which are glycoproteins and induced *in vitro* in the presence of galacturonic acid, the monomer sugar unit from which pectin is formed (Johnston and Williamson, 1992a,b). Since the endo-PGs are constitutively produced and can degrade pectin efficiently in the pH range 4.4–4.6, some type of PG inhibition must be involved in quiescence in fruits which resist attack by this pathogen until ripening occurs. Polygalacturonase-inhibiting proteins (PGIPs) have been detected in fruits of apple (Fielding, 1981; Brown, 1984; Brown and Adikaram, 1984), pear (Abu-Goukh *et al.*, 1983), grape (Prudet *et al.*, 1992) and also in bean tissues (Cervone *et al.*, 1987).

A single wall-bound 38.5 kDa protein with a pI value above 10.0 was obtained from immature raspberry tissues after extraction in 1.0 M NaCl and purification by ion exchange chromatography (Johnston *et al.*, 1993). Kinetic studies suggest that it behaves as a non-competitive inhibitor (attaching to the enzyme in regions other than the active site). This PGIP strongly inhibited both endo-PGs secreted by *B. cinerea* in culture but had no effect on the activity of the exo-PGs, or bacterial endo-PGs or endopectate lyase. The activity of the PGIP was highest in green berries and declined rapidly during fruit ripening. In these studies the inhibitor was not found in flower styles, but this may be because the molecule is highly localized and the assay procedure too insensitive to detect it without further refinement.

It is interesting that the N-terminal sequence of this inhibitory protein from raspberry has close homology with the PGIP from bean (Johnston *et al.*, 1993). The gene encoding the bean PGIP has been cloned (Toubart *et al.*, 1992) and, in our laboratory, progress has been made in cloning the related raspberry gene. Further biochemical, immunological and molecular studies are required to determine if these proteins have a role in pathogenesis, and in quiescence in soft fruits in particular. These fruits of high cash value are difficult to transport to distant markets because of their extreme susceptibility to grey mould. Recombinant DNA technology could perhaps be used to enhance the resistance of these fruits and prolong the quiescent phase of the infection into the post-harvest period by stopping the spread of mycelium from attached dead floral organs.

Colletotrichum *diseases of fruit*

There are several excellent reviews dealing with the infection strategies of *Colletotrichum* spp. and the probable involvement of host antimicrobial substances in quiescence in tropical and subtropical fruits (Swinburne, 1983; Bailey *et al.*, 1992; Prusky and Plumbley, 1992; Prusky and Keen, 1993) and only a brief outline of the common features is warranted here.

The spores of these pathogens germinate on the surface of the fruit and form melanized appressoria which do not begin to produce infection pegs until fruits mature. The appressoria are essential for infection in diseases caused by *Colletotrichum* spp. Mucilaginous material is secreted around the appressoria and this mucilage is likely to be involved in surface adhesion and protection of appressoria from extremes of heat, cold and desiccation (Bailey *et al.*, 1992). In these fungi, successful penetration of the epidermis and ultimate tissue colonization is dictated by conditions stimulating production of melanized appressoria. Spores of *C. musae* are stimulated to form appressoria by anthranilic acid, a compound found in leachates of unripe bananas and rapidly degraded by the pathogen to 2,3-dihydroxybenzoic acid, an iron-chelating agent (Harper and Swinburne, 1979; Swinburne, 1983). Most of the bacteria isolated from banana fruit surfaces produce siderophores, other chelating agents with high affinity for iron (McCracken and Swinburne, 1980; Swinburne, 1986). In this case the competition for this key element and its depletion seems to trigger the differentiation of appressoria, but the formation of appressoria can be seen to be ultimately under the influence of the host and the nature of its exudates which provide essential substrates for the microflora in the carposphere.

The appressorial wall usually consists of two or three carbohydrate layers with melanin deposited in one of these layers (Wolkow *et al.*, 1983). As in other fungal resting bodies, such as sclerotia, the melanin in the appressorial wall may protect the cytoplasm from harmful irradiation and microbial attack during the long quiescent period on the epidermis. The appressorial wall also probably plays a crucial role in infection by providing a rigid wall that can withstand the high internal hydrostatic pressure required for penetration of the cuticle (Kubo *et al.*, 1982; Kubo and Furusawa, 1986). Mechanical force exerted by the infection peg has been shown to be an important factor in penetration, but recent studies indicate that cutinases are also involved (see Bailey *et al.*, 1992). These enzymes have been purified (Dickman *et al.*, 1982), their induction by cutin monomers established (Kolattukudy, 1987) and the gene from *C. capsici* isolated (Soliday *et al.*, 1989).

Little is known about the mechanisms controlling quiescence of appressoria. For example, it is difficult at present to envisage how chemical signalling between host and pathogen could be received through an impervious melanized appressorial wall. At fruit maturation, soluble sugars and ethylene production increase and these may be detected by the fungus, but these changes may also be associated with other, as yet unknown, signal molecules arising through the cuticle. As Swinburne (1983) points out, further information about factors triggering release of appressoria from quiescence could be useful in devising control measures because if the fungal growth resumes prematurely lesion development rarely occurs.

Avocado anthracnose: antimicrobial diene and its removal by a lipoxygenase

Two preformed antifungal compounds are present in the peel of unripe avocado fruits and decline rapidly with fruit maturation as susceptibility to *C. gloeosporioides* increases (see Prusky and Keen, 1993; Prusky and Plumbley, 1992). One of these compounds, 1-acetoxy-2-hydroxy-4-oxo-heneicosa-12,15-diene, accounted for most of the antifungal activity. The concentration of this diene increased in response to inoculation of unharvested or freshly harvested fruits with the pathogen (Prusky *et al.*, 1990). Several lines of enquiry indicate that the degradation of the diene during ripening is associated with lipoxygenase activity and that this enzyme system is under the inhibitory control of epicatechin acting as a trap for free radicals (Prusky and Keen, 1993). The concentration of epicatechin in a very susceptible cultivar decreased in proportion to loss of fruit firmness and the symptoms of disease appeared at the highest concentration of the lipoxygenase inhibitor; in a resistant cultivar epicatechin levels were initially higher and were sustained at moderate levels in the peel of ripe soft fruits (Prusky *et al.*, 1988).

The biochemical basis of quiescence of *C. gloeosporioides* on avocado is therefore understood better than in any other host–pathogen combination and serves as a useful model. There are differences in the detail of the tactics employed by *Colletotrichum* spp. but also many common features. By producing appressoria as a 'time capsule' on the fruit surface, these pathogens seem to have evolved to avoid the toxicity of the preformed antifungal compounds present in unripe fruits and benefit from the underlying substrate when the compounds have been removed at maturity.

Role of phytoalexins in quiescent infections of fruit

Swinburne (1983) explained how the wound pathogen *Nectria galligena* can survive exposure to the phytoalexin benzoic acid in Bramley's Seedling apples. A protease produced by this pathogen, and three other pathogens which establish quiescent infections in apple, is the elicitor of benzoic acid synthesis (Brown and Swinburne, 1973a). In the presence of the phytoalexin, the fungus is capable of only limited colonization of apple tissue before harvest. Benzoic acid is toxic only as the undissociated molecule and, during fruit ripening, slight increases in the pH of cell sap lead to loss of fungitoxicity sufficient to allow the fungus to metabolize any remaining benzoic acid to carbon dioxide (Brown and Swinburne, 1973b).

Unripe bananas inoculated with conidia of *C. musae* produce necrotic spots beneath the inoculum droplets and five fungitoxic compounds were found in solvent extracts which were not present in healthy tissue (Brown and Swinburne, 1980). These unidentified compounds decreased in concentration as fruits ripened and were undetectable when anthracnose lesions started to expand.

Phytoalexins also are involved in the quiescent infections of *C. gloeosporioides* (*Glomerella cingulata*) in anthracnose disease of peppers (*Capsicum annuum*). Capsicannol accumulated within 18 h after inoculation of immature fruits, particularly in superficial tissues, and reached maximum concentration after 4–5 days: as fruits started to ripen c. 10 days after inoculation the capsicannol concentration declined (Adikaram *et al.*, 1982). An elicitor prepared from mycelial cell walls of the pathogen and applied to wounds induced accumulation of capsicannol to 1.25 mg g^{-1} in immature fruits, but with the onset of ripening capsicannol decreased rapidly to 0.1 mg g^{-1} fresh weight in fully ripe fruit. *C. gloeosporioides* metabolized capsicannol to capsenone, a less toxic compound, but the breakdown of capsicannol was much slower than that of capsidiol. The latter is a more toxic molecule than capsicannol, but it was not detected in immature fruit in response to infection by the pathogen. This indicates that capsidiol did not contribute to the quiescence of the fungus in immature fruits (Adikaram *et al.*, 1982).

Proanthocyanidins and quiescence

The group of tannins comprising flavan-3-ol dimers and oligomers called proanthocyanidins combine strongly with proteins and inhibit enzyme activity. Their possible role in quiescence of *Botrytis cinerea* in strawberry fruit tissue was studied by Jersch *et al.* (1989). The pathogen initially infects petals, styles and filaments of stamens, but hyphal growth in the fleshy receptacle is largely inhibited until fruits ripen (Powelson, 1960; Bristow *et al.*, 1986). Jersch *et al.* (1989) found that pelargonidin is the principal proanthocyanidin in strawberry. An inverse relationship between the pelargonidin content of immature fruits and their susceptibility to colonization by *B. cinerea* was found and histological studies showed that phenolic substances were present mainly in the epidermis, in tissue beneath the seeds and in seed traces. Tannins in ripening fruits undergo polymerization (Goldstein and Swain, 1963) and lose their protein-binding activity. Some strawberry cultivars which retain a white collar at the proximal end of the fruit at maturity may be more resistant to the fungus because unpolymerized molecules of the proanthocyanidin are available for cross-linking extracellular fungal enzymes (Jersch *et al.*, 1989).

Grape berries produce two hydroxystilbene phytoalexins, pterostilbene and resveratrol, but *B. cinerea* can detoxify them with an extracellular stilbene oxidase (Pezet *et al.*, 1991) which functions as a laccase. Laccase production in *B. cinerea* is induced by a methylated pectin trimer and by certain phenolic compounds as well as by copper and calcium ions in the medium (Viterbo *et al.*, 1992). Certain tetracyclic triterpenoid-lanosterol derivatives called cucurbitacins which are present in cucumber fruits and other species inhibit the formation of laccase under inducing conditions (Bar Nun and Mayer, 1989). The proanthocyanidins, cyanidin and delphinidin, are present in grape skin and are strong competitive inhibitors of stilbene oxidase; this activity is higher in unripe than in fully ripe berries (Pezet *et al.*, 1992) and it has been suggested that these inhibitors are a factor in the relatively strong resistance of developing grapes to the fungus which is established in dead attached floral organs.

With the increasing evidence that numerous antimicrobial compounds decrease in concentration as fruits of many types ripen, it is perhaps useful to remember that highly coloured and fleshy fruits clearly evolved as a seed dispersal system dependent on birds and mammals which rely on vision and smell when seeking food, but soft texture, sweet taste and fragrant aroma are all important. The fleshy pericarps of these fruits need to retain strong defences against pathogens only until the seeds are fully mature. Post-harvest resistance probably has little survival value during the evolution of these hard-seeded fruits and the need to reduce the concentration and toxicity of many antimicrobial compounds and improve the flavour and colour of the pericarp have led to an inevitable reduction in disease resistance properties. The single exception to this metabolic trend is the fact that among the large number of volatile compounds present in ripe fruits which convey the characteristic aroma 'signatures' there are many which possess potent antimicrobial activity (see Wilson *et al.*, 1987; Avissar *et al.*, 1990).

Fungi which evolved systems to minimize contact with the most toxic plant metabolites, by forming appressoria or releasing detoxifying enzymes, and have adapted to withstand the environmental extremes and forms of microbial competition in the phylloplane or carposphere, are well-placed to utilize eventually the large quantities of nutrients present in highly susceptible fruits and leaves when their extracellular enzymes are no longer inhibited. Ripe fruits provide ideal substrates for rapid growth and sporulation of the pathogens and the prolonged latent period or quiescence represents a sound strategy in terms of pathogen energy conservation, utilization of resources and ecological fitness.

Acknowledgements

I thank Drs G.D. Lyon and J.M. Duncan for helpful criticism of the text and The Scottish Office Agriculture and Fisheries Department for supporting some aspects of the work on fruit described in this chapter.

References

Abu-Goukh, A.A., Greve, L.C. and Labavitch, J.M. (1983) Purification and partial characterization of 'Bartlett' pear fruit polygalacturonase inhibitors. *Physiological Plant Pathology* 23, 111–122.

Adikaram, N.K.B., Brown, A.E. and Swinburne, T.R. (1982) Phytoalexin involvement in the latent infection of *Capsicum annuum* L. fruit by *Glomerella cingulata* (Stonem.). *Physiological Plant Pathology* 21, 161–170.

Anthony, V.M., Shattock, R.C. and Williamson, B. (1985) Life-history of *Phragmidium rubi-idaei* on red raspberry in the United Kingdom. *Plant Pathology* 34, 510–520.

Anthony, V.M., Williamson, B., Jennings, D.L. and Shattock, R.C. (1986) Inheritance of resistance to yellow rust (*Phragmidium rubi-idaei*) in red raspberry. *Annals of Applied Biology* 109, 365–374.

Avissar, I., Droby, S. and Pesis, E. (1990) Characterisation of acetaldehyde effects on *Rhizopus stolonifer* and *Botrytis cinerea*. *Annals of Applied Biology* 116, 213–220.

Bailey, J.A., O'Connell, R.J., Pring, R.J. and Nash, C. (1992) Infection strategies of *Colletotrichum* species. In: Bailey, J.A. and Jeger, M.J. (eds), Colletotrichum: Biology, Pathology and Control. CAB International, Wallingford, UK, pp. 88–120.

Bar Nun, N. and Mayer, A.M. (1989) Cucurbitacins protect cucumber tissue against infection by *Botrytis cinerea*. *Phytochemistry* 29, 787–791.

Bristow, P.R., McNicol, R.J. and Williamson, B. (1986) Infection of strawberry flowers by *Botrytis cinerea* and its relevance to grey mould development. *Annals of Applied Biology* 109, 545–554.

Brown, A.E. (1984) Relationship of endopolygalacturonase inhibitor activity to the rate of fungal rot development in apple fruits. *Phytopathologische Zeitschrift* 111, 122–132.

Brown, A.E. and Adikaram, N.K.B. (1984) The differential inhibition of pectic enzymes from *Glomerella cingulata* and *Botrytis cinerea* by a cell wall protein from *Capsicum annuum* fruit. *Phytopathologische Zeitschrift* 105, 27–38.

Brown, A.E. and Swinburne, T.R. (1973a) Factors affecting the accumulation of benzoic acid in Bramley's seedling apples infected by *Nectria galligena*. *Physiological Plant Pathology* 3, 91–99.

Brown, A.E. and Swinburne, T.R. (1973b) Degradation of benzoic acid by *Nectria galligena* Bres. *in vitro* and *in vivo*. *Physiological Plant Pathology* 3, 453–459.

Brown, A.E. and Swinburne, T.R. (1980) The resistance of immature banana fruits to anthracnose (*Colletotrichum musae* (Berk. & Curt.) Arx.). *Phytopathologische Zeitschrift* 99, 70–80.

Butt, D.J. and Royle, D.J. (1980) The importance of terms and definitions for a conceptually unified epidemiology. In: Palti, J. and Kranz, J. (eds), *Comparative Epidemiology: A Tool for Better Disease Management*. Pudoc Scientific Publishers, Wageningen, pp. 29–45.

Carroll, G.C. (1986) The biology of endophytism in plants with particular reference to woody perennials. In: Fokkema, N.J. and van den Heuvel, J. (eds), *Microbiology of the Phyllosphere*. Cambridge University Press, Cambridge, pp. 205–222.

Cervone, F., De Lorenzo, G., Degra, L., Salvi, G. and Bergami, M. (1987) Purification and characterisation of a polygalacturonase-inhibiting protein from *Phaseolus vulgaris* L. *Plant Physiology* 85, 631–637.

Clay, K. (1986) Grass endophytes. In: Fokkema, N.J. and van den Heuvel, J. (eds), *Microbiology of the Phyllosphere*. Cambridge University Press, Cambridge, pp. 188–204.

Cooper, R.M. and Wood, R.K.S. (1975) Regulation of synthesis of cell wall-degrading enzymes by *Verticillium albo-atrum* and *Fusarium oxysporum* f.sp. *lycopersici*. *Physiological Plant Pathology* 5, 135–156.

Dashwood, E.P. and Fox, R.A. (1988) Infection of flowers and fruits of red raspberry by *Botrytis cinerea*. *Plant Pathology* 37, 423–430.

Dickman, M.B., Patil, S.S. and Kolattukudy, P.E. (1982) Purification, characterization and role in infection of an extracellular cutinolytic enzyme from *Colletotrichum gloeosporioides* Penz. on *Carica papaya* L. *Physiological Plant Pathology* 20, 333–347.

Dubos, B. (1992) Biological control of *Botrytis*: state-of-the-art. In: Verhoeff, K., Malathrakis, N.E. and Williamson, B. (eds). *Recent Advances in* Botrytis *Research*. Pudoc Scientific Publishers, Wageningen, pp. 169–178.

Elad, Y. (1988) Latent infection of *Botrytis cinerea* in rose flowers and combined chemical and physiological control of the disease. *Crop Protection* 7, 361–366.

English, P.D., Jurale, J.B. and Albersheim, P. (1971) Host–pathogen interactions II. Parameters affecting polysaccharide-degrading enzyme secretion by *Colletotrichum lindemuthianum* grown in culture. *Plant Physiology* 47, 1–6.

Fielding, A.H. (1981) Natural inhibitors of fungal polygalacturonases in infected fruit tissues. *Journal of General Microbiology* 123, 377–381.

Goldstein, J.L. and Swain, T. (1963) Changes in tannins in ripening fruits. *Phytochemistry* 4, 371–383.

Gullino, M.L. (1992) Chemical control of *Botrytis* spp. In: Verhoeff, K., Malathrakis, N.E. and Williamson, B. (eds), *Recent Advances in* Botrytis *Research*. Pudoc Scientific Publishers, Wageningen, pp. 217–222.

Hahn, M.G., Bucheli, P., Cervone, F., Doares, S.H., O'Neill, R.A., Darvill, A. and Albersheim, P. (1989) Roles of cell wall constituents in plant–pathogen interactions. In: Nester, E. and Kosuge, T. (eds), *Plant–Microbe Interactions III*. Macmillan Press, New York, pp. 131–181.

Hammer, P.E. and Marois, J.J. (1989) Non-chemical methods for postharvest control of *Botrytis cinerea* on cut roses. *Journal of the American Society for Horticultural Science* 114, 100–106.

Harper, D.B. and Swinburne, T.R. (1979) 2,3-Dihydroxybenzoic acid and related compounds as stimulants of germination of conidia of *Colletotrichum musae* (Berk. and Curt.) Arx. *Physiological Plant Pathology* 14, 363–370.

Hayward, A.C. (1974) Latent infections by bacteria. *Annual Review of Phytopathology* 12, 87–97.

Hirst, J.M. and Schein, R.D. (1965) Terminology of infection processes. *Phytopathology* 55, 1157.

Jarvis, M.C. (1984) Structure and properties of pectin gels in plant cell walls. *Plant, Cell and Environment* 7, 153–164.

Jarvis, W.R. (1962) The infection of strawberry and raspberry fruits by *Botrytis cinerea* Fr. *Annals of Applied Biology* 50, 569–575.

Jersch, S., Scherer, C., Huth, G. and Schlösser, E. (1989) Proanthocyanidins as basis for quiescence of *Botrytis cinerea* in immature strawberry fruits. *Zeitschrift für Pflanzenkrankheiten und Pflanzenschutz* 96, 365–378.

Johnston, D.J. and Williamson, B. (1992a) Purification and characterization of four polygalacturonases from *Botrytis cinerea*. *Mycological Research* 96, 343–349.

Johnston, D.J. and Williamson, B. (1992b) An immunological study of the induction of polygalacturonases in *Botrytis cinerea*. *FEMS Microbiology Letters* 97, 19–24.

Johnston, D.J., Ramanathan, V. and Williamson, B. (1993) A protein from immature raspberry fruits which inhibits endopolygalacturonases from *Botrytis cinerea* and other micro-organisms. *Journal of Experimental Botany* 44, 971–976.

Katan, T. (1982) Resistance to 3,5-dichlorophenyl-*N*-cyclic imide ('dicarboximide') fungicides in the grey mould pathogen *Botrytis cinerea* on protected crops. *Plant Pathology* 31, 133–141.

Knee, M. (1978) Metabolism of polymethylgalacturonate in apple fruit cortical tissue during ripening. *Phytochemistry* 17, 1261–1264.

Knee, M. and Bartley, I.M. (1981) Composition and metabolism of cell wall polysaccharides in ripening fruits. In: Friend, J. and Rhodes, M.J.C. (eds), *Recent Advances in the Biochemistry of Fruits and Vegetables*. Academic Press, New York, pp. 133–148.

Kolattukudy, P.E. (1987) Lipid-derived defensive polymers and waxes and their role in plant–microbe interactions. In: Stumpf, P.K. (ed.), *Biochemistry of Plants, Vol. 7, Lipids: Structure and Function*. Academic Press, New York, pp. 291–314.

Kolattukudy, P.E. (1992) Plant-fungal communications that trigger genes for breakdown and reinforcement of host defensive barriers. In: Verma, D.S. (ed.), *Molecular Signals in Plant–Microbe Communications*. CRC Press, Boca Raton, Florida, pp. 65–83.

Kubo, Y. and Furusawa, I. (1986) Localisation of melanin in appressoria of *Colletotrichum lagenarium*. *Canadian Journal of Microbiology* 32, 280–282.

Kubo, Y., Suzuki, K., Furusawa, I., Ishida, N. and Yamamoto, M. (1982) Relation of appressorium pigmentation and penetration of nitrocellulose membranes by *Colletotrichum lagenarium*. *Phytopathology* 72, 498–501.

Leonard, K.J. and Mundt, C.C. (1984) Methods for estimating epidemiological effects of quantitative resistance to plant disease. *Theoretical and Applied Genetics* 67, 219–230.

Matsuura, Y. (1987) Limit to the de-esterification of citrus pectin by citrus pectinesterase. *Agricultural and Biological Chemistry* 51, 1675–1677.

McClellan, W.D. and Hewitt, W.B. (1973) Early *Botrytis* rot of grapes: time of infection and latency of *Botrytis cinerea* Pers. in *Vitis vinifera* L. *Phytopathology* 63, 1151–1157.

McCracken, A.R. and Swinburne, T.R. (1980) Effect of bacteria isolated from surfaces of banana fruits on germination of *Colletotrichum musae* conidia. *Transactions of the British Mycological Society* 74, 212–213.

McNicol, R.J. and Williamson, B. (1989) Systemic infection of black currant flowers by *Botrytis cinerea* and its possible involvement in premature abscission of fruits. *Annals of Applied Biology* 114, 243–254.

McNicol, R.J., Williamson, B. and Dolan, A. (1985) Infection of red raspberry styles and carpels by *Botrytis cinerea* and its possible role in post-harvest grey mould. *Annals of Applied Biology* 106, 49–53.

McNicol, R.J., Williamson, B. and Young, K. (1989) Ethylene production by black currant flowers infected by *Botrytis cinerea*. *Acta Horticulturae* 262, 209–215.

Millar, C.S. (1970) Role of *Lophodermella* species in premature death of pine needles in Scotland. *Report on Forest Research*, HMSO, pp. 176–178.

Millar, C.S. (1974) Decomposition of coniferous leaf litter. In: Dickinson, C.H. and Pugh, G.J.F. (eds), *Biology of Plant Litter Decomposition*, Vol. 1. Academic Press, London, pp. 105–128.

Mitchell, C.P., Williamson, B. and Millar, C.S. (1976). *Hendersonia acicola* on pine needles infected by *Lophodermella sulcigena*. *European Journal of Forest Pathology* 6, 92–102.

Mitchell, C.P., Millar, C.S. and Minter, D.W. (1978a) Studies on decomposition of Scots pine needles. *Transactions of the British Mycological Society* 71, 343–348.

Mitchell, C.P., Millar, C.S. and Williamson, B. (1978b) The biology of *Lophodermella conjuncta* Darker on Corsican pine needles. *European Journal of Forest Pathology* 8, 108–118.

Nair, N.G. and Parker, F.E. (1985) Midseason bunch rot of grapes: an unusual disease phenomenon in the Hunter Valley, Australia. *Plant Pathology* 34, 302–305.

Pak, H.A., Beever, R.E. and Laracy, E.P. (1990) Population dynamics of dicarboximide-resistant strains of *Botrytis cinerea* on grapevine in New Zealand. *Plant Pathology* 39, 501–509.

Parlevliet, J.E. (1979) Components of resistance that reduce the rate of epidemic development. *Annual Review of Phytopathology* 17, 203–222.

Peace, T.R. (1962) *Pathology of Trees and Shrubs*. Clarendon Press, Oxford.

Petrini, O. (1986) Taxonomy of endophytic fungi of aerial plant tissues. In: Fokkema, N.J. and van den Heuvel, J. (eds), *Microbiology of the Phyllosphere*. Cambridge University Press, Cambridge, pp. 177–187.

Pezet, R., Pont, V. and Hoang-Van, K. (1991) Evidence for oxidative detoxification of pterostilbene and resveratrol by a laccase-like stilbene oxidase produced by *Botrytis cinerea*. *Physiological and Molecular Plant Pathology* 39, 441–450.

Pezet, R., Pont, V. and Hoang-Van, K. (1992) Enzymatic detoxication of stilbenes by *Botrytis cinerea* and inhibition by grape berries proanthocyanidins. In: Verhoeff, K., Malathrakis, N.E. and Williamson, B. (eds), *Recent Advances in* Botrytis *Research*. Pudoc Scientific Publishers, Wageningen, pp. 87–92.

Powelson, R.L. (1960) Initiation of strawberry fruit rot caused by *Botrytis cinerea*. *Phytopathology* 50, 491–494.

Prudet, S., Dubos, B. and LeMenn, R. (1992) Some characteristics of resistance of grape berries to grey mould caused by *Botrytis cinerea*. In: Verhoeff, K., Malathrakis, N.E. and Williamson, B. (eds), *Recent Advances in* Botrytis *Research*. Pudoc Scientific Publishers, Wageningen, pp. 99–103.

Prusky, D. and Keen, N.T. (1993) Involvement of preformed antifungal compounds in the resistance of subtropical fruits to fungal decay. *Plant Disease* 77, 114–119.

Prusky, D. and Plumbley, R.A. (1992) Quiescent infections of *Colletotrichum* in tropical and subtropical fruits. In: Bailey, J.A. and Jeger, M.J. (eds), Colletotrichum: *Biology, Pathology and Control*. CAB International, Wallingford, UK, pp. 289–307.

Prusky, D., Kobiler, I. and Jacoby, B. (1988) Involvement of epicatechin in cultivar susceptibility of avocado fruits to *Colletotrichum gloeosporioides* after harvest. *Phytopathologische Zeitschrift* 123, 140–146.

Prusky, D., Karni, L., Kobiler, I. and Plumbley, R.A. (1990) Induction of the antifungal diene in unripe avocado fruits: effect of inoculation with *Colletotrichum gloeosporioides. Physiological and Molecular Plant Pathology* 37, 425–435.

Salinas, J., Glandorf, D.C.M., Picavet, F.D. and Verhoeff, K. (1989) Effects of temperature, relative humidity and age of conidia on the incidence of spotting on gerbera flowers caused by *Botrytis cinerea. Netherlands Journal of Plant Pathology* 95, 51–64.

Schönbeck, F. and Schroeder, C. (1972) Role of antimicrobial substances (tuliposides) in tulips attacked by *Botrytis* spp. *Physiological Plant Pathology* 2, 91–99.

Sinclair, J.B. (1991) Latent infection of soybean plants and seeds by fungi. *Plant Disease* 75, 220–224.

Soliday, C.L., Dickman, M.B. and Kolattukudy, P.E. (1989) Structure of the cutinase gene and detection of promoter activity in the 5' flanking region by fungal transformation. *Journal of Bacteriology* 171, 1942–1951.

Stone, J.K. (1987) Initiation and development of latent infections by *Rhabdocline parkeri* on Douglas-fir. *Canadian Journal of Botany* 65, 2614–2621.

Stone, J.K. (1988) Fine structure of latent infections by *Rhabdocline parkeri* on Douglas-fir, with observations on uninfected epidermal cells. *Canadian Journal of Botany* 66, 45–54.

Sutton, J.C. (1991) Alternative methods for managing grey mould of strawberry. In: Dale, A. and Luby, J.J. (eds), *The Strawberry into the 21st Century.* Timber Press, Portland, Oregon, pp. 183–190.

Swinburne, T.R. (1983) Quiescent infections in post-harvest diseases. In: Dennis, C. (ed.), *Post-harvest Pathology of Fruits and Vegetables.* Academic Press, London, pp. 1–21.

Swinburne, T.R. (1986) Stimulation of disease development by siderophores and inhibition by chelated iron. In: Swinburne, T.R. (ed.), *Iron, Siderophores and Plant Disease.* Plenum Press, New York, pp. 217–226.

Toubart, P., Desiderio, A., Salvi, G., Cervone, F., Daroda, L., De Lorenzo, G., Bergmann, C., Darvill, A.G. and Albersheim, P. (1992) Cloning and characterisation of the gene encoding the endopolygalacturonase-inhibiting protein (PGIP) of *Phaseolus vulgaris* L. *The Plant Journal* 2, 367–373.

Tronsmo, A., Tronsmo, A.M. and Raa, J. (1977) Cytology and biochemistry of pathogenic growth of *Botrytis cinerea* Pers. in apple fruit. *Phytopathologische Zeitschrift* 89, 208–215.

van der Plank, J.E. (1963) *Plant Diseases: Epidemics and Control.* Academic Press, New York.

Verhoeff, K. (1974) Latent infections by fungi. *Annual Review of Phytopathology* 12, 99–110.

Viterbo, A., Bar Nun, N. and Mayer, A.M. (1992) The function of laccase from *Botrytis cinerea* in host infection. In: Verhoeff, K., Malathrakis, N.E. and Williamson, B. (eds), *Recent Advances in* Botrytis *Research.* Pudoc Scientific Publishers, Wageningen, pp. 76–82.

Williamson, B. and Hargreaves, A.J. (1981) Effects of *Didymella applanata* and *Botrytis cinerea* on axillary buds, lateral shoots and yield of red raspberry. *Annals of Applied Biology* 97, 55–64.

Williamson, B. and Jennings, D.L. (1986) Common resistance in red raspberry to

Botrytis cinerea and *Didymella applanata*, two pathogens occupying the same ecological niche. *Annals of Applied Biology* 109, 581–593.

Williamson, B. and McNicol, R.J. (1986) Pathways of infection of flowers and fruits of red raspberry by *Botrytis cinerea*. *Acta Horticulturae* 183, 137–141.

Williamson, B., Mitchell, C.P. and Millar, C.S. (1976) Histochemistry of Corsican pine needles infected by *Lophodermella sulcigena* (Rostr.) v. Höhn. *Annals of Botany* 40, 281–288.

Williamson, B., McNicol, R.J. and Dolan, A. (1987) The effect of inoculating flowers and developing fruits with *Botrytis cinerea* on post-harvest grey mould of red raspberry. *Annals of Applied Biology* 111, 285–294.

Wilson, C.L., Franklin, J.D. and Otto, B.E. (1987) Fruit volatiles inhibitory to *Monilinia fructicola* and *Botrytis cinerea*. *Plant Disease* 71, 316–319.

Wolkow, P.M., Sisler, H.D. and Vigil, E.L. (1983) Effects of inhibitors of melanin biosynthesis on structure and function of appressoria of *Colletotrichum lindemuthianum*. *Physiological Plant Pathology* 23, 55–71.

13 Aspects of the Autecology of the Take-all Fungus

D. Hornby

AFRC Institute of Arable Crops Research, Rothamsted Experimental Station, Harpenden, Herts AL5 2JQ, UK.

Definitions of Autecology

Autecology has been defined many times in the literature and the substance of most of the definitions is the interrelationships of individual organisms and/or individual species to their environments. Studies of groups of individuals of the same species presumably veer towards population ecology and studies of entire populations of a species are seen by some (e.g. Schmidt, 1978) to have moved into the realms of synecology. Here are two examples of definitions of autecology with emphasis on the individual:

> '... the functioning of any one organism viewed ... in relation to that of other members of its own species but also in relation to the diversity of other organisms, the activities of which it may either affect or be affected by.'
>
> (Cooke and Rayner, 1984, p. 55);
>
> 'the life relations of the individual plant.'
>
> (Braun-Blanquet, 1932, p. 81)

However, the emphasis in this chapter is on autecology as the ecology of individual species (probably the more common view amongst plant pathologists) and with a proviso that this definition does not preclude considerations of populations of the species.

According to Schmidt (1978), the study of disease development concerns primary factors such as the pathogen, that are the realm of aetiology, and secondary environmental factors, which concern the 'ecology of disease', better known as epidemiology. Indeed, within the different levels of investigation of host–parasite relations epidemiology has been considered by others to be within socio-economic and ecological 'space' (Rapilly, 1991).

Autecology of the Take-all Fungus

Overview

In studies of the fungus which causes the take-all disease of wheat, now known as *Gaeumannomyces graminis* var. *tritici*, Garrett (1938) suggested that the alternation between an active parasitic phase and a declining resting phase was characteristic of a highly specialized root-infecting fungus, confined to existence upon its hosts because of microbial competition in the soil. In later works he initially called such an organism a 'semi-obligate parasite', but soon introduced the phrase 'ecologically obligate parasite' (Garrett, 1956). This idea is developed in Fig. 13.1 to show some general areas

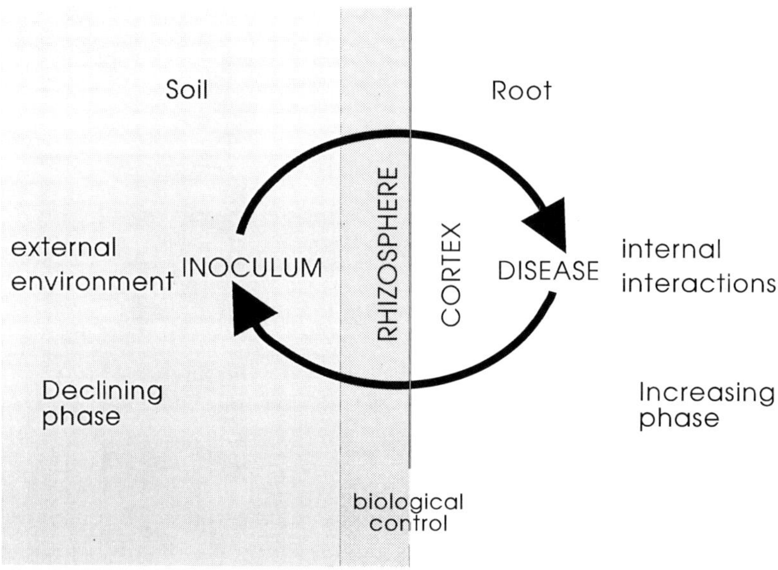

Fig. 13.1. Some general areas of autecological research on the take-all fungus.

of autecological interest. Away from living hosts, therefore, the take-all fungus resides mostly in host residues in the soil in a declining, saprophytic state. Such material constitutes the bulk of inoculum in established arable fields and it has been a conventional focus for autecological studies. Consequently, there is a large body of research on nature, sources, amounts, behaviour, potential, survival and relationships to disease of this inoculum (for reviews and further literature see Asher and Shipton, 1981; Hornby and Bateman, 1991). On and inside hosts, when the take-all fungus is parasitic and/or pathogenic, its biomass increases. The internal environment of the host should be included in the study of autecological relationships, for it is inside the host that interactions operate which influence the extent of disease and the effect that it has on yield. Adjacent to the interface between soil and host are special niches: to the soil side, the rhizosphere (Hornby, 1990) and, to the host side, the outer cells of the root where the phenomenon of root cortex death occurs (Deacon and Henry, 1981). In these niches the study of the autecology of the take-all fungus leads into current research topics on biological control mediated by interactions with other microorganisms. Although much has been written that concerns the autecology of the take-all fungus (see Hornby, 1979; Asher and Shipton, 1981; Kollmorgen, 1985; Hornby, 1990; Hornby and Bateman, 1991), there is still much to learn.

This chapter describes how attempts to simulate patches of take-all disease in winter wheat revealed gaps in our knowledge about the structure of disease patches and their immediate surroundings in the field. It also describes how practical work using ecological techniques to redress this omission unexpectedly provided some insight into unexplained observations within the body of take-all knowledge and posed a new question about inoculum.

Patches of disease in the field

In the absence of uneven growth or other patchiness, it is unlikely that a farmer would be aware of take-all in his wheat crop. Most patches are recognized by poor growth, leaf roll and/or crop discoloration, usually in the form of 'whiteheads' (diseased plants that have died prematurely and have become conspicuous through blanching) or premature ripening. Without such indications it is necessary to dig up plant samples and inspect their roots, or to inspect stem bases for late symptoms just before or after harvest to establish the presence of the disease. Small patches in crops are characteristic of short runs of cereals, or severe take-all infection beginning early. In Fig. 13.2, a small patch reveals itself as small plants; the effect of infection on plant height in a field experiment is shown in Fig. 13.13. In Fig. 13.3, a small patch is conspicuous because of whiteheads, rather than a decrease in plant height. This chapter is mostly about small, distinct patches (less than 1 m in

Fig. 13.2. A small patch (centre) of plants with poor growth in a second winter wheat crop after set-aside (weeds cropped twice) after a first spring wheat. The plants in the patch on Horsepool field at Woburn on 7 June 1991 were at about growth stage 47 (Zadoks *et al.*, 1974).

diameter), because intuitively it seems they will be easier to explain than the larger, irregular patches that develop in crops where take-all is intense (Fig. 13.3).

Patch simulation

Some basic ideas that raise the question of the nature of take-all patches (Fig. 13.8) were used in an attempt to write a computer program (Hornby, unpublished) to simulate the patterns of patches which occur in fields with take-all. The initial selection of a proportion of plants to be foci of slight infections of take-all, occurring at random in an array of healthy plants (the 'field'), started the process. After this the model was deterministic (i.e. having the same outcome for that particular starting point and the same subsequent choices of options), with simple rules for: (i) disease spread from plant to plant; and (ii) changing disease severity on infected plants, either increasing to maximum severity (model default), through concentric zones of slight, moderate and severe disease until, finally, all plants became severely infected, or increasing and then declining to produce the effect in Fig. 13.8. Take-all

Fig. 13.3. A small patch (mostly to the right of a diagonal from the bottom left corner to the top right corner) of plants with whiteheads in a fourth winter wheat crop at growth stage 75 on Great Harpenden field, Rothamsted, 10 July 1992.

Fig. 13.4. Larger patches of take-all in a second winter wheat crop on Dell Piece, Rothamsted at growth stage 75 on 14 July 1987.

Fig. 13.5. Plants dug from a transect being stapled to a wooden batten; crop, site and date as for Fig. 13.2.

decline (Hornby, 1979) rarely, if ever, eliminates the disease, so the decline back to healthy shown in Fig. 13.8 would not be a realistic choice. In the model, spreading patches were allowed to coalesce and patterns could be disrupted at any stage by simulated cultivations between successive cereal crops which displaced and diluted inoculum. Such disruption created distributions of infected plants from which disease would develop in the following crop that were quite different from the initial starting point of a few slightly-infected plants occurring at random. Applying the rules for spread and changing disease severity in following crops extended the range of patterns obtained. Patterns of severely-infected plants emerged in the model that in some cases superficially resembled patterns of patches seen in the field. However, since the model was based on points representing regular plant spacings within and among drilled rows, the patterns that developed

Fig. 13.6. Completed battens representing a transect of 2.5 m and some conventional plant samples in bundles to the right; crop, site and date as for Fig. 13.2. The plants show growth stages 47–55.

Fig. 13.7. A transect through a small patch where plants in the patch are shorter. This was in a winter wheat crop following winter oats in Little Knott I field at Rothamsted at growth stage 73 on 12 July 1991.

```
Disease status

      Build-up                         Decline

Cycle:

0  1     2          3             4            5              6

                                                              1
                                               1            1 2
                                  1          1 2          1 2 3
                    1           1 2        1 2 3        1 2 3 2
          1       1 2 1       1 2 3      1 2 3 2      1 2 3 2 1
0   1   1 2 1   1 2 3 2 1   1 2 3 2    1 2 3 2 1    1 2 3 2 1 0
          1       1 2 1       1 2 3      1 2 3 2      1 2 3 2 1
                    1           1 2        1 2 3        1 2 3 2
                                  1          1 2          1 2 3
                                               1            1 2
                                                              1
```

Fig. 13.8. How disease might develop from a focus of one plant with slight infection in an array of healthy plants, where plant and row spacings are similar. At each step an infected plant changes its disease severity state (initially increasing and then decreasing after peak severity in a crude simulation of take-all decline, which in nature usually requires several consecutive crops to develop) and neighbouring uninfected plants become infected. Disease severity: 0, healthy; 1, slight; 2, moderate; 3, severe. Bold numbers show the centre of the developing patch; after cycle 3 part of the right-hand side of the patch is omitted to save space.

were initially conspicuously angular (as in Fig. 13.8) and unlike the more rounded patches of diseased plant foliage in fields, as seen in Figs 13.2, 13.3 and 13.4. Patchiness in the field is probably partly affected by the lay of the leaves and stems of a plant that 'produces its mainstem leaves to form an overall shape similar to that of a hand-held fan' (Rickman and Klepper, 1991) and the different positions in the drill furrow of seed that gives rise to such plants. Diseased plants may also have a smaller canopy, paler appearance and may be overgrown by healthier plants. The representation of plants by symbols for disease severity in a two-dimensional array with regular spacing is unlikely to result in a realistic picture of patchiness, but it may come close to simulating the zones of disease severities exhibited by the plants that underlie the visible patchiness in the field. Another unrealistic assumption of the model was that plants were in exactly the same positions each year. Since they are not, this also could contribute to modification of pattern development.

The real nature of patches

If a realistic patch program based on the approach outlined above could be created, it would almost certainly include some random element in the disease spread rules. However, without reliable information on the zonation of disease severities in relation to the above-ground signs of take-all, it is not clear what we are trying to simulate. Mapping of take-all in fields generally has been limited in scope and/or detail. Of the few maps (e.g. White, 1945; Wehrle and Ogilvie, 1956; MacNish and Dodman, 1973; Huet and Maumene, 1988) and aerial photographs (e.g. Yarham, 1981; Hornby and Bateman, 1991) published, most show the extent of patches of visibly diseased plants (rarely more frequently than annually) and maps showing the spatial relationships of individual healthy and disease plants are scarce. It is likely that visible patchiness gives no accurate impression of the distribution of all plants with root symptoms within a wheat field. This seems to be one step worse than for aerial diseases, where it is accepted that there is a difference between the visible epidemic (spread of symptoms) and the true epidemic (the increase in infectious area) (Rapilly, 1991).

Zonation in relation to patches arose in the work of Sarniguet and Lucas (1992) on turfgrass infection by *Gaeumannomyces graminis* var. *avenae*. Where there was recolonization by *Festuca* sp. of the centres of diseased patches, the ratios of fluorescent *Pseudomonas* spp. to total bacteria were 1:22 (disease free area outside the patch, where 12–34% of the pseudomonads were antagonistic to the pathogen *in vitro*), 1:15.4 (outer margin of the patch), 1:3.5 (damaged part) and 1:2.9 (recolonized central part, where 44–88% of the pseudomonads were antagonistic to the pathogen *in vitro*).

Similarly, early maps of spatial patterns of Phymatotrichum root rot emphasized radial expansion over a number of seasons and the formation of runs of diseased plants in rows within seasons, but there was little quantitative analysis or interpretation and often assessments were too infrequent to analyse spatial development of the disease (Jeger *et al.*, 1987). Maps are easy to interpret visually. With counts from quadrats, spatial patterns of root diseases and soil-borne plant pathogens may be investigated by: (i) goodness-of-fit techniques for conclusions about randomness or non-randomness; (ii) indices of dispersion to measure departures from randomness; and (iii) techniques using the location of each quadrat for more careful characterization of patterns (Campbell and Noe, 1985).

Studies of disease intensities within and around take-all patches in winter wheat crops

Transects

The sort of transects that have been used commonly in take-all research are exemplified by work at Boxworth Experimental Husbandry Farm, Cambridgeshire to investigate the relationship between infected plants in 1991 and the density of volunteers in 1990. These transects were 24 m long and samples of several plants were taken at *c.* 0.5 m intervals (D.J. Yarham and B. Symonds, personal communication).

At Rothamsted and Woburn during 1991, transects were laid out in and across rows, mostly centred on visible patches, in three different wheat crops (Table 13.1). Wooden battens, 54 cm × 5 cm × 2 cm were placed, end to end, behind the plants at soil level and the positions of the stems marked on the battens by felt pen. The plants were then dug up with a fork, lined up with the marks and stapled to the battens, using binder twine for additional support (Figs 13.5 and 13.6). The battens were then transported to the laboratory and the roots washed clean with the plants still stapled to the battens. Plants were measured and assessed in sequence along the battens and their distances from the origin of the transect noted. In a few cases there was an obvious decrease in height of plants within visible patches (Fig. 13.7). The numbers of root axes and the numbers of root axes infected were counted and an assessment made on a 4-point scale of the percentage of the whole root system that was discoloured on each plant. The scale is one that has been widely used for infection severity categories: healthy (0%), slight (<25%), moderate (25–75%) and severe (>75%) (Clarkson and Polley, 1981). It will be referred to as the DS (disease severity) scale.

Table 13.1. Locations and names of fields from which transects were taken on the sampling dates and at the growth stages (Zadoks *et al.*, 1974) indicated.

	Rothamsted		Woburn
Date (1991)	Little Knott	Osier	Horsepool
24 April			30
7 June			47
20 June		59	
12 July	73	73	
15 July			75

Table 13.2. Clustering of some plant and disease variables along transects.

	Plants		Disease					
			Infrts	%rts	DS			Visible
Field*, (date) and transect	Ht	Rts			Class	Runs†	Mean‡	
Little Knott								
TI1	+++	+	++	+++	+++	5.5	1.35	Y
TA1	–	–	+++	+++	+++	1.0	0.75	Y
TI2	–	–	–	–	+++	5.5	2.47	Y
TA2	–	+	+	+++	+++	1.0	1.25	Y
TI3	–	–	–	+++	+++	12.0	2.79	Y
TA3	–	+	+++	+++	+++	1.0	1.57	Y
Osier								
TI (20 Jun)	+++	–	+++	+++	++	4.0	1.41	Y
TA	++	–	–	+++	+++	1.0	1.58	Y
TI1 (12 Jul)	–	–	+++	+++	+++	8.0	1.65	Y
TA1	–	–	+	++	+++	1.0	1.63	Y
TI2	+	–	++	+++	+++	15.0	2.36	Y
TA2	–	–	+	+++	+++	1.0	1.63	Y
Horsepool								
TI (24 Apr)	–	–	–	+	+	0	1.25	N
TA	–	–	+	+	+	0	1.29	N
TI (7 Jun)	+	+	+	+	+	1.6	1.76	Y
TA	+	–	+	+	+	1.2	1.82	Y
TI (15 Jul)	++	++	++	+++	+++	13.0	2.89	Y
TA	–	–	–	++	+++	4.5	2.75	Y

Ht, plant height; Rts, number of root axes per plant; Infrts, number of infected root axes per plant; %rts, percentage of roots infected; DS, disease severity rating; TI, transect in a crop row; TA, transect across crop rows; –, +, ++, +++ = no, weak, moderate or strong evidence for clustering; Y, transect through a visible patch; N, no above-ground evidence of patches.

*Further details of fields in Table 13.1.

†Mean number of plants in runs in DS class 3 (severe take-all).

‡Mean DS of plants in transect.

Osier field at Rothamsted (Table 13.1) developed larger, widespread patches that continued to expand between June and July, compared to disease development on Little Knott and Horsepool fields, although no stunting was associated with the later occurrence of whiteheads. This contrasts with an observation in France, where two wheat crops had followed 11 years of potatoes and 3 years of lucerne: according to P. Huet (Thiverval-Grignon, 1989, personal communication), the patches occurred unusually early and the outbreak in the second wheat was one of the most severe observed. As early as April there were patches showing decreased tillering, plants with prostrate habit and yellow leaves. In May, plants in patches were shorter and had fewer culms. At flowering, patches contained dwarf plants, fewer heads and yellowed basal leaves and finally, on 24 June, patches also contained whiteheads, but the area of the severe patches did not increase perceptibly during the period of observation.

Although at the time of writing full, detailed analysis is not available for data from the fields at Rothamsted and Woburn, enough has been done to draw attention to certain findings. Ordinary runs analysis, which Madden *et al.* (1982) found better than doublet analysis for determining the randomness of infected plants, was applied to all the two-class samples based on all combinations of the disease severity classes (i.e. 0 *v.* 123, 1 *v.* 023, 2 *v.* 013, 3 *v.* 012, 23 *v.* 01) for each transect. In the first combination it considers spread of disease between adjacent plants in one direction only and is intolerant of missing data. It identifies the occurrence of clusters, but provides no information about their size, shape, number or distance and orientation between clusters (Gray *et al.*, 1986).

Table 13.2, based on ordinary runs analysis of the disease severity data and simple summary statistics of other data, gives an assessment of how different variables were clustered in 18 transects, 16 of which were through visible patches. In general, clustering was most frequently and strongly manifested in the disease variables: particularly assessments using the DS scale, followed closely by percentage of roots infected. Of the disease variables, number of infected roots showed the least tendency to cluster. Clustering of plant heights occurred in a third of the transects and was mostly weak and numbers of roots per plant showed even less clustering in relation to visible patches.

Area samples

It is feasible to sample all the plants in an area including a small patch, or to sample all plants in areas containing successive quadrants of an expanding patch at intervals, using the wooden batten technique. The results presented in Figs 13.9, 13.10 and 13.11 show an area containing a small patch in a first winter wheat crop in France. Figure 13.9 is a scale plan of a 1 m × 1 m square

Fig. 13.9. A plan of plants in a 1 m × 1 m square of a first winter wheat crop at St Gelven, near Rostrenen, Cotes d'Armor, France on 2 July 1992. It contains a small cluster of whiteheads.

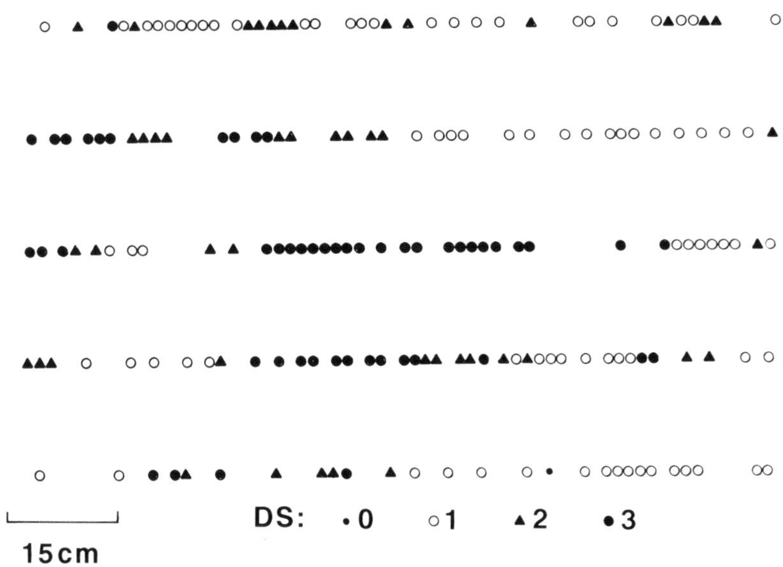

Fig. 13.10. The same plan as in Fig. 13.9, but showing the disease severity scores for each of the plants. The scores are equivalent to healthy (0) and slight (1), moderate (2) and severe (3) take-all.

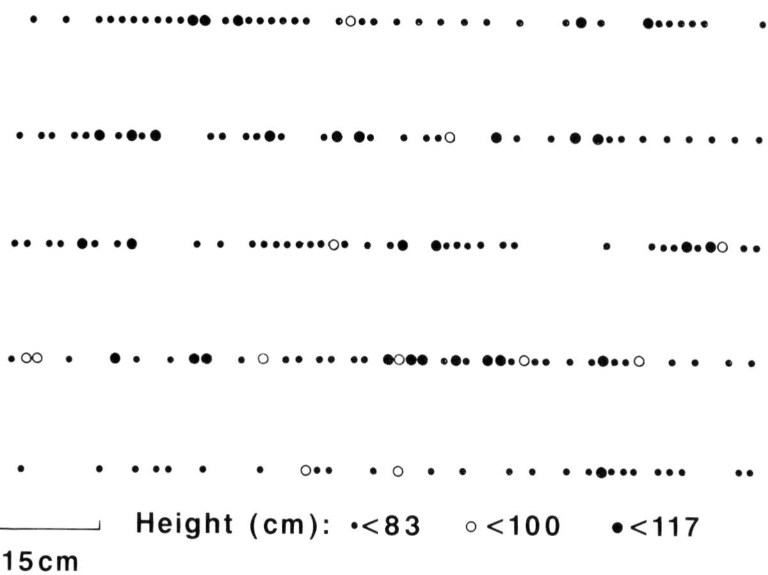

Fig. 13.11. The same plan as in Fig. 13.9, but showing the height categories of the plants.

in the crop showing the positions of nearly all plants (at a few places two or more plants occurred in a line at right-angles to the transect and are represented on the plan as one plant). Gaps in rows of plants, another feature not taken into account in the computer model referred to above, are immediately apparent. Whiteheads form a small cluster rather like that in Fig. 13.3 and they constituted the visible above-ground signs of the patch. Fig. 13.10 shows that the plants with whiteheads, when lifted, were assessed as severely affected with take-all (3 on the DS scale), but also that many other severely affected plants occurred, but did not have the whiteheads symptom. Most of the severely affected plants seem to occur in two major clusters. The distribution of three categories of plant height (Fig. 13.11), however, shows little or no association with DS or whiteheads.

Insights

The data presented demonstrate two aspects of take-all disease that have long been perplexing: (i) disease (DS) on root systems may be as severe outside patches as it is within them; and (ii) association between plant height and DS

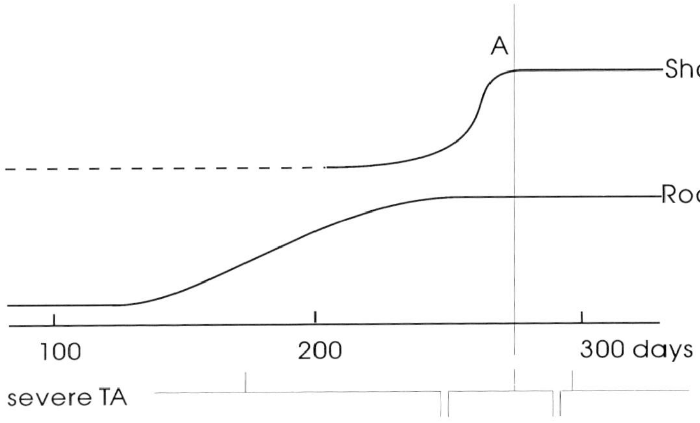

effects:			
patches in crop	Y	?	N
yield decrease	Y	(?)	?
short plants in patches	Y	N	N
farmer aware of TA	Y	?	N
inoculum increase	Y	Y	Y

Fig. 13.12. A sketch suggesting how apparently inconsistent effects of take-all arise. The pattens of shoot heights and root weights are shown against the number of days from sowing winter wheat (x-axis). This time is divided into three approximate periods on the basis of when severe take-all first occurs and some probable effects of the disease in each of these periods are listed. (Further explanation in the text)
A, anthesis; Ht, height; Wt, weight; TA, take-all; Y, yes; N, no; (?), some doubt and ?, considerable doubt that these effects will be apparent.

in the field is variable, often weak when present and frequently absent.

They also suggest an explanation that could accommodate these observations and other observations (Hornby and Bateman, 1991) that: (i) paradoxically, severe disease may occur in years with above average yields; and (ii) the association between disease and yield is strongest at anthesis.

The scheme in Fig. 13.12 is based on the growth of winter wheat in southern England. Root dry weight virtually ceases to increase about anthesis (c. 260 days after sowing), but shoot dry weight continues to increase to beyond 300 days (Barraclough *et al.*, 1991). The crop achieves its maximum height about anthesis (Percival, 1921). This is illustrated in Fig. 13.13 for two sowing dates, with and without the addition of artificial inoculum of the take-all fungus, in conditions where the late sowing produced taller plants than the early sowing. If severe take-all occurs in a period ending a week or so before anthesis (Fig. 13.12), it is likely that the crop will develop visible patches of disease and the farmer would know he has take-all. In such circumstances significant yield decreases are highly likely and diseased plants in the patches would probably be shorter than healthy plants. If severe take-all occurs first about the time of anthesis,

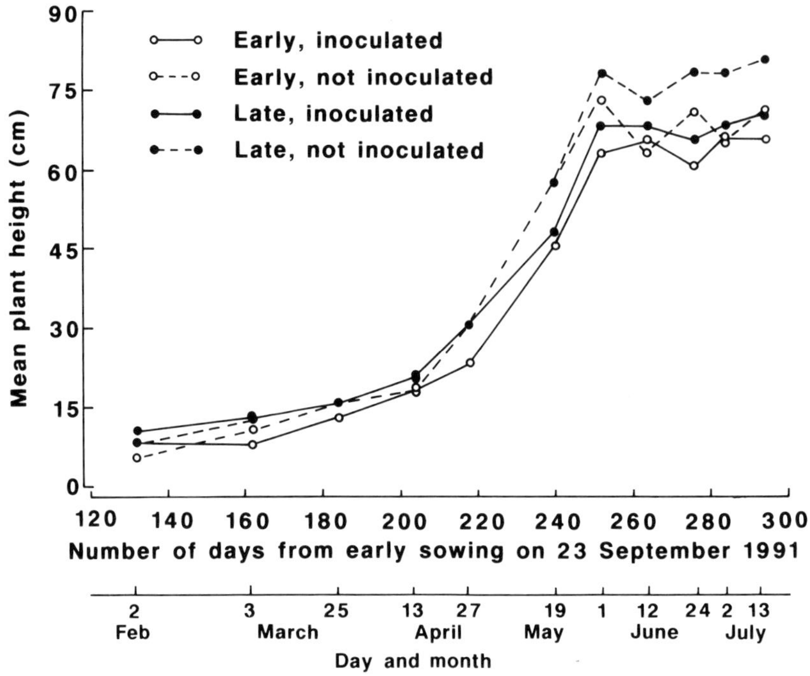

Fig. 13.13. Mean plant height (to the tip of the tallest leaf, or the tip of the ear, whichever was greater) in a first winter wheat crop at Butt Close I field at Woburn, Beds. Late sowing was on 14 October 1991 and inoculation was the addition of 400 kg ha^{-1} of oat grains infected by the take-all fungus. The growth stages at each of the eleven dates were: 13–24, 22–26, 22–24, 23–30, 31, 33–34, 60–61, 69, 73–75, 85–87 and 87–91.

patchiness may not develop and crop height is unlikely to be affected, in which case signs and symptoms of the disease will not be easily detected above ground. It is not clear how much such disease will affect yield, but it will probably be less than when severe infections occur earlier. Severe take-all developing a week or so after anthesis will be even more unlikely to cause height reductions in the crop, or visible patches above ground and the effect on yield may not be great. All these possibilities, however, increase inoculum, but it remains to be discovered which one of these combinations of circumstances is the most damaging for a following susceptible crop and how each of them would affect the onset of take-all decline.

Acknowledgements

Part of this work was funded by MAFF and done as a student project by Michael G. Redfern. The remainder was done during the tenure of my INRA-AFRC Fellowship in the laboratory of Philippe Lucas at Le Rheu (Brittany). Richard J. Gutteridge and Gilles Capron provided much assistance in the field.

References

Asher, M.J.C. and Shipton, P.J. (eds) (1981) *Biology and Control of Take-all*. Academic Press, London.

Barraclough, P.B., Weir, A.H. and Kuhlmann, H. (1991) Factors affecting the growth and distribution of winter wheat roots under UK field conditions. In: McMichael, B.L. and Persson, H. (eds), *Plant Roots and their Environment*. Elsevier, Amsterdam, pp. 410–417.

Braun-Blanquet, J. (1932) *Plant Sociology: the Study of Plant Communities*. Translated, revised and edited by G.D. Fuller and H.S. Conrad. McGraw-Hill, New York.

Campbell, C.L. and Noe, J.P. (1985) The spatial analysis of soilborne pathogens and root diseases. *Annual Review of Phytopathology* 23, 129–148.

Clarkson, J.D.S. and Polley, R.W. (1981) Diagnosis, assessment, crop-loss appraisal and forecasting. In: Asher, M.J.C. and Shipton, P.J. (eds), *Biology and Control of Take-all*. Academic Press, London, pp. 251–269.

Cooke, R.C. and Rayner, A.D.M. (1984) *Ecology of Saprotrophic Fungi*. Longman, London.

Deacon, J.W. and Henry, C.M. (1981) Death of the root cortex of winter wheat in field conditions; effects of break crops and possible implications for the take-all fungus and its biological control agent, *Phialophora radicicola* var. *graminicola*. *Journal of Agricultural Research, Cambridge* 96, 579–585.

Garrett, S.D. (1938) Soil conditions and the take-all disease of wheat. III. Decomposition of the resting mycelium of *Ophiobolus graminis* in infected wheat stubble buried in the soil. *Annals of Applied Biology* 25, 742–766.

Garrett, S.D. (1956) *Biology of Root-infecting Fungi*. Cambridge University Press, Cambridge.

Gray, S.M., Moyer, J.W. and Bloomfield, P. (1986) Two-dimensional distance class model for quantitative description of virus-infected plant distribution lattices. *Phytopathology* 76, 243–248.

Hornby, D. (1979) Take-all decline: a theorist's paradise. In: Schippers, B. and Gams, W. (eds), *Soil-borne Plant Pathogens*. Academic Press, London, pp. 133–156.

Hornby, D. (1990) Root diseases. In: Lynch, J.M. (ed.), *The Rhizosphere*. Wiley and Sons, Chichester, pp. 233–258.

Hornby, D. and Bateman, G.L. (eds) (1991) *Take-all of Cereals*. HGCA Research

Review No. 20. Home-grown Cereals Authority, London.

Huet, P. and Maumene, C. (1988) Nécroses racinaires et élaboration du rendement. *Perspectives Agricoles* 128, 44–50.

Jeger, M.J., Kenerley, C.M., Gerik, T.J. and Koch, D.O. (1987) Spatial dynamics of Phymatotrichum root rot in row crops in the Blackland region of north central Texas. *Phytopathology* 77, 1647–1656.

Kollmorgen, J.F. (1985) Proceedings of the First International Workshop on Take-all of Cereals. In: Parker, C.A., Rovira, A.D., Moore, K.J., Wong, P.T.W. and Kollmorgen, J.F. (eds), *Ecology and Management of Soilborne Plant Pathogens.* The American Phytopathological Society, St. Paul, Minnesota, pp. 291–351.

MacNish, G.C. and Dodman, R.L. (1973) Incidence of *Gaeumannomyces graminis* var. *tritici* in consecutive wheat crops. *Australian Journal of Biological Science* 26, 1301–1307.

Madden, L.V., Louie, R., Abt, J.J. and Knoke, J.K. (1982) Evaluation of tests for randomness of infected plants. *Phytopathology* 72, 195–198.

Percival, J. (1921) *The Wheat Plant. A Monograph.* Duckworth, London.

Rapilly, F. (1991) *L'Épidémiologie en Pathologie Végétale: Mycoses Aériennes.* Institut National de la Recherche Agronomique, Paris.

Rickman, R.W. and Klepper, E.L. (1991) Tillering in wheat. In: Hodges, T. (ed.), *Predicting Crop Phenology.* CRC Press, Boca Raton, Florida, pp. 73–83.

Sarniguet, A. and Lucas, P. (1992) Evaluation of populations of fluorescent pseudomonads related to decline of take-all patch on turfgrass. *Plant and Soil* 145, 11–15.

Schmidt, R.A. (1978). Diseases in forest ecosystems: the importance of functional diversity. In: Horsfall, J.G. and Cowling, E.B. (eds), *Plant Disease. An Advanced Treatise*, Vol II. Academic Press, New York, pp. 287–315.

Wehrle, V.M. and Ogilvie, L. (1956) Spread of take-all from infected wheat plants. *Plant Pathology* 5, 106–107.

White, N.H. (1945) The etiology of take-all disease of wheat. I. A survey of a take-all affected field at Canberra, ACT. *Journal of the Council for Scientific and Industrial Research* 18, 318–328.

Yarham, D.J. (1981) Practical aspects of epidemiology and control. In: Asher, M.J.C. and Shipton, P.J. (eds), *Biology and Control of Take-all.* Academic Press, London, pp. 353–384.

Zadoks, J.C., Chang, T.T. and Konzak, C.F. (1974) A decimal code for the growth stages of cereals. *Weed Research* 14, 415–421.

14 Autecology of Foliar Pseudomonads

S.S. Hirano[1] and C.D. Upper[1,2]

[1]*Department of Plant Pathology and* [2]*US Department of Agriculture, Agricultural Research Service, Plant Disease Resistance Research Unit, University of Wisconsin, Madison, Wisconsin 53706, USA.*

Introduction

For well over a decade, we have used snap bean (*Phaseolus vulgaris*) plants to study the autecology of the foliar bacterium, *Pseudomonas syringae* pv. *syringae* in relation to the epidemiology of bacterial brown spot of bean. Among the questions we have pursued with this system are the role of epiphytic populations of the pathogen in disease development, the role of the physical environment in regulating the dynamics of pathogen population sizes, and the role of lesion formation in the overall ecological success of *P. syringae*. We have assumed that our findings would not only provide an increased understanding of the *P. syringae* pv. *syringae*–bacterial brown spot of bean system, but that these findings should be generally applicable to understanding the ecology of the many strains (or pathovars) of bacteria that comprise the *P. syringae* group and the epidemiology of their associated diseases. Is the *P. syringae* pv. *syringae*–bacterial brown spot system merely a case study or indeed, is it a good model system for understanding the ecology of *P. syringae* and epidemiology of its associated diseases? Some of our findings do not fit accepted paradigms. Is this because our system is unusual (i.e. a case study) or might it be due to the way in which we have approached our system experimentally? It is within the context of these open questions that we discuss the following concepts or hypotheses which are based on findings from the *P. syringae* – snap bean – brown spot system.

1. Epiphytic populations of *P. syringae* are the immediate source of inoculum for disease. Epiphytic population sizes of *P. syringae* are quantitatively related to disease.
2. Rain is the environmental factor that has the greatest effect on the dynamics of *P. syringae* population sizes. Rain with sufficient momentum triggers the onset of growth of *P. syringae*.
3. The selective advantage conferred upon *P. syringae* by its ability to form lesions is in survival, not reproduction.

The *Pseudomonas syringae* Group

We recently raised the question of 'what is a *Pseudomonas syringae*' (Hirano and Upper, 1990) given the extensive phenotypic and genotypic variability among strains that go by the name of *P. syringae*. The use of over 40 different pathovar designations to describe the host specificity of strains of *P. syringae* is but one example of the phenotypic variability within this group of bacteria. Moreover, the use of the pathovar system (and indeed, the description of the species) ignores the existence of those strains which may not be pathogenic on any plant species (Lindow, 1985; Lindemann and Suslow, 1987) and hence does not encompass all strains within the species (Doudoroff and Palleroni, 1974; Dye *et al.*, 1980).

The characteristic symptoms of bacterial brown spot of bean on the foliage are circular brown necrotic lesions, frequently surrounded by a narrow chlorotic ring (Hall, 1991). Necrotic tissues occasionally fall out, giving the leaves a shot-hole appearance. Lesions on pods are small circular necrotic spots. The brown spot pathogen does not cause systemic symptoms as does *P. syringae* pv. *phaseolicola*, causal agent of halo blight of bean. Toxin production is not involved in symptom development, unlike halo blight of bean, wildfire of tobacco (*P. syringae* pv. *tabaci*) or any number of other diseases. Bacterial ooze is not a characteristic of the brown spot disease, as is the case for halo blight of bean. Strains in other pathovars are noted for causing cankers (e.g. *P. syringae* pv. *morsprunorum*) or galls (e.g. *P. syringae* pv. *savastanoi*) of woody tissues (cf. Fahy and Lloyd, 1983). Given this heterogeneity in host range, symptoms expressed, and in any number of bacterial phenotypes such as nutrient utilization patterns, bacteriophage sensitivity, toxin production, or ice nucleation activity (cf. Hirano and Upper, 1990), one may question whether any *P. syringae*–host–disease system can adequately serve as a model system for studies on the ecology and epidemiology of this group of bacteria. At best, any given system should be expected to deviate from a 'model' in some detail.

Epiphytic Populations of *P. syringae* and Disease

Despite the tremendous phenotypic and genotypic diversity within the species, there are certain common features among the many pathogen–host–disease systems. These include their tendency to infect aerial parts of plants and their ability to exist epiphytically on leaves, buds, flowers, etc. of host plants in the absence of disease. Crosse (1959) was the first to report the presence of epiphytic populations of *P. syringae* pv. *morsprunorum* (causal agent of leaf spot and canker of *Prunus* spp.) on healthy leaves of cherry trees. Since the initial report, epiphytic populations of pathogenic *P. syringae* have been found in many pathovar–host–disease systems (Hirano and Upper, 1983). A few examples are shown in Table 14.1. We are not aware of cases where epiphytic populations of *P. syringae* have been rigorously sought and not found.

The concept of Crosse (1959) and Leben (1965) that bacterial pathogens may multiply on the surfaces of healthy plants and thus provide inoculum in the absence of disease was a milestone in the epidemiology of foliar diseases caused by bacterial pathogens, particularly *P. syringae*. Unlike many fungal diseases, inoculum for bacterial foliar diseases need not be produced in lesions. It is well accepted that epiphytic pathogen populations may provide inoculum for disease. However, the extent of the quantitative contribution of these populations to disease development is less clear. This lack of definition may be due to the system examined and/or the experimental approach taken.

Table 14.1. Examples of diseases caused by pathovars of *Pseudomonas syringae* which develop epiphytic populations.

Pathovar of *P. syringae*	Disease	References
pv. *syringae*	Bacterial brown spot of bean	Ercolani *et al.*, 1974; Lindemann *et al.*, 1984a
	Bacterial leaf necrosis of wheat	Fryda and Otta, 1978
	Bacterial canker of *Prunus* spp.	Latorre and Jones, 1979; Sundin *et al.*, 1988
pv. *phaseolicola*	Halo blight of bean	Stadt and Saettler, 1981; Legard and Schwartz, 1987
pv. *glycinea*	Bacterial blight of soybean	Kennedy and Ercolani, 1978; Mew and Kennedy, 1971, 1982
pv. *tomato*	Bacterial speck of tomato	Smitley and McCarter, 1982
pv. *savastanoi*	Knot of olive	Ercolani, 1971

For the *P. syringae*–snap bean–brown spot system, there clearly is such a quantitative relationship. In other systems, for example, *P. syringae* pv. *papulans*–apple–blister spot (Bedford *et al.*, 1988), *P. syringae* pv. *tomato* –tomato–bacterial speck (Smitley and McCarter, 1982), and *P. syringae* pv. *glycinea*–soyabean–bacterial blight (Mew and Kennedy, 1982), increased amounts of disease were associated with larger epiphytic populations of the pathogen. There appears to be a relationship for these systems. It is interesting that the systems for which no quantitative relationship have been found generally involved host plants which are perennials rather than annuals. Examples include *P. syringae* pv. *morsprunorum*–cherry–leaf spot (although the numbers of bacteria recovered from healthy leaves in autumn were related to the incidence of canker infections measured the following May) (Crosse, 1963), *P. syringae* pv. *garcae*–coffee–leaf blight and twig dieback (Ramos, 1979; Ramos and Kamidi, 1981), and *P. syringae* pv. *savastanoi*–olive–olive knot (Ercolani, 1971). Note, however, that the converse is not always true.

The nature of the quantitative relationship which we found for the bacterial brown spot of snap bean system and the way in which this relationship was pursued, both conceptually and empirically, is discussed here to question whether the different findings are associated with the system and/or the experimental approach taken.

The question of whether there is a quantitative relationship between epiphytic pathogen population sizes and disease is a problem of dose versus host response over time. Under controlled conditions, infectivity titrations have shown a clear relationship between the probability of disease on a given plant (or plant part) and the size of the inoculum applied to that plant (or plant part) (Ercolani, 1973, 1984). If we use infectivity titration experiments conducted in the laboratory as a model for field experiments, we need: (i) a set of varying inoculum doses; (ii) a reliable and realistic measure of dose concentrations; (iii) a measure of the responses; and (iv) a method to relate the responses to the given doses. Epiphytic populations of the pathogen become the necessary inoculum dose. To achieve varying inoculum dose, we must manipulate the sizes of these populations. Measures of the sizes of these populations and of disease represent measures of dose and response, respectively. In laboratory experiments, dose is applied and disease estimated on the same site (e.g. leaf). Thus, sampling is not an issue, as it is in the field. Finally, a key issue is how to relate the epiphytic population sizes to resulting disease in a manner which adequately accounts for the natural variability one encounters in the field.

The following approach was taken for the bacterial brown spot system.

Doses

Two different experimental approaches were taken to establish a set of varying inoculum doses on bean plants. In the study of Lindemann *et al.* (1984a,b), bean plots were established at 11 locations on a 64 km east–west transect through the major bean-growing area of central Wisconsin. In general, epiphytic population sizes of *P. syringae* pv. *syringae* were larger on plants from the six plots located within the bean growing area than on plants from the five plots located outside of the area. In the study of Rouse *et al.* (1985), varying pathogen doses were established by several combinations of seed and foliage treatments. Seed treatments included: (i) treatment with a solution of streptomycin; (ii) inoculation of seeds with a powder of brown spot diseased leaves at the time of planting; and (iii) none. Foliage was either treated with a copper bactericide or not treated. All plots were located at a single site.

Measurement of dose

In both studies, careful consideration was given to the observations that the sizes of epiphytic bacterial populations are highly variable from leaf to leaf within a canopy (e.g. plot) and that these populations follow a lognormal distribution (Hirano *et al.*, 1982). Because epiphytic population sizes follow a skewed distribution, estimates of mean population sizes derived from bulk samples (i.e. wherein a few to many individual leaves are combined into one sample) are not reliable and estimates of the variance cannot be obtained from such samples (Hirano *et al.*, 1982; Kinkel, 1992). To relate numbers of epiphytic pathogens to subsequent disease, it is essential to account for the variability in population sizes among leaves. Hence, the ability to calculate meaningful population variances associated with the various dose levels (i.e. epiphytic pathogen population sizes) was critical to the analysis of dose versus response (see below). Thus, the sampling unit for dose measurements was an individual bean leaflet. At each sampling time, a set of leaflets was collected from each plot and processed individually by dilution plating of leaf washings. It is important to note that, in both studies, dose levels were measured more than once during the growing season (five times in the study of Lindemann *et al.* (1984a,b); every 3–4 days for a 5–6 week period in the study of Rouse *et al.* (1985)).

Measurement of response

Response was measured as disease incidence. Lindemann *et al.* (1984a,b) assessed the percentage of diseased bean leaflets in the top third of the plant canopy at pod harvest. Rouse *et al.* (1985) assessed the number of diseased leaflets on each of eight plants per plot in three replicate plots per treatment. Data were expressed as the proportion of diseased leaflets per plot. Disease measurements were made every 3 to 4 days throughout the growing season.

Analysis of dose versus response

Lindemann *et al.* (1984b) found that a simple regression of mean epiphytic pathogen population sizes and disease incidence failed to provide a significant quantitative relationship between dose and response. A quantitative relationship was obtained when consideration was also given to the population variances. The important criterion relating pathogen population sizes and subsequent disease appeared to be the frequency with which pathogen populations exceeded approximately 10^4 colony forming units (CFU) on individual leaflets. This frequency was estimated from the lognormal distribution of epiphytic pathogen population sizes (Hirano *et al.*, 1982). The model used to relate pathogen population sizes and disease by Lindemann *et al.* (1984b) is shown in Equation 14.1,

$$X_2 = X_1 + f_{(p \geq 4)}[1 - (X_1/100)] \tag{14.1}$$

where X_2 and X_1 are percentage disease at times t_2 and t_1, $f_{(p \geq 4)}$ is the frequency (as percentage) with which pathogen populations on individual symptomless bean leaflets exceed 10^4 CFU at t_1 and $[1-(X_1/100)]$ is the proportion of leaflets that are symptomless at t_1 (Lindemann *et al.*, 1984b).

Rouse *et al.* (1985) expanded on the model of Lindemann *et al.* (1984b) as shown in Equation 14.2.

$$PD(\lambda, \tau, \mu, \sigma) = \int_{-\infty}^{\infty} \phi[(y - \lambda)/\tau]\ 1/[\sigma\sqrt{(2\pi)}]\ e^{-(1/2)[(y\ -\ \mu)/\sigma]^2} dy \tag{14.2}$$

PD = probability of disease as a function of four parameters (see below); y = ln of pathogen population size on an individual leaf (leaflet for snap bean).

The model of Rouse *et al.* (1985, Equation 14.2) takes into account both the distribution of sizes of epiphytic pathogen populations (i.e. the lognormal) among individual bean leaflets and the probability of disease given pathogen population sizes on individual leaves (i.e. the probit function). The model has four parameters, two characterizing the dose–response function (λ and τ) and two characterizing the frequency distribution of epiphytic pathogen population sizes on individual leaves (μ and σ). The parameters λ and τ are equivalent to the ED_{50} value and slope of the probit regression equation, respectively and were obtained by a nonlinear regression procedure. Although variability was observed among estimates of the ED_{50}, none of the values were significantly different from approximately 3×10^5 CFU per leaflet for pathogen population sizes determined 4 to 11 days prior to disease assessments. We do not view this ED_{50} value as being absolute for the bacterial brown spot system. The parameters λ and τ may vary with bean cultivar or environmental factors as they, in turn, affect the probability of disease given pathogen population size.

The findings of Lindemann *et al.* (1984b) and Rouse *et al.* (1985) demonstrate that there is a quantitative relationship between epiphytic populations of *P. syringae* pv. *syringae* and brown spot incidence. We developed an assay (tube nucleation test) which takes advantage of the ice nucleation activity of *P. syringae* to rapidly estimate the frequency with which individual leaves or leaflets harbour large epiphytic pathogen populations (Hirano *et al.*, 1985). For both bacterial brown spot of bean and halo blight of oats caused by *P. syringae* pv. *coronafaciens*, these frequencies were predictive of relative disease incidence (Hirano *et al.*, 1987a).

Montesinos and Vilardell (1991) reported that a significant relationship was found between population sizes of *P. syringae* in dormant flower buds in winter and bud blast disease of pear in spring. These authors reported an ED_{50} value of approximately 8.3×10^7 CFU per bud for this disease based on mean values from bulk samples from three cultivars in eight orchards. The very large differences in inoculum levels between orchards probably facilitated analysis based on bulk samples, although the ED_{50} value is probably somewhat overestimated because of the way in which the measurements were made. As they suggested, differences between host, the pathogens, and the interaction between the two may also account for some of the apparent difference between the ED_{50} values for the brown spot and bud blast diseases. Nonetheless, what is significant is the presence of a quantitative relationship between epiphytic pathogen size and subsequent disease incidence for two rather different diseases. For both systems, one involving an annual and the other a perennial host, the probability of disease occurring was related to the establishment of relatively large epiphytic populations of *P. syringae*.

Is the *P. syringae* pv. *syringae* – brown spot of bean system a model or a case study? The case for a quantitative relationship between epiphytic population sizes of the pathogen and disease is well supported for this system. There have, however, been few clear demonstrations that other bacterial systems follow this model. At best, the proven applicability of the model is rather narrow at present. To quote Henis and Bashan (1986) in their review on epiphytic survival of bacterial leaf pathogens '... no relationship was found between *X. campestris* pv. *vesicatoria* number in pepper plants and symptom expression. Therefore, this model (Rouse *et al.*) should be treated at present with some caution'. Clearly the general applicability of models relating sizes of epiphytic populations of bacterial pathogens to amounts of the foliar diseases they cause awaits further experimentation. We would also argue that equal 'caution' and consideration should be given to the empirical approach taken in any such future experimentation. Within that set of diseases for which the relationship between epiphytic pathogen dose and

and McCarter, 1982; Gross *et al.*, 1983; Latorre *et al.*, 1985; Wimalajeewa and Flett, 1985; Bedford *et al.*, 1988). What is not clear from these studies is whether rain is the dominant environmental factor in these systems. For example, on leaves of stonefruit trees grown in Victoria, Australia, high populations of *P. syringae* pv. *syringae* were correlated with periods when 'maximum temperatures ranged from 19 to 25°C and when rainfall was moderately high' (Wimalajeewa and Flett, 1985). On tomatoes grown in Georgia, USA, large populations of *P. syringae* pv. *tomato* were found when 'the temperature was low, generally averaging below 23°C and moisture [rain] levels were high' (Smitley and McCarter, 1982). Bedford *et al.* (1988) suggested that 'seasonal differences in temperature, humidity, and rainfall' influenced the dynamics of *P. syringae* pv. *papulans* on apple leaves in Ontario, Canada.

Roos and Hattingh (1986) examined populations of *P. syringae* on leaves of fruit trees in South Africa and found that population increases were detected even in the absence of rain or mist when presumably dew was present. The authors suggested that 'apparently dew provides conditions suitable for sustained bacterial growth during otherwise dry periods' (Roos and Hattingh, 1986). Because sampling frequencies for measurement of bacterial population sizes and the manner in which weather data were gathered and analysed (e.g. in some cases rainfall data were examined as total volume for a period of up to 2 weeks), it is difficult to evaluate this and other reports to ask if the dynamics of *P. syringae* populations are regulated similarly by the physical environment. It may be that populations of *P. syringae* have evolved different ways to be successful under different environmental conditions. Rain does appear to have an overall positive effect on population sizes of *P. syringae* in all of the systems examined.

In nature, weather variables are frequently correlated. Hence, distinguishing the effects of the various parameters of the physical environment on the population dynamics of *P. syringae* is not easily accomplished in the field. However, from our experiences with the *P. syringae*–snap bean system, we are not strong advocates of the use of controlled environments provided by growth chambers or greenhouses for the examination of the effect of the physical environment on the dynamics of *P. syringae* in the phyllosphere. We have observed a rather curious phenomenon related to growth of *P. syringae* on bean leaves. In the field, rapid growth of *P. syringae* occurs infrequently and is highly dependent on some environmental trigger, most frequently, intense rain. When leaves were kept wet by misting or use of a sprinkler irrigation system that did not deliver water drops with sufficient momentum, growth of *P. syringae* on bean leaves in the field was not detected (Alberga, 1987; Hirano and Upper, unpublished data). However, growth of *P. syringae* does occur on bean plants in a growth chamber maintained at a relatively high humidity, in the absence of rain (cf. Willis *et al.*, 1990; Hirano and Upper, unpublished data). Although we have been unable to relate the behaviour of bacterial populations on plants in the growth chamber to those in the field, the approach taken by Alberga (1987) to modify the microclimate of plants in the field has been useful in our studies of the population dynamics of *P. syringae*.

Rain and disease

Increased amounts of disease caused by pathovars of *P. syringae* have frequently been associated with rain (cf. Hirano and Upper, 1990). We discussed previously our view of the role of rain in the epidemiology of bacterial brown spot of snap bean. The literature would suggest that this is not a prevalent view. Rather, the role of rain in disease development has more frequently been attributed to rain-splash dispersal of *P. syringae*, not necessarily growth of the pathogen (Walker and Patel, 1964; Daft and Leben, 1972; Ercolani *et al.*, 1974; Ramos, 1979, Ramos and Kamidi, 1981; Umekawa *et al.*, 1981; Smitley and McCarter, 1982). In none of these studies was the mechanism by which rain increased disease examined in a way which would separate the role of rain in dispersal from that in promotion of bacterial growth. For example, Umekawa *et al.* (1981) compared the amount of angular leaf spot on cucumber plants grown under a roof of plastic film to plants exposed to rain. The short term dynamics of pathogen population sizes were not followed. The amount of disease was less on plants protected from rain than on plants which were exposed to rain. Their conclusion that the decreased amount of disease was due to decreased dispersal of the pathogen cannot be separated from the alternative hypothesis that pathogen growth was stimulated. Should we continue to assume that the role of rain is primarily related to pathogen dispersal for all of those bacterial diseases for which the effect of rain on pathogen growth has not been examined? Dispersal of pathogen inocula by rain splash plays a major role in the epidemiology of many fungal diseases (cf. Fitt *et al.*, 1989; Madden, 1992). Perhaps it should be asked whether paradigms from studies of the epidemiology of fungal diseases are appropriate for the epidemiology of foliar bacterial diseases.

Lesion Formation in the Ecological Success of *P. syringae*

What role does the ability to form lesions play in the overall ecological success of a bacterial species that can multiply on healthy leaves in the absence of disease? Does the ability to form lesions confer fitness to *P. syringae*? It is interesting to note that there are strains of *P. syringae* for which no host has been found, even after an extensive survey of over 30 to

40 different plant species or cultivars (Lindow, 1985; Lindemann and Suslow, 1987). Of 2080 strains of *P. syringae* isolated from healthy bean leaflets during a growing season, 40% were found to be non-pathogenic to bean in a pod inoculation assay (Hirano *et al*., 1991).

To determine whether the ability to form brown spot lesions confers fitness to *P. syringae*, we compared the population dynamics of a Tn*5*-induced non-lesion forming mutant of *P. syringae* pv. *syringae* strain B728a (Hirano *et al*., 1992; Hirano *et al*., unpublished data). The Tn*5*-mutant NPS3136 was isolated and characterized by Willis *et al*. (1990). The affected gene, *lemA* (for lesion manifestation) has been sequenced and found to have similarity to a family of two-component transmembrane regulators (Hrabak and Willis, 1992). Under controlled environmental conditions, the mutant was similar to the parent in its ability to grow on bean leaves when either misted upon, or infiltrated into, the leaves (Willis *et al*., 1990). Under field conditions, population sizes of the mutant and parent increased following application. However, by 2 weeks post-inoculation, population sizes of the mutant were approximately 50-fold less than those of the parent. The mutant appeared to be less fit than the parent when the strains were applied as single treatments to replicated plots of beans. As the season progressed with few rainfalls, population sizes of the parent and mutant decreased substantially. However, we were able to isolate the parent from brown spot lesions. We hypothesize that the advantage conferred to *P. syringae* by its ability to form brown spot lesions is in survival, not necessarily growth.

An unexpected result was obtained when the mutant and parent strains were simultaneously applied to the same plants (co-inoculated). In this case, population sizes of the parent were not significantly different from those of the mutant and were similar to the densities of the mutant when applied alone. The relative fitness of the parent was decreased in the presence of the mutant. Therefore, a single mutation in a regulatory gene which is required for lesion formation has resulted in a progeny which appears to affect the fitness of the parent. The parent, however, appeared not to have a negative effect on its progeny. Lindow (1986) suggested that competition for limiting resources is a likely mechanism for the types of intraspecific interactions they have observed with strains of *P. syringae*. The resources available to support *P. syringae* in our field experiment should have been equally available on all plants. If availability of resources was the cause of the effect of the mutant on its parent, the total of the two populations (mutant and parent) in the co-inoculation treatment should have approximated that of the parent alone. The sum of population sizes of the mutant plus the parent in the co-inoculation treatment was at least an order of magnitude lower than the densities attained by the parent when it was applied alone. The effect of the non-lesion forming mutant on its wild-type parent is clearly due to some mechanism other than competition for resources. The nature of this mechanism remains to be elucidated.

Summary

Leaves of snap bean plants have provided a useful and interesting host system with which to study the role of epiphytic pathogen populations in disease development, the effect of the physical environment on the dynamics of these populations and the role of lesion formation in the ecological success of *P. syringae*. We raised the question whether our findings with the snap bean system are unique to *P. syringae* on snap bean in Wisconsin, or whether they may be generally applicable to other *P. syringae*–host–disease systems. There are similarities between the *P. syringae*–snap bean–brown spot system and others studied. For example, epiphytic populations of pathogenic *P. syringae* may provide a source of inoculum and larger *P. syringae* populations are found during rainy compared to dry weather. Beyond these generalizations, it is difficult to evaluate whether the quantitative relationship between epiphytic pathogen population sizes and bacterial brown spot incidence, or the role of rain as the primary environmental factor affecting the dynamics of *P. syringae* on bean leaflets may apply to other systems. This may be due partly to differences in experimental approach and design. We note two specific differences between our approach and those of others. The first relates to estimates of bacterial population parameters. In all experiments, means and variances of bacterial populations were calculated from measurements of population sizes on sets of individual leaflets. The second relates to sampling frequency. Daily and even bi-hourly sampling of bean leaflets for estimation of bacterial population sizes showed the inherent dynamics of *P. syringae* populations under field conditions. Changes in population sizes in *P. syringae* can occur rapidly in response to changes in the weather, which in turn, may fluctuate tremendously within a given day and from day to day.

Whether the *P. syringae*–snap bean–brown spot system is a case study or model study remains an open question, we think.

References

Alberga, A.H. (1987) Microclimate influences on *Pseudomonas syringae* bacteria on snap beans. Unpublished PhD Thesis. University of Wisconsin-Madison.

Bedford, K.E., MacNeill, B.H., Bonn, W.G. and Dirks, V.A. (1988) Population dynamics of *Pseudomonas syringae* pv. *papulans* on Mutsu apple. *Canadian Journal of Plant Pathology* 10, 23–29.

Butterworth, J. and McCartney, H.A. (1991) The dispersal of bacteria from leaf surfaces by water splash. *Journal of Applied Bacteriology* 71, 484–496.

Constantinidou, H.A., Hirano, S.S., Baker, L.S. and Upper, C.D. (1990) Atmospheric dispersal of ice nucleation-active bacteria: the role of rain. *Phytopathology* 80, 934–937.

Crosse, J.E. (1959) Bacterial canker of stone-fruits. IV. Investigation of a method for measuring the inoculum potential of cherry trees. *Annals of Applied Biology* 47, 306–317.

Crosse, J.E. (1963) Bacterial canker of stone-fruits. V. A comparison of leaf-surface populations of *Pseudomonas mors-prunorum* in autumn on two cherry varieties. *Annals of Applied Biology* 52, 97–104.

Daft, G.C. and Leben, C. (1972) Bacterial blight of soybeans: epidemiology of blight outbreaks. *Phytopathology* 62, 57–62.

Doudoroff, M. and Palleroni, N.J. (1974) Genus I. *Pseudomonas* Migula. In: Buchanan, R.E. and Gibbons, N.E. (eds), *Bergey's Manual of Determinative Bacteriology*, 8th edn. Williams and Wilkins, Baltimore/London, pp. 217–243.

Dye, D.W., Bradbury, J.F., Goto, M., Hayward, A.C., Lelliott, R.A. and Schroth, M.N. (1980) International standards for naming pathovars of phytopathogenic bacteria and a list of pathovar names and pathotype strains. *Review of Plant Pathology* 59, 153–168.

Ercolani, G.L. (1971) Occurrence of *Pseudomonas savastanoi* (E.F. Smith) Stevens as an epiphyte of olive trees, in Apulia. *Phytopathologia Mediterranea* 10, 130–132.

Ercolani, G.L. (1973) Two hypotheses on the aetiology of response of plants to phytopathogenic bacteria. *Journal of General Microbiology* 75, 83–95.

Ercolani, G.L. (1984) Infectivity titration with bacterial plant pathogens. *Annual Review of Phytopathology* 22, 35–52.

Ercolani, G.L., Hagedorn, D.J., Kelman, A. and Rand, R.E. (1974) Epiphytic survival of *Pseudomonas syringae* on hairy vetch in relation to epidemiology of bacterial brown spot of bean in Wisconsin. *Phytopathology* 64, 1330–1339.

Fahy, P.C. and Lloyd, A.B. (1983) *Pseudomonas*: the fluorescent pseudomonads. In: Fahy, P.C. and Persley, G.J. (eds), *Plant Bacterial Diseases. A Diagnostic Guide*. Academic Press, Sydney, pp. 141–188.

Fitt, B.D.L., McCartney, H.A. and Walklate, P.J. (1989) The role of rain in dispersal of pathogen inoculum. *Annual Review of Phytopathology* 27, 241–270.

Foulkes, J.A. and Lloyd, A.B. (1980) Epiphytic populations of *Pseudomonas syringae* pv. *syringae* and *P. syringae* pv. *morsprunorum* on cherry leaves. *Australasian Plant Pathology* 9, 114–115.

Fryda, S.J. and Otta, J.D. (1978) Epiphytic movement and survival of *Pseudomonas syringae* on spring wheat. *Phytopathology* 68, 1064–1067.

Gross, D.C., Cody, Y.S., Proebsting, E.L., Jr., Radamaker, G.K. and Spotts, R.A. (1983) Distribution, population dynamics, and characteristics of ice nucleation-active bacteria in deciduous fruit tree orchards. *Applied and Environmental Microbiology* 46, 1370–1379.

Hall, R. (ed.) (1991) *Compendium of Bean Diseases*. American Phytopathological Society, St. Paul, Minnesota.

Henis, Y. and Bashan, Y. (1986) Epiphytic survival of bacterial leaf pathogens. In: Fokkema, N.J. and van den Heuvel, J. (eds), *Microbiology of the Phyllosphere*. Cambridge University Press, Cambridge, pp. 252–268.

Hirano, S.S. and Upper, C.D. (1983) Ecology and epidemiology of foliar bacterial plant pathogens. *Annual Review of Phytopathology* 21, 243–269.

Hirano, S.S. and Upper, C.D. (1989) Diel variation in population size and ice nucleation activity of *Pseudomonas syringae* on snap bean leaflets. *Applied and Environmental Microbiology* 55, 623–630.

Hirano, S.S. and Upper, C.D. (1990) Population biology and epidemiology of *Pseudomonas syringae*. *Annual Review of Phytopathology* 28, 155–177.

Hirano, S.S. and Upper, C.D. (1992) Population dynamics of *Pseudomonas syringae* in the phyllosphere. In: Galli, E., Silver, S. and Witholt, B. (eds), Pseudomonas: *Molecular Biology and Biotechnology*. American Society for Microbiology, Washington, D.C., pp. 21–29.

Hirano, S.S., Nordheim, E.V., Arny, D.C. and Upper, C.D. (1982) Lognormal distribution of epiphytic bacterial populations on leaf surfaces. *Applied and Environmental Microbiology* 44, 695–700.

Hirano, S.S., Baker, L.S. and Upper, C.D. (1985) Ice nucleation temperature of individual leaves in relation to population sizes of ice nucleation active bacteria and frost injury. *Plant Physiology* 77, 259–265.

Hirano, S.S., Rouse, D.I. and Upper, C.D. (1987a) Bacterial ice nucleation as a predictor of bacterial brown spot disease on snap beans. *Phytopathology* 77, 1078–1084.

Hirano, S.S., Tanner, C.B. and Upper, C.D. (1987b) Rain-triggered multiplication of *Pseudomonas syringae* on snap bean leaflets. *Phytopathology* 77, 1694.

Hirano, S.S., Baker, L.S. and Upper, C.D. (1991) Intraspecific variation in the population dynamics of *Pseudomonas syringae* on snap bean (*Phaseolus vulgaris*) leaflets. In: Durbin, R.D., Surico, G. and Mugnai, L. (eds), Pseudomonas syringae *Pathovars*, Proceedings of the 4th International Working Group on *Pseudomonas syringae* pathovars. Stamperia Granducale, Florence, Italy, pp. 214–217.

Hirano, S.S., Willis, D.K. and Upper, C.D. (1992) Population dynamics of a Tn*5*-induced non-lesion forming mutant of *P. syringae* pv. *syringae* on bean plants in the field. *Phytopathology* 82, 1067.

Hrabak, E.M. and Willis, D.K. (1992) The *lemA* gene required for pathogenicity of *Pseudomonas syringae* pv. *syringae* on bean is a member of a family of two-component regulators. *Journal of Bacteriology* 174, 3011–3020.

Kennedy, B.W. and Ercolani, G.L. (1978) Soybean primary leaves as a site for epiphytic multiplication of *Pseudomonas glycinea*. *Phytopathology* 68, 1196–1201.

Kinkel, L. (1992) Statistical consequences of combining population samples. *Phytopathology* 82, 1168.

Latorre, B.A. and Jones, A.L. (1979) *Pseudomonas morsprunorum*, the cause of bacterial canker of sour cherry in Michigan, and its epiphytic association with *P. syringae*. *Phytopathology* 69, 335–339.

Latorre, B.A., Gonzalez, J.A., Cox, J.E. and Vial, F. (1985) Isolation of *Pseudomonas syringae* pv. *syringae* from cankers and effect of free moisture on its epiphytic populations on sweet cherry trees. *Plant Disease* 69, 409–412.

Leben, C. (1965) Epiphytic microorganisms in relation to plant disease. *Annual Review of Phytopathology* 3, 209–230.

Legard, D.E. and Schwartz, H.F. (1987) Sources and management of *Pseudomonas syringae* pv. *phaseolicola* and *Pseudomonas syringae* pv. *syringae* epiphytes on dry beans in Colorado. *Phytopathology* 77, 1503–1509.

Lindemann, J. and Suslow, T.V. (1987) Characteristics relevant to the question of environmental fate of genetically engineered INA$^-$ deletion mutant strains of *Pseudomonas*. In: Civerolo, E.L., Collmer, A., Davis, R.E. and Gillaspie, A.G.

(eds), *Plant Pathogenic Bacteria*. Martinus Nijhoff, Dordrecht, pp. 1005–1012.

Lindemann, J. and Upper, C.D. (1985) Aerial dispersal of epiphytic bacteria over bean plants. *Applied and Environmental Microbiology* 50, 1229–1232.

Lindemann, J., Arny, D.C. and Upper, C.D. (1984a) Epiphytic populations of *Pseudomonas syringae* pv. *syringae* on snap bean and nonhost plants and the incidence of bacterial brown spot disease in relation to cropping patterns. *Phytopathology* 74, 1329–1333.

Lindemann, J., Arny, D.C. and Upper, C.D. (1984b) Use of an apparent infection threshold population of *Pseudomonas syringae* to predict incidence and severity of brown spot of bean. *Phytopathology* 74, 1334–1339.

Lindow, S.E. (1985) Ecology of *Pseudomonas syringae* relevant to the field use of Ice$^-$ deletion mutants constructed *in vitro* for plant frost control. In: Halvorson, H.O., Pramer, D. and Rogul, M. (eds), *Engineered Organisms in the Environment, Scientific Issues*. American Society for Microbiology, Washington, D.C., pp. 23–35.

Lindow, S.E. (1986) Construction of isogenic ice strains of *Pseudomonas syringae* for evaluation of specificity of competition on leaf surfaces. In: Megusar, F. and Gantar, M. (eds), *Perspectives in Microbial Ecology*. Slovene Society for Microbiology, Ljubljana, pp. 509–515.

Madden, L.V. (1992) Rainfall and the dispersal of fungal spores. *Advances in Plant Pathology* 8, 39–79.

Mew, T.W. and Kennedy, B.W. (1971) Growth of *Pseudomonas glycinea* on the surfaces of soybean leaves. *Phytopathology* 61, 715–716.

Mew, T.W. and Kennedy, B.W. (1982) Seasonal variation in populations of pathogenic pseudomonads on soybean leaves. *Phytopathology* 72, 103–105.

Montesinos, E. and Vilardell, P. (1991) Relationships among population levels of *Pseudomonas syringae*, amount of ice nuclei, and incidence of blast of dormant flower buds in commercial pear orchards in Catalunya, Spain. *Phytopathology* 81, 113–119.

Ramos, A.H. (1979) Bacterial blight of coffee: the inoculum supply and avenues of infection. *Plant Disease Reporter* 63, 6–10.

Ramos, A.H. and Kamidi, R.E. (1981) Seasonal periodicity and distribution of bacterial blight of coffee in Kenya. *Plant Disease* 65, 581–584.

Roos, I.M.M. and Hattingh, M.J. (1986) Resident populations of *Pseudomonas syringae* on stone fruit tree leaves in South Africa. *Phytophylactica* 18, 55–58.

Rouse, D.I., Nordheim, E.V., Hirano, S.S. and Upper, C.D. (1985) A model relating the probability of foliar disease incidence to the population frequencies of bacterial plant pathogens. *Phytopathology* 75, 505–509.

Smitley, D.R. and McCarter, S.M. (1982) Spread of *Pseudomonas syringae* pv. *tomato* and role of epiphytic populations and environmental conditions in disease development. *Plant Disease* 66, 713–717.

Stadt, S.J. and Saettler, A.W. (1981) Effect of host genotype on multiplication of *Pseudomonas phaseolicola*. *Phytopathology* 71, 1307–1310.

Sundin, G.W., Jones, A.L. and Olson, B.D. (1988) Overwintering and population dynamics of *Pseudomonas syringae* pv. *syringae* and *P. s.* pv. *morsprunorum* on sweet and sour cherry trees. *Canadian Journal of Plant Pathology* 10, 281–288.

Umekawa, M., Watanabe, Y. and Inomata, Y. (1981) Facilitative effect of rainfall on the transmission of the pathogen and the development of angular leaf spot of cucumber. *Annals of the Phytopathological Society of Japan* 47, 346–351.

Walker, J.C. and Patel, P.N. (1964) Splash dispersal and wind as factors in epidemiology of halo blight of bean. *Phytopathology* 54, 140–141.

Willis, D.K., Hrabak, E.M., Rich, J.J., Barta, T.M., Lindow, S.E. and Panopoulos, N.J. (1990) Isolation and characterization of a *Pseudomonas syringae* pv. *syringae* mutant deficient in lesion formation on bean. *Molecular Plant–Microbe Interactions* 3, 149–156.

Wimalajeewa, D.L.S. and Flett, J.D. (1985) A study of populations of *Pseudomonas syringae* pv. *syringae* on stonefruits in Victoria. *Plant Pathology* 34, 248–254.

15 Autecology and Evolution of the Witches' Broom Pathogen (*Crinipellis perniciosa*) of Cocoa

G.W. Griffith,[1] E. Bravo-Velasquez,[2] F.J. Wilson, D.M. Lewis and J.N. Hedger

Department of Biological Sciences, University College of Wales, Aberystwyth, Dyfed SY23 3DA, UK; [1] Present address – School of Biological Sciences, University College of North Wales, Bangor, Gwynedd LL57 2UW, UK; [2] Present address – Department of Biology, Catholic University, Quito, Ecuador.

Introduction and History of the Disease

The fungus *Crinipellis perniciosa* (Agaricales, Tricholomataceae, Marasmieae), causal agent of witches' broom disease (WBD) of cocoa (*Theobroma cacao*), is an important and destructive hemibiotrophic pathogen of the crop throughout tropical South America (Lass, 1985; Wheeler, 1985). The disease first became known to science at the beginning of this century when symptoms of an epiphytotic in Surinam in 1895 were described by Went (1904). Some years later the pathogen was described as *Marasmius perniciosus* (Stahel, 1915) but was moved to the genus *Crinipellis* by Singer (1942). However, it has recently come to light, through the diaries of Alexandre Rodrigues Ferreira from 1785–1787, that WBD, known then as 'lizard', was present on cocoa in Amazonia at least a century earlier (Silva, 1987).

During the first half of this century the disease spread, with disastrous economic consequences, to Guyana (1906), Ecuador (1918), Trinidad (1928), Colombia (1929) and Grenada (1948) (Pound, 1943; Baker and Holliday, 1957; Barros, 1979). More recently, WBD has become rife in Amazonian

South America, notably the Brazilian Province of Rondonia (Thorold, 1975; Rudgard, 1986), where it had been hoped that cocoa as a tree crop might provide an ecologically acceptable alternative to cattle ranching and arable farming.

The most recent stage in the spread of WBD was its appearance in Bahia Province on the Atlantic coast of Brazil (separated from Amazonia by the 1000 km wide and semi-arid Caatinga region) in 1989 (Pereira *et al.*, 1990). Still the most important cocoa-growing area in South America, the future of the industry in this region seems bleak, given the absence of either resistant cultivars or cost-effective methods of control (Rudgard and Andebrhan, 1988).

The centre of evolution of *T. cacao*, and probably related species, is believed to be the upper Amazon Basin (Cuatrecasas, 1964), although it is endemic to much of South and Central America. In nature it grows as understorey trees in moist tropical and subtropical forest (Allen, 1987) but although cultivation of cocoa as a plantation crop became widespread after the 17th century, historical records of the use of cocoa beans by Aztecs suggests that cultivation may have begun in Pre-Colombian times (Toxopeus, 1985; Wood, 1985).

C. perniciosa has generally been assumed to have co-evolved with hosts belonging to the genera *Theobroma* and *Herrania* (Sterculiaceae) in Amazonia. Pound (1943) suggested that the disease originated in western Amazonia where he found a high proportion of apparently resistant trees. Spread of the disease beyond Amazonia is believed to have occurred by the distribution of diseased pods for use as planting material (Baker and Holliday, 1957).

This chapter aims to provide a consideration of the autecology of this fungus and its population structure in South America. By interpreting the ecology of this pathogen, it is possible to suggest an alternative route by which it may have evolved and indeed that it may be of comparatively recent origin. In other words, an ecological perspective may alter the views of plant pathologists by viewing the 'pathogen' in the context of the wider ecology of the genus or group of fungi to which it belongs.

Biotypes of *C. perniciosa*

Host specificity

In addition to causing WBD in cocoa and its wild relatives, *C. perniciosa* has also been associated with WBD symptoms in *Solanum rugosum* and several other solanaceous hosts (Bastos and Evans, 1985), as well as the shrub *Bixa orellana* (Bixaceae) (Bastos and Andebrhan, 1986), in Brazil (Table 15.1). The fact that *C. perniciosa* from cocoa is able to cause WBD symptoms in *B.*

Table 15.1. A summary of the different characteristics of the three biotypes of *Crinipellis perniciosa*.

Biotype	Host range	WBD symptoms	Breeding strategy	Population structure	Genetic variability
Cocoa (C) biotype	*Theobroma/Herrania* spp. (Sterculiaceae) *Bixa orellana* (Bixaceae)	+	1° Homothallic	Clonal (few widespread SCGs)	Low
Liana (L) biotype	*Arrabidaea verrucosa* (Bignoniaceae)	–	Bifactorial heterothallic	Outcrossing (many local SCGs)	High
Solanum (S) biotype	*Solanum* spp. (Solanaceae)	+	1° Homothallic	Clonal (few widespread SCGs)	Moderate

Fig. 15.1. Basidiocarps of the L-biotype of *C. perniciosa* on a living liana of *Arrabidaea verrucosa* at Pichilingue, coastal Ecuador.

orellana (Bastos and Andebrhan, 1986), together with both cultural and molecular evidence (Griffith, 1989; F.J. Wilson, unpublished data), suggests that the *Bixa* biotype is identical to the C-biotype. However, cross-infection experiments with basidiospores of the *Solanum* (S) biotype and the C-biotype on cocoa and tomato respectively resulted in the production of only slight symptoms (i.e. slight swelling rather than broom formation) (Bastos and Evans, 1985; Bastos *et al.*, 1988). Furthermore, C-biotype and S-biotype isolates were clearly differentiated by both isozyme profiles (Griffith, 1989) and mitochondrial DNA (mtDNA) restriction patterns (F.J. Wilson, unpublished data).

Basidiocarps of *C. perniciosa* have also been found on both live and dead liana vines, as well as associated plant debris, in Amazonian Ecuador (Desrosiers and van Buchwald, 1949), coastal Ecuador (Evans, 1977, 1978; Hedger *et al.*, 1987; Griffith, 1989) and Brazil (Bastos *et al.*, 1981). However, unlike the other hosts, WBD symptoms have never been observed on lianas. Attempts have been made to infect cocoa plants with basidiospores from basidiocarps formed on lianas (Evans, 1977, 1978; Hedger *et al.*, 1987; Bastos *et al.*, 1988; Purdy and Dickstein, 1990). Only slight symptoms (slight swelling and abnormal bark formation) were occasionally observed, although Evans (1978) was able to re-isolate the fungus from some infected plants.

A more detailed study by Griffith (1989) in Ecuador established that basidiocarps of the L-biotype were consistently found on, or closely associated with, living and dead stems of the liana species *Arrabidaea verrucosa* (Bignoniaceae) (Fig. 15.1). As with earlier studies, examination of green shoots and woody tissues of this liana gave no indication of symptoms reminiscent of WBD on cocoa. The liana species associated with *C. perniciosa* basidiocarps in Brazil is not known, since *A. verrucosa* is limited to the foothills of the Andes (A. Gentry, Missouri Botanical Gardens, 1989, personal communication). Given the broad host range of the other two biotypes of *C. perniciosa*, *A. japurensis* (Gentry, 1979), which is both closely related to *A. verrucosa* and widely distributed in central Amazonia, may represent an alternative host for the fungus.

Subspecies taxonomy

Stahel (1924) erected the variety *ecuadoriensis* (of what is now *C. perniciosa*), having compared the slightly darker crimson pileal colour of basidiocarps on cocoa brooms from Ecuador to those from Surinam. This subclassification was verified by Pegler (1978), who further defined var. *perniciosa*, typified by C-biotype basidiocarps from Brazil and Surinam, and var. *citriniceps*, which has a yellow pileus and was found once in coastal Ecuador and is probably a mutant variety deficient in pileal pigment synthesis. The existence of two races of *C. perniciosa* on cocoa is further suggested by inoculation experiments with cocoa seedlings, in which isolates from Ecuador and Colombia were found to be significantly more pathogenic than those from Brazil and Trinidad (denoted pathotypes A and B respectively) (Wheeler, 1985; Wheeler and Mepsted, 1988).

Although Pegler (1978) made no mention of significant morphological differences between herbarium specimens of basidiocarps found on liana and cocoa material from Ecuador, Hedger *et al.* (1987) were able to distinguish basidiocarps of *C. perniciosa* from different hosts. They noted that L-biotype basidiocarps from Ecuador, in addition to being significantly larger (20–40 mm cap diameter, as opposed to 5–20 mm for C-biotype basidiocarps) and more darkly pigmented, had both stouter basidiospores and cheilocystidia than basidiocarps of the C-biotype. Bastos and Evans (1985) were unable to place S-biotype basidiocarps into any of Pegler's varietal classes, since pileal pigmentation was variable. However, data presented for cap diameter and basidiospore shape indicates greater similarity to C- rather than L-biotype basidiocarps.

Related species

A number of other *Crinipellis* species have been recorded as minor pathogens, including *C. stipitaria* on rye and *C. siparunae* on a woody tropical shrub (*Siparuna* sp.; Monimiaceae) (Singer, 1942). In Subsection Iopodinae of the genus *Crinipellis* (which includes *C. perniciosa*), all seven species have purple-red pilei and are associated with woody plants in neotropical forests (Singer, 1942). These include *C. eggersii*, an aerial litter saprotroph and *C. trinitatis*, which was found growing on a vine belonging to the genus *Vitex* in Trinidad (Pegler, 1983).

Ecophysiology

Biotrophic establishment

In the S- and C-biotypes, the biotrophic phase of the life cycle is readily apparent, since infection quickly results in disruption of development of normal host tissue and premature necrosis. The L-biotype produces no symptoms in its host but basidiospores are able to form a biotrophic infection of potato callus in which a mycelium identical to that produced by the C-biotype in both callus cultures and cocoa meristems proliferates (Griffith, 1989). The fungus has yet to be isolated from meristematic tissues of *A. verrucosa*, but given its host specificity and known occurrence on the bark of living lianas, it can be argued that infections are asymptomatic and that only very sparse mycelial systems occur in living tissues. In this respect there are parallels with a number of other fungal taxa (often closely related to known pathogens), where latent infections become visible only when the host becomes stressed or necrosed (Boddy and Rayner, 1983; Rayner and Boddy, 1986; Rayner, 1991). A growing body of evidence suggests that such endophytic species, which become established by a biotrophic but asymptomatic pathway, are present in most plant species hitherto examined (Petrini, 1986; Petrini and Müller, 1986; Carroll, 1988; Clay, 1990).

Examination of the distribution of mycelia of the L-biotype (Griffith, 1989) suggests that establishment via basidiospores may be a rare event. Compared to plantation cocoa which is usually grown without shade and consequently produces extensive flushes of meristematic tissue, green shoots were only rarely observed on *A. verrucosa* stems in the forest understorey. The fact that lianas did not appear to be adversely affected by the presence of the fungus suggests that the putative biotrophic phase of the life cycle may be chronic.

Saprotrophic growth

In the broom-forming biotypes, a metamorphosis occurs from a swollen, flexuose (5–20 µm) mycelium in green tissues, to a narrower (1–3 µm) clamped form in dead brooms. The broad terms 'biotrophic' and 'saprotrophic' are used to describe these two mycelial types and are preferred here, since terminology distinguishing the two forms on the basis of nuclear condition (i.e. monokaryotic vs. dikaryotic and primary vs. secondary) could now be misleading given recent information about the breeding biology of *C. perniciosa* (see below) and observations of multinucleate hyphae in biotrophic infections of potato callus (Griffith, 1989).

It is not known whether the occurrence of the secondary mycelium is a cause or a consequence of host tissue necrosis. On agar media, the hyphae of germling colonies initially resemble the biotrophic mycelium but over the course of 1–2 weeks the saprotrophic mycelium proliferates (Evans, 1980; Griffith, 1989). Evans (1980) proposed the existence in meristematic host tissues of a 'modifier' substance, which arrests development in the juvenile biotrophic phase and suppresses growth of the saprotrophic mycelium. In this model, progressive hypertrophic and hyperplasic growth of the host meristem combined with proliferation of the biotrophic mycelium leads to diminished modifier concentrations and the development of the more invasive saprotrophic mycelium. Investigations of the effects of cocoa extracts on basidiospore germlings indicate that the 'modifier' may be a polymeric procyanidin (condensed tannin) (Brownlee *et al.*, 1990).

The switch from biotrophic to saprotrophic growth also represents a change in resource utilization. Resource capture is achieved by basidiospore infection and biotrophic intercellular growth prior to resource exploitation by the saprotrophic mycelium. The saprotrophic mycelium is known to produce a range of degradative extracellular enzymes, including cellulases (Bravo, 1989), pectinases (Griffith, 1989) and phenoloxidases (Krupasagar and Sequeira, 1969). Hedger (1985) recorded that the degradation of sterile cocoa leaves inoculated with C- and L-biotype isolates of *C. perniciosa* was rapid compared to degradation by several other tropical agarics, and associated with bleaching or partial destruction of leaf tissues.

A key feature of the saprotrophic mycelium of all biotypes is the ability to colonize contiguous fragments of plant debris (secondary resource capture) by means of distinctive dark crimson pseudosclerotial pads (PSPs) (Fig. 15.2). The PSPs, which are both sturdy and persistent, can lead to the formation of 'nets' of debris, which, with progressive efficiency, intercept plant material falling from the canopy (Hedger, 1990). This process of saprotrophic colonization appears to be of greater importance for the longer-lived mycelial systems of the L-biotype. It was found that pieces of debris colonized from living lianas and subsequently detached by physical disturbance represented an important mechanism for short-distance dispersal (Griffith, 1989).

Fig. 15.2. A pseudosclerotial pad (PSP) of *Crinipellis perniciosa* connecting two pieces of woody debris and a partly decomposed cocoa leaf. Arrows indicate the position of the dark crimson PSPs.

A similar role for PSPs in mycelial expansion following latent establishment has been proposed for the temperate polypore *Hymenochaete corrugata* (Ainsworth and Rayner, 1990). Hedger (unpublished data) observed that 20–30% of sterilized twig fragments, placed on the living stems of several liana species, were colonized via PSPs within 1 month. The fungi involved were not identified, although his results suggest that latent endophytic species, occupying a similar niche to the L-biotype, occur in other liana species. It is noteworthy that several small agaric species belonging to the genera *Marasmius* and *Marasmiellus*, which are closely related to *Crinipellis* spp., also colonize aerial debris in neotropical rainforests, although colonization of newly arrived debris is achieved by threadlike rhizomorphs as well as PSPs in some species (Hedger *et al.*, 1993).

Competition strategy

Initial establishment of *C. perniciosa* from basidiospores and resource capture by the biotrophic mycelium is achieved in the absence of competition

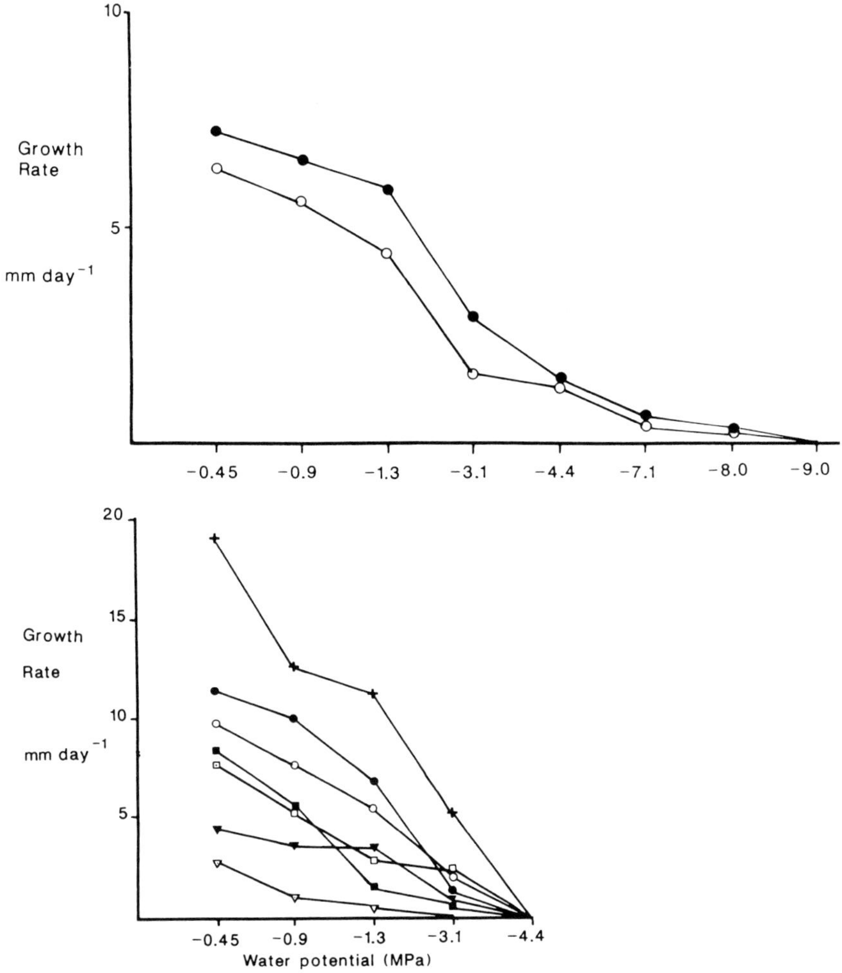

Fig. 15.3. The effect of water potentials on the growth rates of isolates of the C- (●) and L-(○) biotypes of *C. perniciosa* (upper figure) and seven other agarics: +, *Coprinus jamaicenis*; ●, *Clitopilus rhodotrama*; ○, *Melanotus alpiniae*; □, *Crepitodus cuneiformis*; ▼, *Hohenbuehelia barbatula*; ▽, *Mycena theobromicola*; □, *Pluteus* sp. (lower figure). Reproduced by permission of the Royal Society of Edinburgh and E. Bravo and J.N. Hedger from Proceedings of the Royal Society of Edinburgh, Section B: Biological Sciences, volume 94 (1988), pp. 159–166. Isolates were grown on a basal medium (2% glucose, 2% malt extract, 0.1% peptone, 1.5% agar) and a range of water potentials was achieved by addition of potassium chloride. Cultures were incubated at 25°C and radial growth rate was measured daily.

from other microbes. Its ability to persist in the previously colonized host tissues, so that maximal resource exploitation can be achieved, and to escape via basidiospores to colonize new substrates are also important.

Several lines of evidence suggest that survival of *C. perniciosa* in dead brooms and other plant material is critically influenced by substrate moisture conditions. Basidiocarp production by *C. perniciosa* is optimal when brooms are subjected to drastic daily fluctuations in moisture level (Rocha and Wheeler, 1985) and survival of *C. perniciosa*, as assessed by basidiocarp production, in dead brooms, is significantly higher in the canopy compared to brooms on the ground (Hedger, 1988). Although tropical moist forests are associated with high rainfall, significant fluctuations in humidity occur and these might reasonably be expected to cause similar moisture fluctuations in aerial debris. Bravo and Hedger, (1988a, b) have suggested that *C. perniciosa* and other canopy-inhabiting fungi are adapted to water stress, but are unable to compete effectively with other fungi under the conditions of higher water activity found on the ground. They found that *C. perniciosa* was able to grow at water potentials as low as –8.0 MPa and fared best under conditions of low water activity in mycelial competition experiments with other broom-colonizing fungi (Fig. 15.3).

Population Structure and Genetic Variability

Breeding biology

Delgado and Cook (1976) found that basidiospores of the C-biotype are uninucleate and formed in tetrads on basidia following meiotic nuclear division. Mycelia derived from single basidiospores were dikaryotic with clamp connections and single spore isolates (SSIs) derived from the same basidiocarp were morphologically identical (Griffith, 1989). These lines of evidence, along with the ability of SSIs to produce basidiocarps (Purdy *et al.*, 1983), indicate that the C-biotype of *C. perniciosa* is non-outcrossing with a primary homothallic (homomictic) breeding strategy. The results of a more limited investigation indicate that the S-biotype also exhibits the same breeding strategy (Griffith, 1989).

The growth of L-biotype basidiospores during the first week after germination was identical to that of the C- and S-biotypes. However, most SSIs of the L-biotype did not produce clamp connections, even after repeated subculture, and sibling SSIs were morphologically variable. The absence of clamp connections on the uninucleate hyphae of SSIs of the L-biotype and their presence on the binucleate mycelia derived from basidiocarp stipe tissues (Griffith, 1989) suggested that the L-biotype might be outcrossing (i.e. heterothallic) and that mating between compatible unclamped primary

	A_xB_x						A_yB_y				A_xB_y			A_yB_x	A_yB_z
	1	2	6	15	17	18	4	8	9	11	10	14	16	3	12
1	-	-	-	-	-	-	+	+	+	+	-	-	-	f	-
2		-	-	-	-	-	+	-	+	+	-	-	-	f	+
6			-	-	-	-	+	+	+	+	-	-	-	-	+
15				-	-	-	+	+	+	+	-	-	-	-	+
17					-	-	+	-	+	+	-	-	-	f	-
18						-	+	+	+	+	-	-	-	-	+
4							-	-	-	-	-	f	f	-	-
8								-	-	-	f	f	f	-	-
9									-	-	-	-	f	-	-
11										-	f	-	f	-	-
10											-	-	-	+	+
14												-	-	+	+
16													-	+	+
3														-	-
12															-

Fig. 15.4. The results of a representative sib-mating of single (basidio)spore isolates (SSIs) of the L-biotype of *C. perniciosa*. The results are arranged on the hypothesis that mating compatibility is under the control of two, unlinked mating factors (A and B; i.e. bifactorial or tetrapolar heterothallism; Burnett, 1975). Fifteen sibling single spore isolates (1 to 15) derived from a basidiocarp of the L-biotype isolate LC3 (from Pichilingue, coastal Ecuador) were paired in all combinations on agar media (3% malt extract–0.5% yeast extract). The presence (+) or absence (–) of fused clamp connections on mycelia from the interaction zone (after incubation for 4–6 weeks at 25°C) was recorded. The occurrence of false (i.e. unfused) clamp connections (f) permitted the identification of some common-B factor (hemi/compatible) pairings. Most SSIs could be assigned to one of four mating genotypes (A_xB_x, A_yB_y, A_xB_y or A_yB_x). The mating phenotype of one SSI (LC3.12) was interpreted as being the result of intra-B-factor recombination (A_yB_z).

mycelia is a prerequisite for the formation of secondary clamped mycelia (Burnett, 1975).

Mating pairings conducted with sets of sibling SSIs from basidiocarps collected at sites in coastal and Amazonian Ecuador and Brazil revealed that ca. 25% of sib matings were compatible, giving rise to mycelia bearing clamp connections (Fig. 15.4). The results of 20 separate sib-matings were consistent with the hypothesis that mating compatibility in the L-biotype is under the control of two unlinked mating factors (diaphoromixis or bifactorial heterothallism) (Burnett, 1975). Most pairings between non-sibling SSIs were compatible, demonstrating that multiple alleles at both loci are present with L-biotype populations.

A small proportion of SSIs (6% overall) exhibited a mating phenotype indicating that recombination of the component loci of the B mating factor had occurred. Intrafactor recombination, which results in a reduction in inbreeding bias, has been observed at similar frequencies in a number of bifactorial heterothallic agarics (e.g. *Coprinus cinereus*) (Casselton and Economou, 1985) and is consistent with current models of the molecular structure of basidiomycete mating type factors (Kües *et al.*, 1992).

Fused and unfused clamp connections were observed in 19% of SSI (these were not included in sib-matings). It is possible that these were the result of inaccurate microdissection of basidiospore germlings, but examination of the nuclear condition of basidiospores revealed that a significant proportion were multinucleate (7.6% binucleate and 1.6% trinucleate in an isolate from Amazonian Ecuador). This indicates that the L-biotype exhibits a limited degree of secondary heterothallism. A similar phenomenon also occurs in *Agaricus subrufescens* and is thought to provide a degree of reproductive assurance, for instance when the probability of encountering other compatible primary mycelia is low (Kerrigan and Ross, 1987).

There was some evidence that isolates from Amazonian Ecuador might have a greater inbreeding bias, since the frequency of multinucleate basidiospores and intrafactor recombination was greater than that observed for coastal Ecuadorian isolates. It is known that both these processes are influenced by environmental conditions, particularly temperature (Schaap and Simchen, 1971; Kerrigan and Ross, 1987). SSIs used in this study were obtained both from basidiocarps collected in the field and produced under artificial conditions (by the bran-vermiculite method; Griffith and Hedger, 1993), so it remains to be established whether L-biotype populations do actually differ in the fine-tuning of outcrossing mechanisms under natural conditions.

Compatible mating reactions were also observed in most non-sib matings between SSIs from coastal Ecuador, Amazonian Ecuador and the progeny of a single L-biotype isolate from Brazil. The mating compatibility of geographically-distant L-biotype isolates (ca. 3000 km apart) and the fact that some of the resulting synthetic dikaryons were induced to produce basidiocarps suggests that no physiological barrier to gene flow exists between populations of the L-biotype.

Population structure

Examination of somatic incompatibility reactions provides a simple method for identifying genets (see Chapter 1 by Andrews and Chapter 2 by Shaw and Peters) in mycelial fungi. Genetical control of the expression of somatic incompatibility, leading to the production of barrages and pigment production between incompatible mycelia, is known to be complex (Croft and Dales, 1984; Dales and Croft, 1990; Kay and Vilgalys, 1992), so compatibility (indicated by free intermingling of hyphae from different isolates) provides a reliable indication of genetic similarity or identity between mycelia.

Pairings between isolates of the C-biotype from several South American countries (McGeary and Wheeler, 1988; Wheeler and Mepsted, 1988; Griffith, 1989) have shown that populations of this biotype consist of comparatively few and geographically widespread somatic compatibility groupings (SCGs). For instance, over 200 C-biotype isolates obtained throughout the coastal regions of Ecuador and Colombia were found to belong to a single SCG (Griffith, 1989). Indeed, only 15 SCGs in total were identified among all the C-biotype isolates examined by Griffith (1989). A separate study by McGeary and Wheeler (1988) found the C-biotype to be similarly homogeneous. Although such a clonal population structure is consistent with a non-outcrossing breeding strategy, the significant morphological variation on agar media of ramets (i.e. isolates belonging to the same SCG) suggested that these clones may be long-established.

Whereas most sympatric C-biotype isolates were fully compatible, a more complex situation was observed in pairings between 79 isolates from Amazonian Ecuador (Griffith, 1989). The majority of isolates belonged to two SCGs which were clearly incompatible with each other but pairings with 14 isolates produced less clear and often intransitive results (i.e. isolate x was compatible with isolates y and z but isolates y and z were incompatible with each other). Isolates belonging to one of the predominant SCGs were compatible with C-biotype isolates from coastal Ecuador, possibly introduced to Amazonian Ecuador during the recent extensive colonization of this region. The occurrence of several sympatric but indistinct SCG genotypes further suggests that recombination between the introduced and indigenous populations of the pathogen may have occurred.

The field distribution of L-biotype SCGs was assessed by Griffith (1989), using secondary mycelial isolates from L-biotype basidiocarp stipe tissues and liana bark cores. A large number of SCGs were identified (Fig. 15.5) and in contrast to the situation observed in the other biotypes, L-biotype SCGs were highly local in distribution, extending at most 10 m. Among 69 mycelial isolates of the L-biotype obtained from a 1 km^2 forest site at Pichilingue in Ecuador, 21 SCGs were identified. A similar level of diversity was found at two other forest sites. Isolates belonging to the same SCG were morphologically identical in culture and always contained identical complements of mating type factors. These lines of evidence suggested that isolates belonging to the same SCG were genetically identical and part of the same mycelial system.

Several examples were found of SCGs confined to a single liana stem, in some cases occupying substantial lengths of stem. Such a distribution could have resulted from a prolonged period of latent growth in meristematic tissue

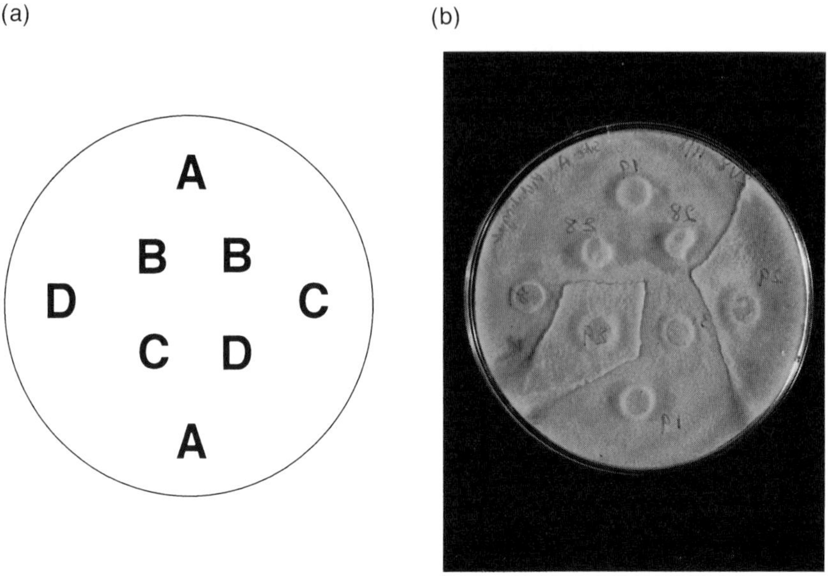

Fig. 15.5. Somatic incompatibility reactions between isolates obtained from the stipe tissues of four L-biotype basidiocarps of *C. perniciosa* from Jauneche, coastal Ecuador (Griffith, 1989). 9 cm Petri dishes containing 20% V8 juice medium were inoculated with plugs of the four isolates (denoted A, B, C and D), as shown in the schematic diagram (a) and incubated for 4 weeks at 25°C. The antagonistic and sometimes pigmented reaction zones (b) clearly demarcated incompatible isolates. The interpretation from this experiment is that isolates A, B and D are compatible and are therefore grouped in the same SCG.

and subsequent proliferation into developing woody tissues. In other cases, isolates belonging to the same SCG were obtained from substrates connected by PSPs or substrates in the same vertical plane, consistent with the mechanisms of saprotrophic spread described earlier.

Correlations between the distribution of mycelial genets (SCGs) and mating factors gave some indication of patterns of mycelial establishment and spread. In two cases, adjacent genets were found to share two mating type factors. This is consistent with the interpretation that they were sib-related and that the two secondary mycelia resulted from two separate mating events both involving the same primary mycelium (Fig. 15.6). If this is the case, then the primary mycelia of the L-biotype may be capable of substantial growth prior to mating. A further possibility is that mating and secondary mycelial development occurs only during the saprotrophic phase of growth. Attempts to clarify this situation with potato callus dual cultures were inconclusive. A schematic representation of the life cycle of the L-biotype is shown in Fig. 15.7.

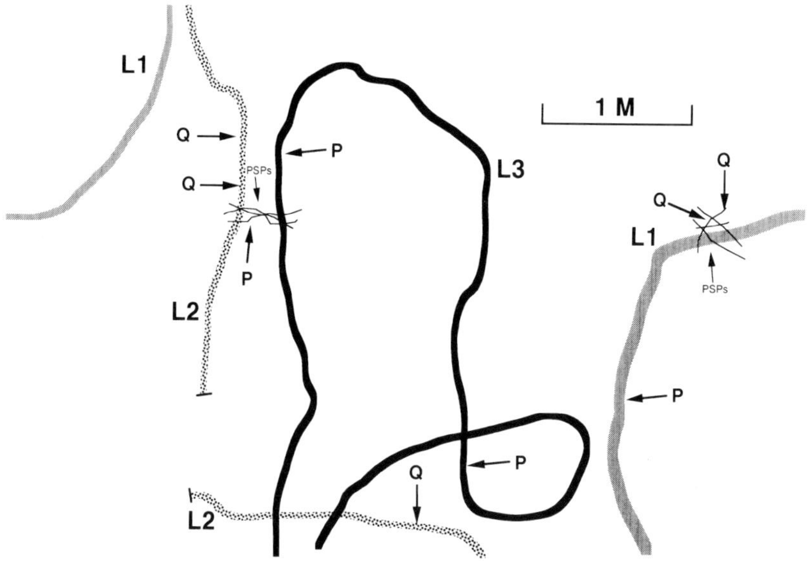

Fig. 15.6. A schematic representation of the distribution of genets (SCGs) of the L-biotype at Pichilingue, coastal Ecuador. Nine isolates were obtained from the stipe tissues of L-biotype basidiocarp collected from a thicket of *A. verrucosa* lianas which were entwined around a large tree trunk. Three liana stems (5–8 cm diam.), which appeared to be unconnected, were present. The approximate position of pseudo-sclerotial pads (PSPs) which connected pieces of debris to the lianas is shown. Pairings between the isolates established that two SCGs (denoted P and Q) were present. Matings between SSI progeny of isolates belonging to each of the genets showed that two mating factors were common to both, suggesting that the two genets are possibly sib-related. This situation could have arisen by two independent mating events between a pre-established primary (and probably extensive) mycelium with two other primary mycelia.

Genetic variation within and between biotypes

The somatic compatibility reactions observed during pairings between isolates from the same and different biotypes suggested that the three biotypes were genetically distinct (i.e. reactions between isolates from different biotypes were more intense). Furthermore, reactions between incompatible C-biotype isolates were less distinct than reactions between L- and S-biotype isolates, suggesting that the C-biotype isolates were genetically more homogeneous.

These suspicions were confirmed by investigations of isozyme (Griffith, 1989) and mitochondrial DNA (mtDNA) polymorphisms (F.J. Wilson, unpublished data). Staining for 18 enzyme systems and the use of two mitochondrial probes (from *Coprinus cinereus*, supplied by Professor L.A.

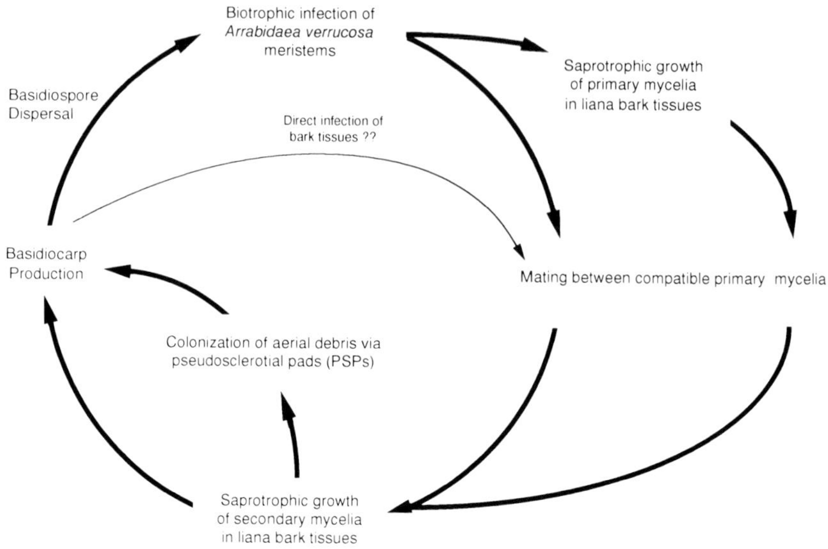

Fig. 15.7. A schematic representation of the life cycle of the L-biotype of *C. perniciosa*.

Casselton) did not reveal any polymorphisms between C-biotype isolates from throughout South America (Fig. 15.8). In contrast, polymorphisms were identified among L-biotype isolates with five enzyme systems and one of the mtDNA probes. No mtDNA polymorphisms were detected among 30 S-biotype isolates belonging to three different SCGs. The data obtained from these studies was not detailed enough to permit the construction of a phylogeny tree but the banding patterns of L- and S-biotype isolates were more similar to each other than to those of C-biotype isolates.

Conclusions

The evidence presented above shows that substantial differences exist between the L-biotype and the other biotypes of *C. perniciosa* and that many of these differences stem from differences in breeding strategy. The breeding biology, population structure and ecological niche of the L-biotype is similar in many respects to that observed in a number of host specific agaric species associated with the aerial tissues of woody hosts in temperate habitats. Rayner and Boddy (1986) have also proposed that the occurrence of single genets occupying large volumes of wood is indicative of prolonged periods

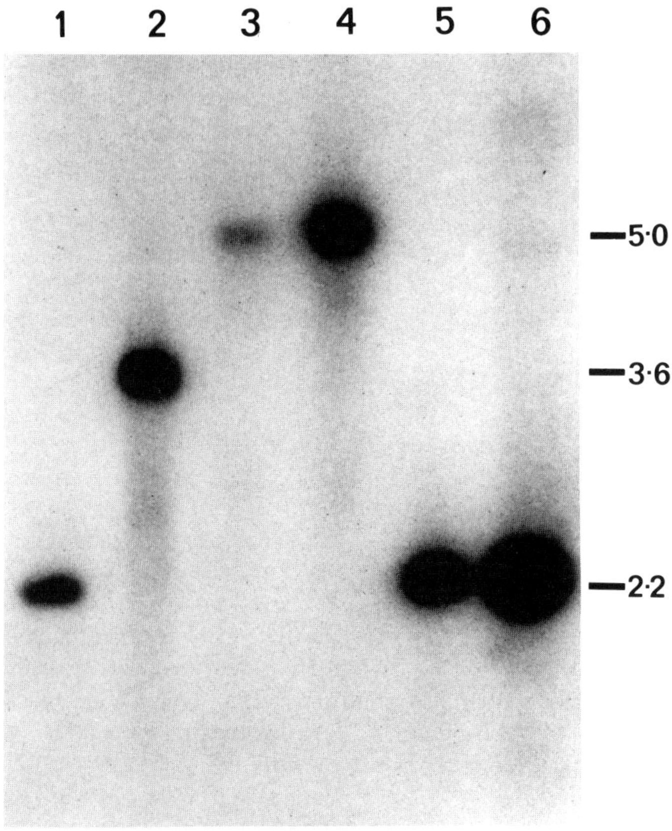

Fig. 15.8. Southern analysis of total DNA of *C. perniciosa* digested with *Bgl* II and hybridized with a radiolabelled fragment of mitochondrial DNA from *Coprinus cinereus* (J variable region; *Pst*I-*Bam*HI digest, 6.57 kb; Economou and Casselton, 1989). Lanes 1 and 2: L-biotype, Pichilingue, Ecuador. Lane 3: C-biotype, Manaus, Brazil. Lane 4: C-biotype, Bahia, Brazil. Lane 5: S-biotype, Bahia, Brazil. Lane 6: S-biotype, Manaus, Brazil. Fragment sizes in kilobase pairs (kb) are shown to the right of the figure.

of latent infection and that establishment via infection of living tissues may have occurred.

In comparison with the L-biotype, the C-biotype is *r*-selected (Andrews, 1992) with a brief but spectacular biotrophic phase, followed by limited saprotrophic exploitation of host tissues and rapid exit to new substrates via basidiospores. Mating is not a prerequisite to basidiocarp formation, so every successful infection is potentially fertile. Since conditions required for basidiocarp production are similar to those which induce flushing in cocoa, rapid spread of the pathogen is possible, as evidenced by the WBD epiphytotics of the past century.

In nature cocoa grows as an understorey tree at low density in the rainforest. Young shoots, flowers and pods are only rarely seen, usually when the collapse of a canopy tree has allowed higher light levels to reach the forest understorey (Allen, 1987). Therefore, the amount of meristematic cocoa tissue in a given area of rainforest can reasonably be assumed to be several orders of magnitude lower than that found in a cocoa plantation. Given these facts, it is difficult to conceive how the C-biotype, in its present form, could have existed prior to the advent of cocoa cultivation. Indeed, the low level of genetic variability among geographically widespread isolates of the C-biotype (compared to the L-biotype) suggest that it may be recent in origin and that new selective pressures associated with the monoculture of cocoa without shade may have precipitated its evolution. One is reminded of Yarwood's (1968) sentiment that man's mismanagement has turned 'the innocent elements of balanced systems' into pathogens.

Outcrossing mechanisms, controlled by either one or two mating type factors with multiple alleles, are widespread among the basidiomycetes (Whitehouse, 1949) and are considered to be an ancestral feature of this taxon (Koltin *et al.*, 1972). Indeed, Lemke (1969) has estimated that only 1% of basidiomycetes exhibit a primary homothallic non-outcrossing breeding strategy, as found in the C- and S-biotypes. There are several other examples of outcrossing species which are closely related to, or taxonomically conspecific with, primary homothallic taxa (Kemp, 1975; David and Déquatre, 1984; Ainsworth, 1987) and it is possible that comparatively small changes at the genetic level may result in the evolution of a non-outcrossing breeding strategy.

The fact that the S-biotype was found to be genetically more similar to the L-biotype than the C-biotype suggests that, despite similarities in WBD symptom production, the two non-outcrossing biotypes evolved separately. It is tempting to speculate that the common ancestor was the L-biotype. Similarities between all three biotypes in basidiocarp morphology, isozyme and mtDNA profiles, as well as ecological niche (i.e. as xerotolerant and host specific colonizers in the rainforest understorey) add weight to this argument.

An alternative hypothesis, given the wide range of host species susceptible to infection by *C. perniciosa* (Table 15.1), is that other biotypes, as yet undiscovered but more similar to the L-biotype than the C-biotype, exist in the neotropical rainforests and that the pathogenic biotypes evolved from these. The other *Crinipellis* species briefly described by Singer (1942) and Pegler (1983) may therefore merit further attention.

If either hypothesis is correct, then the widespread assumption of WBD researchers that the pathogen has coevolved with cocoa in the rainforests (Baker and Holliday, 1957; Wheeler, 1987) may require reassessment. Our scenario, which is not inconsistent with the apparent absence of effective resistance to WBD among cocoa germplasm collections (despite extensive international efforts; Anon., 1987), implies that cocoa may be a very recent host for *C. perniciosa* and that it has 'jumped' (*sensu* Hijwegen, 1988; i.e. a sudden non-coevolutionary change in host range) from another host. Leppik (1967) has proposed that similar jumps have played an important role in the evolution of rust fungi. The symptoms which characterize the disease in cocoa could be the fortuitous (though not for cocoa growers) result of the interaction of *C. perniciosa* and its new host.

Acknowledgements

The authors would like to express their thanks to the staff of INIAP, Pichilingue, Ecuador, and CEPLAC, Belem, Brazil for their assistance and hospitality, in particular Dr Alan Maddison, Ing. Jaime Aragundi, Dr Cleber Bastos and Dr Teklu Andebrhan. This research was funded by the UK Science and Engineering Research Council (G.W.G.), the Scottish Office Agriculture and Fisheries Department (F.J.W.) and the British Council (E.B.V.). Field visits were made possible by the award of Wain Fellowships to G.W.G. and F.J.W.

References

Ainsworth, A.M. (1987) Occurrence and interactions of outcrossing and non-outcrossing populations in *Stereum*, *Phanerochaete* and *Coniophora*. In: Rayner, A.D.M., Brasier, C.M. and Moore, D. (eds), *Evolutionary Biology of the Fungi. British Mycological Society Symposium, Volume 12*. Cambridge University Press, Cambridge, pp. 286–299.

Ainsworth, A.M. and Rayner, A.D.M. (1990) Aerial mycelial transfer by *Hymenochaete corrugata* between stems of hazel and other trees. *Mycological Research* 94, 263–266.

Allen, J.B. (1987) London Cocoa Trade Amazon Project. Final Report Phase Two, *Cocoa Growers' Bulletin* 39 (special issue), 96 pp.

Andrews, J.H. (1992) Fungal Life-History Strategies. In: Carroll, G.C. and Wicklow, D.T. (eds), *The Fungal Community: its Organization and Role in the Ecosystem*, 2nd edn. *Mycology Series, vol. 9*. Marcel Dekker, New York, pp. 119–145.

Anon. (1987) Managing Witches' Broom Disease of Cocoa. Lass, R.A. and Rudgard, S.A. (eds), Report of the 4th IWBP Workshop, IOCCC, Brussels.

Baker, R.E.D. and Holliday, P. (1957) Witches' broom disease of cocoa (*Marasmius perniciosus* Stahel). *Phytopathological Papers* 2, Commonwealth Mycological Institute, Kew, 42 pp.

Barros, O. (1979) Algunos aspectos de la 'escoba de bruxa' en Colombia. *El Cacaotero Colombiano* 7, 43–46.

Bastos, C.N. and Andebrhan, T. (1986) Urucu (*Bixa orellana*): Nova espécie hospedera da vassoura de bruxa (*Crinipellis perniciosa*) do cacueiro. *Fitopatologia Brasiliera* 11, 963–965.

Bastos, C.N. and Evans, H.C. (1985) A new pathotype of *Crinipellis perniciosa* (witches' broom disease) on solanaceous hosts. *Plant Pathology* 34, 306–312.

Bastos, C.N., Andebrhan, T. and de Almeida, L.C. (1988) Comparação morfológica de isolados de *Crinipellis perniciosa*. *Fitpatologia Brasiliera* 13, 202–206.

Bastos, C.N., Evans, H.C. and Samson, R.A. (1981) A new hyperparasitic fungus, *Cladobotryum amazonense*, with potential for control of the fungal pathogens of cocoa. *Transactions of the British Mycological Society* 77, 273–278.

Boddy, L. and Rayner, A.D.M. (1983) Origins of decay in living deciduous trees: the role of moisture content and a reappraisal of the expanded concept of tree decay. *New Phytologist* 94, 623–641.

Bravo, E. (1989) Biological Control of *Crinipellis perniciosa* (Stahel) Sing. by Fungi Isolated from Cocoa Brooms. Unpublished PhD Thesis, University of Wales.

Bravo, E. and Hedger, J.N. (1988a) The effect of ecological disturbance on competition between *Crinipellis perniciosa* and other tropical fungi. *Proceedings of the Royal Society of Edinburgh, Series B* 94, 159–166.

Bravo, E. and Hedger, J.N. (1988b) Microflora associated with "witches' broom" in cocoa and its potential role in the biological control of the pathogen *Crinipellis perniciosa*. *Proceedings of the 10th International Cocoa Research Conference, 1987, Santo Domingo, Dominican Republic.* Cocoa Producers Alliance, Lagos, Nigeria, pp. 345–348.

Brownlee, H.E., McEuen, A.R., Hedger, J.N. and Scott, I.M. (1990) Anti-fungal effects of cocoa tannin on the witches' broom pathogen *Crinipellis perniciosa*. *Physiological and Molecular Plant Pathology* 36, 39–48.

Burnett, J.H. (1975) *Mycogenetics*. John Wiley and Sons, London.

Carroll, G.C. (1988) Fungal endophytes in stems and leaves: from latent pathogen to mutualistic symbiont. *Ecology* 69, 2–9.

Casselton, L.A. and Economou, A. (1985) Dikaryon formation. In: Moore, D., Casselton, L.A., Wood, D.A. and Frankland, J.C. (eds), *Developmental Biology of Higher Fungi. British Mycological Society Symposium, Volume 10.* Cambridge University Press, Cambridge, pp. 213–229.

Clay, K. (1990) Fungal endophytes of grasses. *Annual Reviews of Ecology and Systematics* 21, 275–297.

Croft, J.H. and Dales, R.B.G. (1984) Mycelial interactions and mitochondrial inheritance in *Aspergillus*. In: Jennings, D.H. and Rayner, A.D.M. (eds), *Ecology and Physiology of the Fungal Mycelium. British Mycological Society Symposium 8*. Cambridge University Press, Cambridge, pp. 433–450.

Cuatrecasas, J. (1964) Cocoa and its allies: a taxonomic revision of the genus *Theobroma*. *Contributions from the US National Herbarium* 35, 375–614.

Dales, R.B.G. and Croft, J.H. (1990) Investigations of the *het* genes that control heterokaryon incompatibility between members of the heterokaryon-compatibility (h-c) groups A and G1 of *Aspergillus nidulans*. *Journal of General Microbiology* 136, 1717–1724.

David, A. and Déquatre, B. (1984) Deux ultraspecies: *Antrodia malicola* (Berk. et

Curt.) Donk et *Antrodia ramentacea* (Berk. et Br.) Donk (Basidiomycete, Aphyllophorales). *Cryptogamie, Mycologie* 5, 293–300.

Delgado, J.C. and Cook, A.A. (1976) Nuclear condition of the basidia, basidiospores and mycelia of *Marasmius perniciosus*. *Canadian Journal of Botany* 54, 66–72.

Desrosiers, R. and van Buchwald, A. (1949) Report of a trip to the Napo river. Estacion Experimental Agricola, Pichilingue, Ecuador, pp. 1–11.

Economou, A. and Casselton, L.A. (1989) Polymorphisms of the mitochondrial L-RNA gene of the basidiomycete fungus *Coprinus cinereus*. *Current Genetics* 16, 41–46.

Evans, H.C. (1977) The occurrence of pathotypes of *Crinipellis perniciosa* (Stahel) Singer in the tropical forest ecosystem. *Proceedings of the 6th International Cocoa Research Conference, 1977, Caracas, Venezuela*. Cocoa Producers Alliance, Lagos, Nigeria, pp. 166–170.

Evans, H.C. (1978) Witches' broom disease of cocoa (*Crinipellis perniciosa*) in Ecuador. I. The fungus. *Annals of Applied Biology* 89, 185–192.

Evans, H.C. (1980) Pleomorphism in *Crinipellis perniciosa*, causal agent of witches' broom disease of cocoa. *Transactions of the British Mycological Society* 74, 515–523.

Gentry, A.H. (1979) Distribution patterns of neotropical *Bignoniaceae*: some phytogeographical considerations. In: Larsen, K. and Holm-Nielsen, L.B. (eds), *Tropical Botany*. Academic Press., London, pp. 337–354.

Griffith, G.W. (1989) Population structure of the cocoa pathogen *Crinipellis perniciosa* (Stahel) Sing. Unpublished PhD Thesis, University of Wales.

Griffith, G.W. and Hedger, J.N. (1993) A novel method for producing basidiocarps of the cocoa pathogen *Crinipellis perniciosa* using a bran-vermiculite medium. *Netherlands Journal of Plant Pathology* 99, 227–230.

Hedger, J.N. (1985) Tropical agarics: Resource relations and fruiting periodicity. In: Moore, D., Casselton, L.A., Wood, D.A. and Frankland, J.C. (eds), *Developmental Biology of Higher Fungi. British Mycological Society Symposium 10*. Cambridge University Press, Cambridge, pp. 41–86.

Hedger, J.N. (1988) Decomposition and basidiocarp production by witches' broom of cocoa under field conditions in Ecuador. *Proceedings of the 10th International Cocoa Research Conference, 1987, Santo Domingo, Dominican Republic*. Cocoa Producers Alliance, Lagos, Nigeria, pp. 319–323.

Hedger, J.N. (1990) Fungi in the tropical forest canopy. *The Mycologist* 4, 200–202.

Hedger, J.N., Pickering, V. and Aragundi, J.A. (1987) Variability of populations of the witches' broom disease of cocoa (*Crinipellis perniciosa*). *Transactions of the British Mycological Society* 88, 533–546.

Hedger, J.N., Lewis, P.A. and Gitay, H. (1993) Litter trapping by fungi in moist tropical forest. In: Isaac, S., Frankland, J.C., Watling, R. and Whalley, A.J.S. (eds), *Aspects of Tropical Mycology, British Mycological Society Symposium*, Cambridge University Press, Cambridge, pp. 15–36.

Hijwegen, T. (1988) Coevolution of flowering plants with pathogenic fungi. In: Pirozynski, K.A. and Hawksworth, D.L. (eds), *Coevolution of Fungi with Plants and Animals*. Academic Press, London, pp. 63–78.

Kay, E. and Vilgalys, R. (1992) Spatial distribution and genetic relationships among individuals in a natural population of the oyster mushroom *Pleurotus ostreatus*. *Mycologia* 84, 173–182.

Kemp, R.F.O. (1975) Breeding biology of *Coprinus* species in the section *Lanatuli*. *Transactions of the British Mycological Society* 65, 375–388.

Kerrigan, R.W. and Ross, I.K. (1987) Dynamic aspects of basidiospore number in *Agaricus*. *Mycologia* 79, 204–215.

Koltin, Y., Stamberg, J. and Lemke, P.A. (1972) Genetic structure and evolution of the incompatibility factors in higher fungi. *Bacteriological Reviews* 36, 156–171.

Krupasagar, V. and Sequeira, L. (1969) Auxin destruction by *Marasmius perniciosus*. *American Journal of Botany* 56, 390–397.

Kües, U., Richardson, W.V.J., Tymon, A.M., Mutasa, E.S., Gottgens, B., Gaubatz, S., Gregoriades, A. and Casselton, L.A. (1992) The combination of dissimilar alleles of the Aα and Aβ gene complexes, whose proteins contain homeo domain motifs, determines sexual development in the mushroom *Coprinus cinereus*. *Genes and Development* 6, 568–577.

Lass, R.A. (1985) Diseases. In: Wood, G.A.R. and Lass, R.A. (eds), *Cocoa. Tropical Agriculture Series*, 4th edn. Longman, London, pp. 265–365.

Lemke, P.A. (1969) A reevaluation of homothallism, heterothallism and the species concept in *Sistotrema brinkmannii*. *Mycologia* 61, 57–76.

Leppik, E.E. (1967) Some viewpoints on the phylogeny of rust fungi. VI. Biogenic radiation. *Mycologia* 59, 568–579.

McGeary, F.M. and Wheeler, B.E.J. (1988) Growth rates of, and mycelial interactions between, isolates of *Crinipellis perniciosa* from cocoa. *Plant Pathology* 37, 489–498.

Pegler, D.N. (1978) *Crinipellis perniciosa* (Agaricales). *Kew Bulletin* 32, 731–736.

Pegler, D.N. (1983) *The Agaric Flora of the Lesser Antilles*, Kew Additional Bulletin Series 9. HMSO, London.

Pereira, J.L., Ram, A., de Figueiredo, J.M. and de Almeida, L.C.C. (1990) First occurrence of witches' broom disease in the principal cocoa-growing region of Brazil. *Tropical Agriculture* 67, 188–189.

Petrini, O. (1986) Taxonomy of endophytic fungi of aerial plant tissues. In: Fokkema, N.J. and van den Huevel, J. (eds), *Microbiology of the Phyllosphere*. Cambridge University Press., Cambridge, pp. 175–187.

Petrini, O. and Müller, E. (1986) Research on endophytes. *Swiss Biotechnology* 4, 8–10.

Pound, F.J. (1943) *Cacao and Witches' Broom Disease* (Marasmius perniciosus). *A report on a recent visit to the Amazon region of Peru, September 1942–February 1943*, Department of Agriculture of Trinidad and Tobago. Yuille's Printerie, St. Augustine, Trinidad, 14 pp.

Purdy, L.H. and Dickstein, E.R. (1990) Basidiocarp development on mycelial mats of *Crinipellis perniciosa*. *Plant Disease* 74, 493–496.

Purdy, L.H., Trese, A.T. and Aragundi, J.A. (1983) Proof of pathogenicity of *Crinipellis perniciosa* to *Theobroma cacao* by using basidiospores produced in *in vitro* cultures. *Revista Theobroma (Brazil)* 13, 157–163.

Rayner, A.D.M. (1991) The phytopathological significance of mycelial individualism. *Annual Review of Phytopathology* 29, 305–323.

Rayner, A.D.M. and Boddy, L. (1986) Population structure and infection biology of wood-decay fungi in living trees. *Advances in Plant Pathology* 5, 119–160.

Rocha, H.M. and Wheeler, B.E.J. (1985) Factors influencing the production of basidiocarps and the deposition and germination of basidiospores of *Crinipellis perniciosa*, the causal agent of witches' broom disease on cocoa (*Theobroma cacao* L.). *Plant Pathology* 34, 319–328.

Rudgard, S.A. (1986) Witches' broom disease on cocoa in Rondonia, Brazil. Basidiocarp production on detached brooms in the field. *Plant Pathology* 35, 434–442.

Rudgard, S.A. and Andebrhan, T. (1988) Predicting the cost-benefits of sanitation pruning for the management of witches' broom disease. *Proceedings of the 10th International Cocoa Research Conference, 1987, Santo Domingo, Dominican Republic.* Cocoa Producers Alliance, Lagos, Nigeria, pp. 341–344.

Schaap, T. and Simchen, G. (1971) Genetic control of recombination affecting mating factors in populations of *Schizophyllum commune* and its relation to inbreeding. *Genetics* 68, 67–75.

Silva, P. (1987) Cocoa and 'lizard' or witches' broom: records made by Alexandre Rodrigues Ferreira in 1785 and 1787 in Amazonia. *Boletim Tecnico, Centro de Pesquisas do Cacau, Brazil*, Vol. 146, 21 pp.

Singer, R. (1942) A monographic study of the genera *Crinipellis* and *Chaetocalathus*. *Lilloa* 8, 441–534.

Stahel, G. (1915) *Marasmius perniciosus (nov. spec.)*, the cause of Krulloten disease of cocoa in Suriname. *Dept. van den Landbouw in Suriname*, Bulletin No. 33 (September 1915), 5–49.

Stahel, G. (1924) The name *Marasmius perniciosus* var. *ecuadoriensis* proposed. *Report Departement van den Landbouw in Suriname* Verslag over het jaar 1923. 2. Plantzenziekten, 30–32.

Thorold, C.A. (1975) *Diseases of Cocoa*. Clarendon Press, Oxford.

Toxopeus, H. (1985) Botany, types and populations. In: Wood, G.A.R. and Lass, R.A. (eds), *Cocoa. Tropical Agriculture Series*, 4th edn. Longman, London, pp. 11–37.

Went, F.A.F.C. (1904) Krulloten en versteende vruchten van de cacao in Suriname. *Verhandelingen der K. Akademie van Wetenschappen, Amsterdam, 2 Sect.* 10, 1–40.

Wheeler, B.E.J. (1985) The growth of *Crinipellis perniciosa* in living and dead cocoa tissue. In: Moore, D., Casselton, L.A., Wood, D.A. and Frankland, J.C. (eds), *Developmental Biology of Higher Fungi. British Mycological Society Symposium, Volume 10.* Cambridge University Press, London, pp. 103–116.

Wheeler, B.E.J. (1987) Plant pathology in a developing world. *Plant Pathology* 36, 430–437.

Wheeler, B.E.J. and Mepsted, R. (1988) Pathogenic variability amongst isolates of *Crinipellis perniciosa* from cocoa (*Theobroma cacao*). *Plant Pathology* 37, 475–488.

Whitehouse, H.L.K. (1949) Multiple allelomorph heterothallism in the fungi. *New Phytologist* 48, 212–244.

Wood, G.A.R. (1985) History and development. In: Wood, G.A.R. and Lass, R.A. (eds), *Cocoa. Tropical Agriculture Series*, 4th edn. Longman, London, pp. 1–10.

Yarwood, C.E. (1968) Rust infection protects beans against heat injury. *Nature* 220, 813.

16 The Ecology of African Cassava Mosaic Geminivirus

D. Fargette[1] and J.M. Thresh[2]

[1]*ORSTOM, Laboratoire de Phytovirologie, CIRAD, BP 5035, Montpellier 34032, France;* [2]*Natural Resources Institute, Chatham Maritime, Kent ME4 4TB, UK.*

Introduction

In studying the epidemiology of vector-borne virus diseases and in developing control measures it is advantageous to adopt an ecological approach to the complex interactions between viruses, vectors and their host plants. This became apparent from some of the earliest studies on sugarbeet curly top and other viruses in the 1920s and 1930s, as discussed by Carter (1973) and Thresh (1981). However, ecological studies are currently neglected in developed countries because of the increasing preoccupation of virologists with the biochemical features of viruses. There are different problems in sub-Saharan Africa where the dearth of trained personnel restricts the study of even the most important virus diseases (Thresh, 1991).

The lack of adequate ecological information is a serious obstacle in developing effective virus disease control measures as discussed here in relation to African cassava mosaic disease (ACMD) which is caused by a whitefly-borne geminivirus (ACMV). ACMD is a striking example of a disease that is prevalent every year and on a continental scale (Fauquet and Fargette, 1990). This reflects the efficient dual mode of dispersal by the whitefly vector (*Bemisia tabaci*) and in the stem cuttings which are the usual means by which cassava is propagated.

In this text, it is argued that the current prevalence of ACMD in Africa is relatively recent and avoidable and that it masks contrasting situations in different agroecological zones. We recapitulate the key landmarks which have led to the present state of knowledge and discuss the main factors influencing the incidence of ACMD based on the evidence available. The complexity of the problem and the paucity of data from many cassava growing countries explains why there is no general agreement on the most effective means of disease control in the different areas and an appreciation of the key determinants will facilitate the development of appropriate strategies for each agroecological zone.

Cassava in Africa and the Appearance and Spread of ACMD

Cassava was introduced by the Portuguese from South America to West Africa at the end of the 16th century and to East Africa in the 17th (Carter *et al.*, 1992). However, cassava seems to have been grown on a limited scale until the 19th century and only became widely cultivated at the beginning of the 20th. It is now grown extensively in many parts of Africa and in very diverse agroecological conditions. These include upland and lowland areas of long or relatively short growing season with single or double seasonal peaks of rainfall. Carter *et al.* (1992) present a detailed map of the distribution of cassava in Africa and categorize the range of environments in which the crop is grown. Individual farms are usually small and cassava is often interplanted with one or more other crops, which include maize, sorghum, sweet potato, beans, groundnuts and cotton. There are few large mechanized farms for commercial production of the tuberous roots for export as food for livestock, or for processing to produce starch or alcohol.

ACMD is not known to occur in South America and it was first observed in 1894 in what is now Tanzania. It has since been recorded in virtually all parts of Africa where cassava is grown and in the islands of Madagascar, Réunion and Mauritius. There are no reliable estimates of the losses caused by ACMD, but they are known to be large (Fauquet and Fargette, 1990), and ACMV was regarded as the most important vector-borne pathogen of any African crop in a recent economic assessment (Geddes, 1990).

Little information is available about the early history and progress of ACMD in Africa, but it was noted as destructive in West Africa in the 1920s and 1930s, with an apparent progression of the disease northwards from the coastal regions (Guthrie, 1988). The relative importance of natural spread by whiteflies and dissemination by man through the movement of infected cuttings is uncertain. However, the increase in importance of ACMD in recent decades has been associated with the intensification of crop production. These trends are likely to continue as cassava cultivation increases in response to human population pressure and as the crop is introduced to new areas.

Additional assumptions can be made on the likely sequence of events which led to the rapid spread of ACMV in Africa. The virus must have pre-existed in some indigenous natural hosts and have spread to cassava after the crop was introduced. It is also likely that different strains of ACMV occurred in Africa before cassava was introduced, as serological studies have clearly distinguished isolates from Madagascar and eastern Africa from those obtained elsewhere (Harrison *et al.*, 1991). Indeed, some wild plants are now known to be reservoirs of ACMV, but their current role in the ecology of ACMV is uncertain as it cannot be determined whether they are the original primary hosts or merely secondary ones contaminated from cassava. Whatever the initial situation, epidemiological studies indicate that cassava is now the main source of ACMV from which spread occurs and possibly also the main host of the whitefly vector. Thus wild plants play, at most, a marginal role in ACMV epidemiology (see below).

Symptom Expression and Virus Incidence

The symptoms of ACMV in cassava are usually conspicuous and easy to diagnose and much of the evidence on the incidence and spread of ACMV is based on visual observations. However, symptoms are sometimes indistinct and virus content seems to be low, especially in dry conditions when vegetative growth is restricted, or when plants develop symptoms of mineral deficiency, or are severely attacked by cassava green mites (*Mononychellus tanajoa*) or cassava mealybug (*Phenacoccus manihoti*). This indicates the limitations of relying solely on symptom expression in ecological studies and such evidence should be treated with caution.

The problems that can arise are apparent from experience in a survey of northern areas of the Ivory Coast, in which symptomless cassava plants were sampled during the dry season and cuttings were grown on in insect-proof glasshouses. All eventually developed clear symptoms of infection with ACMV (D. Fargette and C. Fauquet, ORSTOM, unpublished results). From this and other observations, it is apparent that the health status of cassava cannot be assessed satisfactorily from the presence of symptoms on mature, or slowly growing or badly-infested plants, even if such observations are supplemented by virus detection tests. The simplest and most reliable way to assess the presence of ACMV in such plants is to take cuttings from suspected stems and to follow symptom expression soon after planting when leaves develop rapidly and show conspicuous symptoms if the plants are infected.

Production and Maintenance of ACMV-Free Cassava

One of the first major advances in understanding the ecology of ACMV was to show that infection is not inevitable and that cassava fields can be maintained free, or largely free, of infection. This was established when ACMV-free cassava was selected and propagated in Tanzania in upland conditions where there was little or no spread by whiteflies (Storey, 1936). From this experience it was concluded that 'it was possible to set healthy plots and to maintain them virus-free through survey and eradication'. This finding was a major breakthrough, but only limited attempts were made at the time to exploit the benefits of phytosanitation on a large scale in Tanzania or elsewhere and the main attention of Storey and his collaborators turned to breeding for resistance to ACMV. Nevertheless, phytosanitation measures involving ACMV-free planting material and roguing were practised widely in Uganda during the 1940s and 1950s and achieved considerable success (Jameson, 1964).

The scope for sanitation was demonstrated elsewhere in the 1970s, when ACMV-free stocks of cassava of several varieties were established in Kenya and maintained over successive years (Bock, 1983). This was done in both coastal and western areas of the country where there was little spread by whiteflies and most contamination originated from infected cuttings. In these circumstances, a simple combination of selection and propagation of symptomless cuttings and eradication of any infected plants that occurred was highly effective. Such measures were not widely adopted in Kenya but they have since been used successfully in parts of Malawi and Uganda where overall inoculum pressure is low (R.F. Sauti and G.W. Otim-Nape, unpublished information).

Whitefly vs. Cutting Transmission and the Role of Cassava as a Source of Infection

High rates of spread of ACMV by whiteflies are a feature of the lowland forest and transitional zones of West Africa, as established near Ibadan in Nigeria (Leuschner, 1977) and Adiopodoumé near Abidjan in the Ivory Coast (Fargette, 1985). However, initial generalizations regarding an apparent difference in ACMV ecology between the rapid spread in West Africa and limited spread in Tanzania and Kenya (Bock, 1983) were discarded when it was shown that ACMV-free cassava could also be cultivated at Toumodi, 200 km north of Abidjan, in the savannah region of the Ivory Coast (Fauquet *et al.*, 1988a). Little spread occurred there and ACMV-free cassava of a wide range of cultivars (varying from susceptible to resistant) were cultivated over areas of several hectares in each of several successive years. The contrast between high rates of contamination at Abidjan and low rates at Toumodi was attributed to a difference between the rainforest and savannah environments, but no explanation was provided at the time on the precise factors and underlying mechanisms involved.

The crucial role of cassava in the epidemiology of ACMV as the major virus reservoir and possibly also the main host of whitefly vectors was suspected from various results obtained in the Ivory Coast. It was shown that infected cassava, by its prevalence and its virus content, was the most important virus source (Fargette, 1985). Moreover, host range studies indicated that other crops or wild species were unlikely to be involved in the spread of the disease. Because of their limited distribution they would play, at most, a marginal role (Fargette, 1985). Furthermore, studies suggested that there were different *Bemisia tabaci* biotypes, the one found on cassava in the Ivory Coast being largely restricted to this host and characterized by a specific electrophoretic pattern (Burban *et al.*, 1992). By contrast, a much more polyphagous *B. tabaci* biotype was unable to colonize cassava. It was also apparent that whiteflies are carried by the prevailing wind and can spread ACMV over distances of several kilometres downwind from cassava fields (Fargette, 1985). Finally, the role of cassava as the main major virus and vector reservoir was established through multilocational trials near Abidjan in the lowland rain forest zone.

In these trials, differences in rates of spread between sites were associated with the presence or absence of infected cassava fields upwind (Fauquet *et al.*, 1988b). The role of such fields as sources of infection was reinforced by observations made in other parts of the Ivory Coast. For instance, much spread occurred at Tontonou in the savannah region, c. 15 km from Toumodi, at a site where the experiments were surrounded by diseased cassava fields (Fargette, 1985). Similarly, considerable spread occurred in Kenya in plots adjacent to much diseased cassava (Bock, 1988). Therefore, it is apparent that it is not a difference between savannah and forest environments *per se* but rather differences in the amount of infected cassava upwind and close to the trials, which best explain the differences observed in rates of spread by whiteflies. Cassava tends to be more widely grown in forest areas than in savannah, where the distance between plantings is greater and the opportunity for spread is usually much less.

Complexity of ACMV Ecology

Spread by whiteflies is not only dependent on the cropping system adopted, but also on seasonal factors, the host plant and vector characteristics.

Sequential monthly planting of susceptible cultivars over 2 years at Kiwanda near Amani in Tanzania and over a 6 year period near Abidjan showed big seasonal differences in virus spread (Storey and Nichols, 1938b; Fargette, 1985). Such differences have also been reported in Nigeria (Leuschner, 1977) and coastal Kenya (Bock, 1988; Robertson, 1987, 1988). The respective role of radiation and rainfall-associated parameters on cassava growth, whitefly populations and ACMV spread is discussed below. However, despite the rapid spread and the seasonal variation observed near Abidjan, cassava remained largely free of infection whatever the month of planting when very resistant cultivars were grown (D. Fargette and C. Fauquet, ORSTOM, unpublished results).

There are indications that dissemination by both whitefly and by cuttings is not as straightforward as hitherto assumed. Whitefly species other than *B. tabaci* may be involved in the transmission of ACMV. In particular, the role of *B. afer* has not been determined and yet it occurs widely on cassava and predominates at some periods of the year or in some areas, as in Kenya, Uganda and Malawi. This may further affect the variation and complexity of ACMV epidemiology.

Moreover, the significance of the failure of ACMV to become completely systemic in cassava has not been fully appreciated or exploited. This phenomenon is termed 'reversion' and one of the most important consequences is that a proportion of the cuttings collected from infected sources are free of ACMV and grow into uninfected plants. Reversion has been known since the early work of Storey and Nichols (1938a) in Tanzania, but it has not been fully studied or documented. It is clearly linked to varietal characteristics and is most marked in highly resistant cultivars (Fauquet *et al.*, 1988a). It is also possibly dependent on environmental conditions, as considerable variation in reversion rates was observed from year to year in the Ivory Coast and preliminary experiments suggested that reversion is greater at high temperatures than during relatively cool periods (C. Fauquet and D. Fargette, ORSTOM, unpublished results).

Collectively, the various findings made over many years have gradually revealed the complexity of ACMV ecology and the various interactions between virus, host and vector, and with the environment. It is now apparent that the disease can be controlled by sanitation in at least some circumstances, that the respective role of cuttings and whiteflies in spread differs between ecological regions and that environmental factors play a crucial role. What is not yet clear is the relative importance of biotic factors (species and biotype of *Bemisia*, cassava variety and growth rates) and abiotic ones (temperature, rainfall, etc.). In the following sections we discuss some of the data on the interactions between these factors and propose a model of spread to account for their effects.

Relationships Between Climatic Factors, Cassava Growth, Whitefly Numbers and ACMV Spread

Data sets

ACMV ecology was studied in monthly plantings at Adiopodoumé over a period of 6 years in which disease incidence, whitefly numbers, cassava growth and climatic data were recorded (Fargette, 1985). Based on this comprehensive set of data, hypotheses have been developed on the main features of ACMV ecology and on the key factors influencing virus spread. ACMV epidemiology has also been studied in some detail at Kiwanda near Amani in Tanzania (Storey and Nichols, 1938b), at Ibadan in Nigeria (Leuschner, 1977), in the coastal and western parts of Kenya (Bock, 1983, 1988; Robertson, 1987, 1988), at Toumodi in the Ivory Coast (Fauquet *et al.*, 1988b) and in various parts of Uganda (G.W. Otim-Nape, Kampala, unpublished) and Malawi (Nyirenda *et al.*, 1993). There is also information on whitefly population dynamics on cassava from Togo (Dengel, 1981) and Malawi (Nyirenda *et al.*, 1993).

Some of the results obtained have not been published in detail. Others are not sufficiently comprehensive because information is lacking on one or more components of the pathosystem, or for some periods of the year. Furthermore, the experimental systems and cultivars used differed widely according to the priorities and purpose of the studies. Although this makes it difficult to make a comprehensive comparative analysis of the results, some appropriate comparisons help to validate, refine and set the limits of the proposed ACMV ecological model based on Adiopodoumé data so that it can be adapted to other regions.

Adiopodoumé and Kiwanda: the role of radiation-associated parameters

Some of the most important results obtained in the monthly plantings at Adiopodoumé are illustrated in Fig. 16.1. Spread of ACMV varied widely over the year and differences in rate were closely associated with cassava growth and whitefly numbers. On average, high rates of spread were associated with rapid vegetative growth and high whitefly numbers recorded 1 month earlier, when infection is likely to have occurred. Conversely, periods of little spread were associated with slow growth and low whitefly populations. Virus spread, cassava growth and whitefly populations were also dependent on climatic factors. The climatogram (Fig. 16.2) exhibits for each month the relationship between temperature and rainfall at Adiopodoumé. Rapid virus spread, quick cassava growth and high whitefly numbers occurred soon after the start of the rainy season (March), when temperatures increased. Virus spread, whitefly numbers and cassava growth decreased

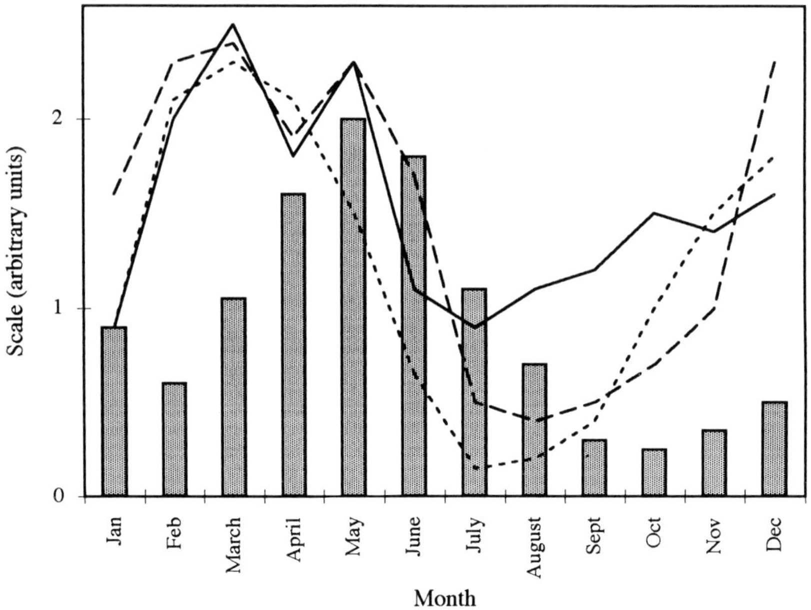

Fig. 16.1. Monthly spread of ACMV at Adiopodoumé, Ivory Coast (histogram), average monthly maximum temperature (dotted line), whitefly numbers (dashed line), monthly cassava growth expressed as the leaf area index (solid line); arbitrary scale.

when temperature and rainfall decreased to a minimum in July.

These results suggest that, for much of the year at Adiopodoumé, the same climatic factors determine cassava growth, whitefly numbers and ACMV spread. There is evidence that whitefly developmental rates and flight activity are linked to temperature (Leuschner, 1977; Butler *et al.*, 1986). It is also known that cassava growth is closely dependent on radiation, provided that soil moisture is not limiting (Sylvestre and Arradeau, 1983). Furthermore, rapidly growing cassava supports high whitefly populations (Dengel, 1981). It is also likely to be more susceptible to virus infection and to multiply the virus more efficiently, as indicated by field observations, although this remains to be confirmed under controlled conditions. Thus high radiation and temperature would be expected to favour ACMV spread directly and also indirectly through effects on all three components of the pathosystem involving virus, vector and host.

Statistical analysis (including non-linear regression between spread and month and stepwise regression between monthly virus spread and average monthly climatic factors) indicate that rates of ACMV spread at Adiopodoumé follow a sinusoidal pattern and are closely associated with radiation-associated parameters (Fargette *et al.*, 1993). A similar relationship between

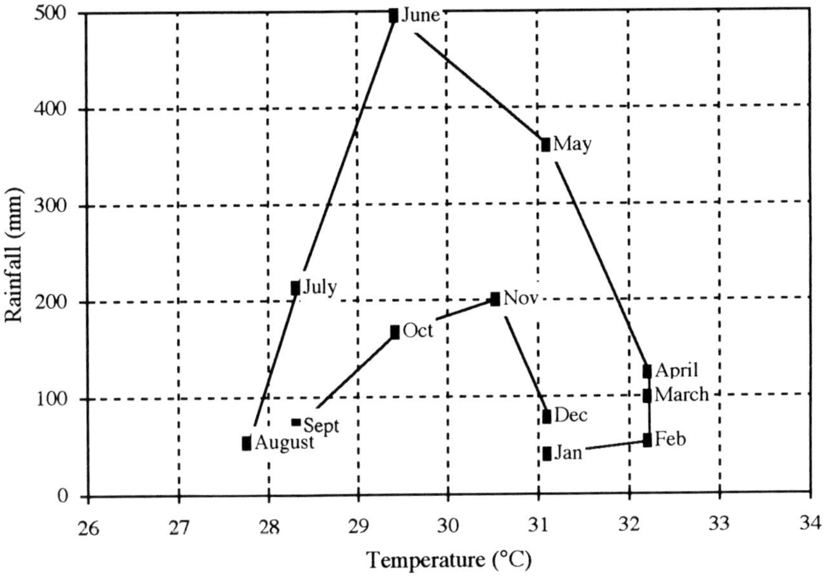

Fig. 16.2. Climatogram giving the monthly rainfall and mean maximum temperatures at Abidjan, Ivory Coast.

ACMV spread and mean maximum temperatures was also found in monthly plantings over a 2 year period at Kiwanda in Tanzania, which is the only other site where comprehensive data are available on monthly spread of ACMV (Storey and Nichols, 1938b). However, there were no observations at Kiwanda on whitefly numbers or cassava growth.

ACMV spread at Kiwanda fluctuated widely over the year with a pattern similar to that observed at Adiopodoumé. Maximum spread occurred between March and May and least between August and November, with intermediate values during the rest of the year. Comprehensive climatic data are not available for Kiwanda, only the maximum and minimum temperature over the year being quoted. However, by using data from a nearby site with comparable temperature and rainfall fluctuations, temperature was shown to be the main factor influencing spread, which was greatest at high temperature and least at low temperature, whereas there was no apparent relationship with rainfall (Fargette *et al.*, 1993). Similar results showing temperature as the key determinant and the non-significant effect of rainfall have also been obtained for tobacco leaf curl disease in India. This is caused by another whitefly-transmitted geminivirus (Valand and Muniyappa, 1992), indicating that this relationship applies to other whitefly-transmitted viruses and to other regions (Fargette *et al.*, 1993).

Parameters in the Ivory Coast, Nigeria and Kenya

It is not possible in the humid conditions of Adiopodoumé or Kiwanda to determine the effects of rainfall on spread of ACMV, as soil moisture deficits are not a factor limiting cassava growth for much of the year at these sites. Furthermore, the planting date experiments at Adiopodoumé were watered during the short relatively dry season between November and February. Toumodi, by contrast, is located in a drier savannah region of the Ivory Coast where ACMV spread was more limited, but also showed some seasonal periodicity. Most rapid spread occurred in April/May during the main rainy season, at a time of high temperatures and mean rainfall exceeding 100 mm per month (Fauquet *et al.*, 1988b). Relatively little spread occurred in July/August which is a cooler period of limited rainfall. In Ibadan (Nigeria) seasonal differences in rates of spread are linked to whitefly numbers. Maximum spread occurred in April, May and June in the rainy season, a hot period characterized by monthly rainfall exceeding 100 mm (Leuschner, 1977). By contrast, spread was low in August, September and October with lower temperatures and limited rainfall. The available data are not comprehensive and information on monthly spread is lacking for some periods of the year so as to preclude statistical analyses similar to those conducted on the Adiopodoumé and Kiwanda data. Nevertheless, the results from Toumodi and Ibadan suggest that rainfall-associated factors influence ACMV spread.

Other evidence on the role of rainfall has been obtained in experiments and observations in coastal Kenya where spread was greatest at sites in areas where mean annual rainfall exceeded 1200 mm and least where it did not exceed 1000 mm. However, rainfall is unlikely to have been the only factor involved because the area of low rainfall was also one of limited cassava production, where there was considerable separation between plantings (Bock, 1988).

In other Kenyan trials spread of ACMV was greatest between May and August and least between September and March (Robertson, 1988). Effects on whitefly numbers and cassava growth are difficult to interpret, as data are lacking for the crucial period between May and July. However, the period of low spread is associated with poor cassava growth and high whitefly numbers. Thus, the results obtained suggest that alternating periods of high and low virus spread do not primarily reflect differences in whitefly populations, but are associated with the cassava growth pattern. In these trials, the pattern of ACMV spread was associated with rainfall, possibly because rain and not temperature is the main limiting factor for cassava growth in the hot conditions of coastal Kenya where the rainfall is seasonal and less than in coastal Ivory Coast.

Conclusions

Knowledge of ACMV ecology has accumulated erratically over the last 60 years reflecting the lack of continuity and coordination between the few research projects. These have been mounted at different times and places in Africa and with different cassava varieties. Information has accumulated through the classical sequence of observations, hypotheses, experiments and results, leading to new or refined hypotheses. However, much further work is required to obtain a more detailed understanding of the situation in the many different environments in which cassava is grown.

Despite the limitations of the available data, generalizations on ACMV ecology are now possible, in particular on the crucial role of cassava as the main source of infection. The information is entirely consistent on this point and cassava may also be the main host of the whitefly vector. Moreover, seasons of fast spread coincide with periods of rapid cassava growth, high susceptibility to virus infection and favourable conditions for virus multiplication and spread, whether growth is primarily linked to radiation-associated parameters in humid environments or to rain-associated ones in drier conditions.

Thus, it is tempting to speculate that in areas where growing conditions are generally favourable and cassava is cultivated intensively, conditions facilitate spread by whiteflies because infection sources are abundant, vectors are numerous and plants are extremely vulnerable to infection. This occurs where mean annual rainfall exceeds 1500 mm and the length of the crop growing period exceeds 270 days (Anon., 1978, 1989; Geddes, 1990) – conditions likely to be optimal for cassava growth. In such areas methods of control by sanitation are unlikely to be successful unless very resistant varieties are used (Bock, 1983; Marquette, 1988; D. Fargette and C. Fauquet, ORSTOM, unpublished results). The situation is completely different in areas where cassava is little grown and growth is curtailed at periods of the year when conditions are too dry or too cold. In such areas, spread by whiteflies is restricted because of limited inoculum (small, scattered and remote virus sources of low potency), small numbers of whiteflies and reduced plant susceptibility during the dry and/or cool seasons (Bock, 1983). In these circumstances, infection is largely due to the use of infected cuttings and control by sanitation is feasible and achieved readily.

Much additional information is required from many other cassava growing areas of Africa before there can be any real understanding of the ecology of ACMV and the most appropriate means of control. Such information will not be obtained unless there is a greatly increased commitment of manpower and resources. This is because ACMV and other important viruses of African food crops are inadequately studied and receive totally inadequate attention, especially when considered in relation to viruses of temperate crops in developed countries (Thresh, 1991).

In these circumstances, there is obvious scope for collaboration to make the best possible use of the resources available and to develop a coordinated research programme. The immediate aim should be to expose virus-free material at a wide range of sites using a standard experimental design, the same set of varieties and uniform recording procedures to follow rates of infection. In the longer term there is scope for developing a multi-million dollar international programme of the type mounted in recent years against the threat posed by cassava mealybug and green mite (Herren and Neuenschwandar, 1991).

References

Anon. (1978) Report on the agro-ecological zones project, Vol. 1, Methodology and results for Africa. *World Soil Resources Report* No. 48. FAO, Rome.

Anon. (1989) *Crop Production Yearbook*. FAO, Rome, p. 141.

Bock, K.R. (1983) Epidemiology of cassava mosaic disease in Kenya. In: Plumb, R.T. and Thresh, J.M. (eds), *Plant Virus Disease Epidemiology*. Blackwell Scientific Publications, Oxford, pp. 337–347.

Bock, K.R. (1988) Some aspects of the epidemiology of African cassava mosaic virus in coastal districts of Kenya. In: Fauquet, C. and Fargette, D. (eds), *Proceedings of the International Seminar on African Cassava Mosaic Disease and Its Control*, Yamoussoukro, 4–8 May 1987. CTA/FAO/ORSTOM/IITA/IAPC, Wageningen, pp. 125–130.

Burban, C., Fishpool, L.D.C., Fauquet, C., Fargette, D. and Thouvenel, J.-C. (1992) Host-associated biotypes within West-African populations of the whitefly *Bemisia tabaci* (Homoptera: Aleyrodidae). *Journal of Applied Entomology* 113, 416–423.

Butler, G.D., Henneberry, T.J. and Hutchison, W.D. (1986) Biology, sampling and population dynamics of *Bemisia tabaci*. *Agricultural Zoology Reviews* 1, 167–195.

Carter, S.E., Fresco, L.O., Jones, P.G. and Fairbairn, J.N. (1992) *An Atlas of Cassava in Africa: Historical Agroecological and Demographic Aspects of Crop Distribution*. Centro Internacional de Agricultura Tropical, Cali, Colombia.

Carter, W. (1973) *Insects in Relation to Plant Disease*, 2nd edn. John Wiley and Sons, New York.

Dengel, H.J. (1981) Untersuchungen über das Auftreten der Imagines von *Bemisia tabaci* (Genn.) auf verschiedenen Manioksorten. *Zeitschrift für Pflanzenkrankheiten und Pflanzenschutz* 88, 355–366.

Fargette, D. (1985) Epidémiologie de la mosaïque africaine du manioc en Côte d'Ivoire. Editions de l'ORSTOM, Paris.

Fargette, D., Muniyappa, V., Fauquet, C., N'Guessan, P.N. and Thouvenel, J.-C. (1993) Comparative epidemiology of three tropical whitefly-transmitted geminiviruses. *Biochimie* 75, 547–554.

Fauquet, C. and Fargette, D. (1990) African cassava mosaic virus: etiology, epidemiology and control. *Plant Disease* 74, 404–411.

Fauquet, C., Fargette, D. and Thouvenel, J.-C. (1988a) Selection of healthy cassava plants obtained by reversion in cassava fields. In: Fauquet, C. and Fargette, D. (eds), *Proceedings of the International Seminar on African Cassava Mosaic Disease and Its Control*, Yamoussoukro, 4–8 May 1987. CTA/FAO/ORSTOM/IITA/IAPC, Wageningen, pp. 146–149.

Fauquet, C., Fargette, D. and Thouvenel, J.-C. (1988b) Some aspects of the epidemiology of African cassava mosaic virus in Ivory Coast. *Tropical Pest Management* 34, 92–96.

Geddes, A.M. (1990) The relative importance of crop pests in sub-saharan Africa. *Natural Resources Institute Bulletin*, No. 36.

Guthrie, E.J. (1988) African cassava mosaic disease and its control. In: Fauquet, C. and Fargette, D. (eds), *Proceedings of the International Seminar on African Cassava Mosaic Disease and Its Control*, Yamoussoukro, 4–8 May 1987. CTA/FAO/ORSTOM/IITA/IAPC, Wageningen, pp. 1–9.

Harrison, B.D., Swanson, M.M., McGrath, P. and Fargette, D. (1991) Patterns of antigenic variation in whitefly-transmitted geminiviruses. *Report of the Scottish Crop Research Institute for 1990*, pp. 88–90.

Herren, H.R. and Neuenschwandar, P. (1991) Biological control of cassava pests in Africa. *Annual Review of Entomology* 36, 257–283.

Jameson, J.D. (1964) Cassava mosaic disease in Uganda. *East African Agriculture and Forestry Journal* 29, 208–213.

Leuschner, K. (1977) Whiteflies; biology and transmission of African mosaic disease. In: Bellotti, A. and Lozano, J.E. (eds), *Proceedings of Cassava Protection Workshop*. CIAT, Cali, Colombia, pp. 51–58.

Marquette, J. (1988) Search for and dissemination of a cassava clone that is relatively tolerant to African cassava mosaic. In: Fauquet, C. and Fargette, D. (eds), *Proceedings of the International Seminar on African Cassava Mosaic Disease and Its Control*, Yamoussoukro, 4–8 May 1987. CTA/FAO/ORSTOM/IITA/IAPC, Wageningen, pp. 270–276.

Nyirenda, G.K.C., Munthali, D.C., Phiri, G.S.N., Sauti, R.F.N. and Gerling, D. (1993) Integrated pest management of *Bemisia* ssp. whiteflies, in Malawi. Unpublished Report, Makoka Research Station, Thondwe, Malawi.

Robertson, I.A.D. (1987) The whitefly, *Bemisia tabaci* (Gennadius) as a vector of African cassava mosaic virus at the Kenya coast and ways in which the yield losses in cassava, *Manihot esculenta* Cranz, caused by the virus can be reduced. *Insect Science Applicata* 8, 797–801.

Robertson, I.A.D. (1988) The role of *Bemisia tabaci* Gennadius in the epidemiology of ACMV in East Africa. In: Fauquet, C. and Fargette, D. (eds), *Proceedings of the International Seminar on African Cassava Mosaic Disease and Its Control*, Yamoussoukro, 4–8 May 1987. CTA/FAO/ORSTOM/IITA/IAPC, Wageningen, pp. 57–63.

Storey, H.H. (1936) Virus diseases of East African plants; VI. A progress report on studies of the diseases of cassava. *East African Agricultural Journal* 2, 34–39.

Storey, H.H. and Nichols, R.F.W. (1938a) Studies of the mosaic disease of cassava. *Annals of Applied Biology* 25, 790–806.

Storey, H.H. and Nichols, R.F.W. (1938b) Virus diseases of East African plants. VII.

A field experiment in the transmission of cassava mosaic. *The East African Agricultural Journal* 3, 446–449.

Sylvestre, P. and Arradeau, M. (1983) *Le Manioc*. Maisonneuve and Larose, Paris.

Thresh, J.M. (1981) The role of weeds and wild plants in the epidemiology of plant virus diseases. In: Thresh, J.M. (ed.), *Pests, Pathogens and Vegetation*. Pitman, London, pp. 53–70.

Thresh, J.M. (1991) The ecology of tropical plant viruses. *Plant Pathology* 40, 324–339.

Valand, G.B. and Muniyappa, V. (1992) Epidemiology of tobacco leaf curl virus in India. *Annals of Applied Biology* 120, 257–267.

IV MICROBIAL INTERACTIONS

17 Straw Disposal and Cereal Diseases

J.F. Jenkyn, R.J. Gutteridge and M. Jalaluddin

AFRC Institute of Arable Crops Research, Rothamsted Experimental Station, Harpenden, Herts AL5 2JQ, UK.

Introduction

Direct action by farmers and growers to control diseases, for example by using fungicides, is now commonplace. However, many other agricultural practices have the potential to influence pathogens directly or indirectly by altering the physical and perhaps microbiological environment in which they exist. Such effects are, therefore, very relevant to a discussion on the ecology of plant pathogens. This chapter focuses on one aspect of cereal husbandry, the disposal of straw.

Production of cereal straw in the UK greatly exceeds current demand. This is partly because much less straw is now used than formerly (especially for animal feed and bedding) but also because the area and yield of cereals (and hence cereal straw) have greatly increased (Prew and Lord, 1988). During the 1970s and early 1980s it became increasingly common to dispose of the surplus by burning it in the field.

Straw burning has much to commend it. It is a cheap and effective way of disposing of a waste product and, after burning, minimal cultivation is usually sufficient to prepare seed beds for succeeding crops. Reductions in weeds, pests and diseases are among the other proven or assumed benefits of straw burning although fire will usually affect harmful and beneficial organisms equally.

Despite the benefits of burning there were problems. Fires occasionally got out of control and caused damage but more commonly it was the smoke and smuts that burning generated that led to increasing criticism of the practice, especially where farms bordered urban areas. This led to increasing restrictions on burning and ultimately to the almost complete ban that now exists.

Straw has the potential to be used for a variety of industrial uses but most are, at present, uneconomic. For most farmers, therefore, the only practical alternative to burning is to incorporate the straw in the soil. In the mid 1980s, the prospects of having to do this on a large scale caused some concern. Firstly, there were fears that incorporating large quantities of straw, and achieving satisfactory seedbeds, would be difficult, especially on the heavier soils. There were further fears that this would, in turn, make it difficult to complete field work in autumn, which is a busy time of the year and one when weather conditions are usually deteriorating. Decomposition of the straw by microorganisms requires nitrogen (N) and it was expected that additional fertilizer N would need to be applied to compensate for this but just how much and when was uncertain. Increasing problems with pests, diseases and weeds were also anticipated. As far as diseases are concerned this seemed a reasonable fear because cereal straw has often been used in experiments as an effective source of inoculum (e.g. Jenkyn and King, 1977; Jenkyn *et al.*, 1989).

Against this background, a multidisciplinary co-ordinated programme, principally involving scientists in the Agricultural and Food Research Council (AFRC) and the Agricultural Development and Advisory Service (ADAS), was initiated in 1984 to study the problems and consequences of incorporating cereal straw instead of burning it (Prew and Lord, 1988). In some circumstances, farmers may find it necessary to make changes to their primary cultivation systems if they are to incorporate straw successfully. Therefore, some of our experiments also compared ploughing with non-inversion cultivations, usually using discs or tines. In this chapter, we restrict our attention to effects on diseases (of both straw and cultivation treatments) mostly using data from experiments grown on our Rothamsted and Woburn Farms in 1985–1991. Our conclusions, although broadly similar to those described in an earlier report (Jenkyn *et al.*, 1988), have been modified in the light of the further evidence obtained since that time.

Experimental Details

Most of the results that we describe are from continuing long-term experiments on winter wheat, begun in 1985, that test the same treatments applied to the same plots in successive years. Two of them test the effects of burnt versus incorporated straw in factorial combination with non-inversion cultivations (discing or tining) versus ploughing. One of these experiments is on the Rothamsted Farm, where the soils are mostly clay loams over clay with flints, and the other is on the contrasting sandy loam at Woburn. Both experiments receive basal fungicides.

Another experiment on winter wheat at Rothamsted, also started in 1985, was established to test the effects of autumn N and different pesticide treatments (including a fungicide programme) in factorial combination on plots where the straw is burnt, baled or chopped and spread, before cultivating, without inversion, to *c.* 10 cm.

Data for winter barley are mostly from an experiment at Rothamsted in 1985–1989 (now finished) in which the straw was either chopped before incorporating by plough, rotary digger or deep tines, or burnt and the plots then shallow cultivated to provide a seed bed. Basal fungicides were applied to this experiment in 1985 but not thereafter. Reference is also made to an experiment on winter barley at Whaddon in Buckinghamshire in 1986. This was on a very heavy clay soil and tested the effects of burnt versus incorporated straw in factorial combination with tine cultivations versus ploughing; fungicides were applied to all plots.

Full details of all the above experiments, including effects on growth and yield, will be published shortly but further information on the treatments applied to, and the husbandry of, experiments at Rothamsted and Woburn can be obtained from *Yields of the Field Experiments* published annually by Rothamsted Experimental Station.

In 1985–1987, diseases were also monitored in a winter wheat experiment at Faringdon in Oxfordshire which tested different amounts of straw, up to 20 t ha^{-1}, versus none, and received basal fungicides.

In some experiments, diseases affecting the leaves, stem bases and roots were assessed in autumn, spring and summer. In others, monitoring was less frequent and, in some, attention was confined to those diseases affecting the stem bases and roots. A logit transformation was used for the statistical analyses of disease data expressed as percentages and the percentage values quoted were obtained by back transformation. Some variates provided evidence for differences between treatments that were relatively consistent but not always significant in individual experiments in individual years. To test the consistency of these effects, analysis of variance was applied to means of treatments from each of the individual experiments in each year. Where possible the data from the winter wheat experiments at Rothamsted and Woburn were combined.

Effects on Leaf Diseases

Samples taken from the winter wheat experiments in autumn occasionally showed increases in the severity of septoria (usually *S. tritici*) where straw had been incorporated instead of burnt. Samples taken in spring and summer, however, provided no evidence of any consistent effect of the straw treatments on *Septoria* spp. (i.e. *S. tritici* and *S. nodorum*). As an example, Fig. 17.1 compares average amounts of disease in summer in plots with burnt or incorporated straw. Cultivation treatments sometimes had significant effects on septoria in individual experiments and years but they were inconsistent. When averaged over sites and years there was little difference between plots that were ploughed and those that were tined and disced.

In the winter barley experiment at Rothamsted, net blotch (*Drechslera teres*) in autumn was invariably more severe where straw had been incorporated using non-inversion cultivations than where it had been incorporated by ploughing or had been burnt (Table 17.1). However, by spring there was no evidence of significant or consistent differences between treatments, and by summer the disease was always more severe where the

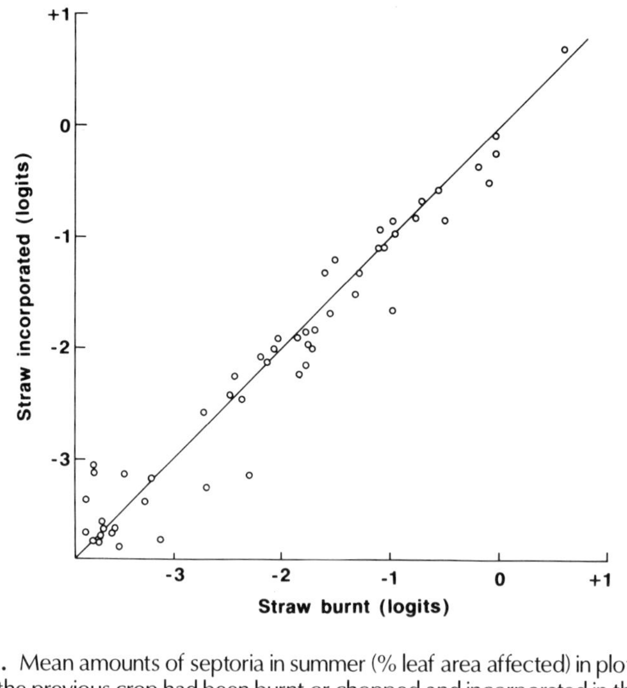

Fig. 17.1. Mean amounts of septoria in summer (% leaf area affected) in plots where the straw of the previous crop had been burnt or chopped and incorporated in the same experiment. Data are for shallow-cultivated (i.e. disced or tined) plots only and were obtained from three sites over 7 years. The sloping line represents equal severities.

Table 17.1. Mean amounts of net blotch in winter barley grown at Rothamsted in 1985–1989 in plots where straw of the previous crop had been burnt or incorporated using different cultivation systems.[1]

	Straw burnt	Straw incorporated			
	Discs or tines	Plough	Rotary digger	Deep tines	S.E.D. (11 d.f.)
Autumn	−2.45 (0.7)	−2.56 (0.6)	−2.06 (1.6)	−1.95 (2.0)	0.140
Spring	−2.44 (0.8)	−2.25 (1.1)	−2.48 (0.7)	−2.30 (1.0)	0.131
Summer	−1.90 (2.2)	−2.22 (1.2)	−2.21 (1.2)	−2.18 (1.3)	0.105

[1]Data, which are for the first seedling leaf in autumn and second youngest leaves in spring and summer, are logit transform values with, in parentheses, the corresponding percentage values obtained by back transformation.

straw had been burnt than where it had not (Table 17.1).

Leaf blotch (*Rhynchosporium secalis*) in spring was, like net blotch in autumn, usually more severe where straw had been incorporated using non-inversion cultivations than where it had been incorporated by ploughing or had been burnt but only at Whaddon in 1986 did these effects persist until summer. At Rothamsted, leaf blotch in summer was, like net blotch, usually most severe where the straw had been burnt.

These mostly small and short-lived effects on leaf diseases were initially unexpected but they can probably be explained because even where straw is burnt much inoculum remains. This, together with the potential that most of these diseases have to multiply very rapidly, means that even quite large differences in amounts of primary inoculum will often be relatively unimportant. The sometimes greater severity of leaf diseases in summer where straw has been burnt than where it has been incorporated can probably be explained by better crop growth (but not necessarily better yield) after burning and consequently a more favourable microclimate for disease development.

Effects on Stem Base Diseases

Few of the plants sampled in autumn showed symptoms of pathogens infecting the stem base.

In spring, winter wheat sampled from plots that had been disced or tined (i.e. not ploughed) commonly had less eyespot (*Pseudocercosporella herpotrichoides*) where the straw had been incorporated than where it had been

Table 17.2. Mean amounts of eyespot and sharp eyespot in winter wheat sown where straw of the previous crop had been burnt or chopped and incorporated. (Data used were for shallow-cultivated plots only and were from three sites over 7 years.)

	Straw[1]		
	Burnt	Incorporated	S.E.D.
Eyespot			
Spring (% shoots)	−1.14 (9.2)	−1.28 (7.1)	0.058 } 18 d.f.
Summer (% straws)	−0.80 (16.8)	−1.08 (10.3)	0.053 } 18 d.f.
Sharp eyespot			
Spring (% shoots)	−2.29 (1.0)	−2.47 (0.7)	0.073 } 26 d.f.
Summer (% straws)	−1.69 (3.3)	−2.01 (1.8)	0.077 } 26 d.f.

[1]Data are logit transform values with, in parentheses, the corresponding percentage values obtained by back transformation.

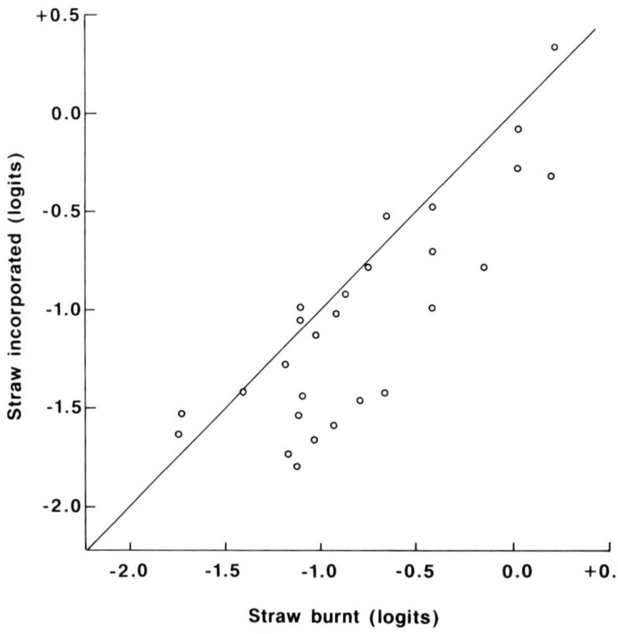

Fig. 17.2. Mean amounts of eyespot in summer (% straws) in plots where the straw of the previous crop had been burnt or chopped and incorporated in the same experiment. Data are for shallow-cultivated (i.e. disced or tined) plots only and were obtained from three sites over 7 years. The sloping line represents equal severities.

burnt. Effects in individual site × year combinations were not always large but they occurred relatively consistently and in an analysis of variance combining treatment means for all sites and years this effect of the straw treatments was significant (P<0.05; Table 17.2). A similar effect was apparent in summer (Fig. 17.2) and was apparently larger and even more significant than in spring (P<0.001; Table 17.2). Incorporating straw instead of burning it similarly decreased sharp eyespot (*Rhizoctonia cerealis*) in both spring and summer (Table 17.2).

Analyses of the combined data from the winter wheat experiments that also compared cultivation methods showed that incorporating straw decreased eyespot and sharp eyespot only where it was done using discs or tines and not where the straw had been buried by ploughing. The same analyses also showed that the cultivation treatments themselves had relatively consistent effects on sharp eyespot which, in both spring and summer, was significantly more severe in the ploughed plots than in the disced or tined. Overall, the cultivation treatments had much less effect on eyespot but this is misleading because there were apparently systematic differences between sites and seasons. Thus at Rothamsted in summer, the disease was less severe in the ploughed plots than in the disced or tined in the first two years (averaging 3.8 and 6.6% straws, respectively), but thereafter the reverse was generally true (averaging 19.4 and 7.8%, respectively, in 1987–1991); the effects of cultivations were very similar whether the straw was burnt or incorporated. At Woburn, effects of the cultivation treatments were more variable. Where straw was incorporated there tended to be more eyespot in the ploughed plots than in the disced or tined (averaging 15.3 and 10.1% straws affected, respectively), but where straw was burnt there was usually least in the ploughed plots (11.5 and 15.6%, respectively).

In the winter barley experiment at Whaddon in 1986, eyespot in summer was less severe where the straw had been chopped and incorporated by tining than where it had been burnt or incorporated by ploughing. In contrast, eyespot in the winter barley at Rothamsted in spring was usually less severe where the straw had been burnt than where it had been incorporated. In summer, however, burning the straw instead of incorporating it had no consistent affect. Eyespot at Rothamsted, in both spring and summer, tended to be more severe where straw was incorporated by ploughing than where it was incorporated using non-inversion cultivations but the mean effects, averaged over years, were not significant.

Brown foot rot (*Fusarium* spp.) in summer was prevalent only in the wheat experiments in 1989–1991. In only two of the individual site × year combinations was it significantly affected by the straw treatments, being less common where the straw was burnt than where it was incorporated. It was almost invariably less severe after ploughing than after discing or tining, and usually significantly so. On average, 27.5 and 41.5% of straws were affected after ploughing and non-inversion cultivations, respectively.

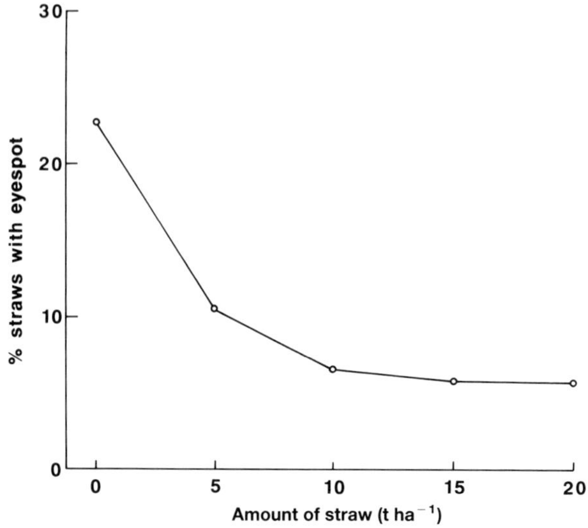

Fig. 17.3. Mean effects on eyespot of winter wheat of incorporating different amounts of straw at Faringdon in Oxfordshire in 1985–1987.

Why eyespot and sharp eyespot should often be less severe where straw is incorporated than where it is burnt is uncertain. However, similar effects on eyespot of incorporating straw were seen in the experiment at Faringdon (Fig. 17.3) in which the 'nil' treatment was achieved by removing the straw rather than burning it. This suggests that the effects, at least on eyespot, are associated with the presence and absence of straw and not burning *per se*. The effects on eyespot confirm limited evidence obtained from other experiments at Rothamsted in the 1970s (Prew and Read, 1978) and more recent evidence obtained elsewhere (e.g. Schulz *et al.*, 1990). Other experiments have also shown that there is often less eyespot where wheat is direct drilled after cereals than where it is sown after conventional cultivations (Brooks and Dawson, 1968; Prew, 1976). Yarham and Norton (1981) suggested that this might be explained by differences in growth habit, because direct-drilled seedlings are typically very prostrate, and they demonstrated correlations that were consistent with this view. However, differences in amounts of debris (assuming it had not been destroyed by burning) might also provide at least a partial explanation because direct drilling leaves most of it on the surface whereas cultivations bury it to a greater or lesser extent.

Any effects of the debris might be purely physical. It might, for example, interfere with the dispersal of spores by rain splash. Another possibility is that microorganisms growing on the decomposing straw might compete with the eyespot fungus to affect its ability to survive, sporulate or infect. Some evidence that would appear to support this was obtained by Prew (1977) who

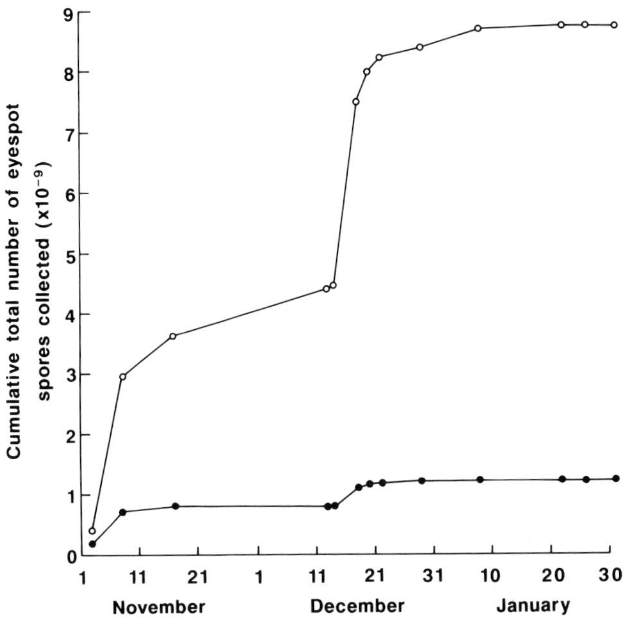

Fig. 17.4. Production of spores of *Pseudocercosporella herpotrichoides* on infected stem bases in the presence (●) or absence (○) of chopped straw.

Table 17.3. Incidence of eyespot in winter wheat at Rothamsted in 1990 where straw of the previous crop had been burnt, baled or chopped and incorporated.[1]

	Eyespot on exposed seedlings[2] (% shoots)	Eyespot on the sown wheat		
		28 March (% shoots)	27 June (% straws)	27 June (% straws; mod. + sev.)
Straw				
Burnt	−1.28 (7.2)	−0.22 (39.4)	+0.27 (63.2)	−0.31 (34.8)
Baled	−1.09 (10.2)	−0.07 (46.3)	+0.12 (55.9)	−0.69 (20.1)
Incorporated	−1.07 (10.5)	−0.25 (37.8)	+0.12 (56.0)	−0.67 (20.7)
S.E.D. (14 d.f.)	0.061	0.135	0.128	0.110

[1]Data are logit transform values with, in parentheses, the corresponding percentage values obtained by back transformation.
[2]Healthy seedlings were exposed at intervals in the plots between 1 November 1989 and 9 May 1990.

sampled cultivation experiments and showed that while more stem bases remained on the surface in direct-drilled plots than in cultivated plots, a smaller percentage of them produced spores. In more recent experiments, stem bases with symptoms of eyespot were placed in mesh bags supported in large plastic funnels so that rain water passing over them could be collected. Results showed that mixing the infected stem bases with chopped straw greatly decreased sporulation (Fig. 17.4). However, other results cast some doubt on differences in sporulation being responsible for the differences in eyespot. In a 1 year winter wheat experiment at Rothamsted in 1990, which compared different straw treatments, pots containing healthy wheat seedlings were exposed at intervals in all plots between November 1989 and May 1990 to monitor inoculum of *P. herpotrichoides*. Differences in amounts of eyespot that developed on the seedlings exposed in individual test periods were mostly small and not significant. However, mean infection over the whole period of study was significantly less on seedlings exposed in plots where straw had been burnt than in those where it had not (Table 17.3). Plants sampled from the plots themselves showed no significant differences in the incidence of eyespot in spring but in summer there tended to be more disease, and significantly more straws with moderate or severe symptoms, where straw had been burnt than where it had not (Table 17.3). In the Rothamsted winter barley experiment, described above, burning the straw decreased eyespot in spring but not in summer. In the long-term wheat experiments, the straw treatments had similar effects on eyespot in both spring and summer but the effects were apparently smaller in spring than subsequently. Collectively, these results seem to suggest that burning does, perhaps, decrease the production of inoculum (as one might expect) and that straw has its effect, by whatever mechanism, at a relatively late stage in crop growth and disease development.

If the effects on eyespot of incorporating straw are a result of competition from other microorganisms growing on that straw then it is conceivable that the two common pathotypes (the R(rye) and the W(wheat) types; Hollins *et al.*, 1985) might not be equally sensitive. If so, changes in the relative frequencies of the two types might occur especially after repeatedly incorporating straw for some years. However, samples taken from one of the Rothamsted wheat experiments in 1989–1991 provided no evidence for differences in the relative frequencies of the two types as a consequence of different straw treatments (Table 17.4).

The different effects of the cultivation treatments on eyespot at Rothamsted in the first 2 years of the experiment and subsequently are, to some extent, understandable. Initially, ploughing would have buried infected debris and might, therefore, have been expected to decrease the disease. In later years it would still have been expected to bury the most recently-produced debris but would, at the same time, bring previously buried debris to the surface. However, if taken at face value, these later results for eyespot

Table 17.4. Mean frequencies of isolation, in summer 1989–1991, of the W and R types of *Pseudocercosporella herpotrichoides* from winter wheat grown where the straw of the previous crop had been burnt, baled or chopped and incorporated.[1]

	W type (% straws)	R type (% straws)
Straw		
Burnt	−0.25 (38.0)	+0.80 (83.1)
Baled	−0.48 (27.9)	+0.72 (80.8)
Incorporated	−0.46 (28.5)	+0.62 (77.5)
S.E.D. (4 d.f.)	0.118	0.056

[1]Data are logit transform values with, in parentheses, the corresponding percentage values obtained by back transformation.

at Rothamsted and, even more consistently, those for sharp eyespot, seem to suggest that debris that has been buried is a more potent source of inoculum of both *P. herpotrichoides* and *R. cerealis* than is fresh debris. We are not aware of any evidence to support this for these pathogens but there may be an interesting parallel in the case of *Gibellina cerealis* in which inoculum sources apparently show an increase in potency during periods of fallow lasting up to several years (Glynne *et al.*, 1985). Ploughing does, of course, bury debris apart from that bearing eyespot and sharp eyespot lesions, and, as mentioned earlier, it is conceivable that cultivations have their effect by altering the amounts of such debris that remain on the surface. The eyespot results from our Woburn experiment seem to support this hypothesis because ploughing on that site did not increase the disease where the straw had been burnt but tended to do so where the straw was incorporated. However, eyespot at Rothamsted and sharp eyespot at both Rothamsted and Woburn were increased by ploughing, compared to disc/tine cultivations, whether the straw was incorporated or burnt.

Effects on Take-all

In our winter wheat experiments, effects of the straw treatments on take-all (*Gaeumannomyces graminis* var. *tritici*) have generally been small. Nevertheless, when averaged over sites and years, the disease was, in summer, slightly, but significantly, more severe where straw had been incorporated than where it had been burnt (84.3 and 77.3% plants affected, respectively).

Effects of cultivation treatments were detected more frequently than effects of straw but they were different in different years so, when averaged over sites and years, there was little difference between plots that were

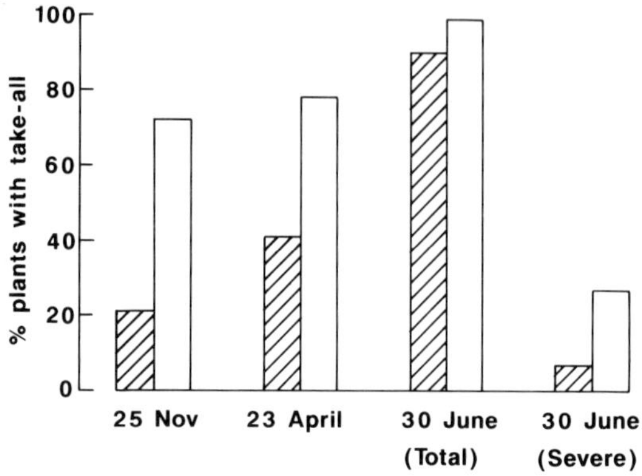

Fig. 17.5. Incidence of take-all in winter barley sown after ploughing (▨) or tine cultivations (□) at Whaddon in Buckinghamshire in 1986.

ploughed and those that were disc or tine cultivated. However, the different effects of the cultivation treatments did not occur randomly and they provide a good illustration of the dangers of reaching hasty conclusions and of the need for a long-term perspective in agricultural research. Thus in the first 3 years of the wheat experiments at Rothamsted and Woburn (1985–1987) and in the experiment at Whaddon in 1986, whenever an effect of ploughing was detected, it was to decrease the severity of take-all in comparison with amounts of disease in the tined plots. Effects were particularly large in the experiment at Whaddon (Fig. 17.5). Similar effects of ploughing have been seen in many of the straw incorporation trials done by ADAS (D.J. Yarham, personal communication). These decreases can probably be explained because inoculum of the take-all fungus is concentrated in the upper layers of the soil (Hornby, 1975). During ploughing this is buried and deeper, relatively less infective soil is brought to the surface. On the basis of available data up until the end of 1987, we tentatively concluded that straw residues *per se* were unlikely to have much effect on take-all. If, however, farmers ploughed-in straw instead of using shallow cultivations, there were likely to be useful decreases in the disease. Perhaps we should not have been surprised that in 1988 cultivations had the opposite effect! However, with the 7 years' data now available, the Rothamsted results at least do seem to make reasonable sense (Fig. 17.6). Thus, it now appears that ploughing did decrease the disease in 1986 and 1987 but this slowed the epidemic so that maximum take-all (and the subsequent, and expected, decline) occurred 1 year later in these plots than in those that were disced or tined. These results agree with others from Rothamsted which show that take-all decline is a consequence of take-all and not cropping history, and that severe disease is necessary for decline to occur. As emphasized above, it needed a relatively long run of data

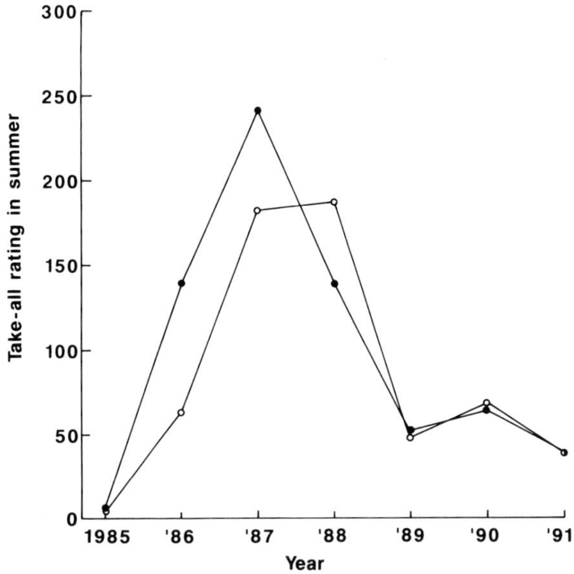

Fig. 17.6. Severity of take-all in continuous winter wheat sown after ploughing (○) or shallow cultivations (discs or tines; ●) at Rothamsted.

to demonstrate (and understand) the changing effects of cultivations on take-all. It also seems certain that they would not have been detected had we accumulated the same volume of data but from a series of experiments in successive years each grown on a different site.

The Rothamsted winter wheat data (Fig. 17.6) suggest that there is little effect of cultivations on take-all once take-all decline has established. However, data from the winter barley experiment (which followed many years of cereals and was well into the decline phase) tended to contradict this so further work is needed.

Conclusions

The experimental evidence so far obtained shows that the effects on cereal diseases of incorporating straw are generally much smaller than originally feared. Indeed there now seems little doubt that some diseases, especially eyespot and sharp eyespot, are quite consistently decreased by incorporating straw instead of burning it. However, for those diseases that tend to be increased or are little affected by the presence of straw, it is important to

Table 17.5. Diseases affecting the leaves (% leaf area affected) and stem bases (% straws) of winter wheat and winter barley, measured during grain filling, in crops where the straw of the previous crop was burnt or not burnt, and after 'reduced' cultivations or ploughing.[1]

	Straw		Cultivations	
	Burnt	Not burnt	'Reduced'	Plough
Wheat[2]				
Septoria spp.	6.9	5.7	5.8	6.5
Eyespot	4.2	3.5	3.8	3.8
Sharp eyespot	2.5	2.6	2.3	2.7
Barley[3]				
Leaf blotch	2.3	3.1	2.5	2.9
Net blotch	4.4	4.5	4.1	4.4
Eyespot	5.3	5.1	5.6	5.0

[1]Data, which are reproduced from Jenkyn *et al.* (1988), are derived from ADAS surveys.
[2]Mean for 1975–1987, excluding 1983 and 1984 when no surveys were done.
[3]Mean for 1981–1987, excluding 1984 and 1985 when no surveys were done.

Table 17.6. Percentages of winter barley crops grown where the straw of the previous crop had been burnt (total number = 435) or not burnt (total number = 870) that received different numbers of fungicide sprays and, in parentheses, the percentages that would have been predicted if there were no association between burning and spraying.[1]

	Number of fungicide sprays			
Straw	0	1	2	3
Burnt	13.1 (17.2)	47.8 (52.9)	32.6 (25.4)	6.4 (4.5)
Not burnt	19.2 (17.2)	55.4 (52.9)	21.8 (25.4)	3.6 (4.5)

[1]Data, which are reproduced from Jenkyn *et al.* (1988), are derived from ADAS surveys in 1981–1987, excluding 1984 and 1985.

consider whether results from plots, albeit relatively large plots in our experiments, are reliable. They might, for example, be affected by interplot interference (Jenkyn and Dyke, 1985), in which case larger effects might be predicted in commercial practice (i.e. farmers' fields) where interactions between differently treated crops should be relatively small.

To examine this possibility, Jenkyn *et al.* (1988) examined crop survey data collected by ADAS, and compared average amounts of disease in crops where straw of the previous crop had been burnt or not burnt, and where the soil had been prepared for sowing by ploughing or using reduced cultivations. Their data, which are summarized in Table 17.5, provide little evidence that the principal trash-borne diseases of wheat or barley are decreased by burning, with the possible exception of leaf blotch on barley. Indeed, *Septoria* spp. and eyespot on wheat both tended to be more severe where the straw had been burnt than where it had not. Cultivation systems also had mostly small effects on these diseases.

However, contrary to the general trends indicated above, farmers might conceivably have observed more disease where straw was incorporated and then responded by applying more fungicides. Reassuringly, however, data for winter barley from the same ADAS surveys provide no evidence for this practice (Table 17.6). Indeed the tendency was for more fungicide sprays to be applied to crops of both barley (Table 17.6) and wheat where the straw of the previous crop had been burnt than where it had not. These survey data are largely supported by anecdotal evidence from the few farmers who have been incorporating straw for many years, which generally confirms that diseases have not increased as a result.

There is, therefore, little if any evidence at present to suggest that diseases are an important factor limiting the success of straw incorporation systems. Most of our experiments are planned to continue for several more years, mainly to study the long-term effects of straw on nitrogen cycling, biomass and soil structure, but it is considered important to also continue studying diseases to guard against the possible emergence of 'new' problems that might develop after straw has been repeatedly incorporated for many years. Further work to explain the decreases in eyespot and sharp eyespot where straw is incorporated should improve our understanding of the ecology of the causal fungi.

References

Brooks, D.H. and Dawson, M.G. (1968) Influence of direct-drilling of winter wheat on incidence of take-all and eyespot. *Annals of Applied Biology* 61, 57–64.

Glynne, M.D., Fitt, B.D.L. and Hornby, D. (1985) *Gibellina cerealis*, an unusual pathogen of wheat. *Transactions of the British Mycological Society* 84, 653–659.

Hollins, T.W., Scott, P.R. and Paine, J.R. (1985) Morphology, benomyl resistance and pathogenicity to wheat and rye of isolates of *Pseudocercosporella herpotrichoides*. *Plant Pathology* 34, 369–379.

Hornby, D. (1975) Inoculum of the take-all fungus: nature, measurement, distribution and survival. *EPPO Bulletin* 5, 319–333.

Jenkyn, J.F. and Dyke, G.V. (1985) Interference between plots in experiments with plant pathogens. *Aspects of Applied Biology* 10, *Field Trials Methods and Data Handling*, 75–85.

Jenkyn, J.F. and King, J.E. (1977) Observations on the origins of *Septoria nodorum* infection of winter wheat. *Plant Pathology* 26, 153–160.

Jenkyn, J.F., Gutteridge, R.J. and Thomas, M.R. (1988) Effects of straw incorporation and cultivations on cereal diseases. *Aspects of Applied Biology* 17, *Environmental Aspects of Applied Biology*, 181–189.

Jenkyn, J.F., Stedman, O.J., Dyke, G.V. and Todd, A.D. (1989) Effects of straw inoculum and fungicides on leaf blotch (*Rhynchosporium secalis*), growth and yield of winter barley. *Journal of Agricultural Science* 112, 85–95.

Prew, R.D. (1976) Diseases in reduced cultivation systems. Winter wheat. *Report of the Rothamsted Experimental Station for 1975*, Part 1, pp. 256–257.

Prew, R.D. (1977) Diseases in reduced cultivation systems. Eyespot. *Report of the Rothamsted Experimental Station for 1976*, Part 1, p. 264.

Prew, R.D. and Lord, E.I. (1988) The straw incorporation problem. *Aspects of Applied Biology* 17, *Environmental Aspects of Applied Biology*, 163–171.

Prew, R.D. and Read, P.J. (1978) Diseases in reduced cultivation systems. Eyespot (*Pseudocercosporella herpotrichoides*) and straw burning. *Report of the Rothamsted Experimental Station for 1977*, Part 1, pp. 217–218.

Schulz, H., Bødker, L., Jørgensen, L.N. and Kristensen, K. (1990) Influence of different cultural practices on distribution and incidence of eyespot (*Pseudocercosporella herpotrichoides*) in winter rye and winter wheat. *Danish Journal of Plant and Soil Science* 94, 211–221.

Yarham, D.J. and Norton, J. (1981) Effects of cultivation methods on disease. In: Jenkyn, J.F. and Plumb, R.T. (eds), *Strategies for the Control of Cereal Disease*. Blackwell Scientific Publications, Oxford, pp. 157–166.

18 The Cereal *Fusarium* Complex

D.W. Parry, T.R. Pettitt, P. Jenkinson and A.K. Lees

Crop and Environment Research Centre, Harper Adams Agricultural College, Newport, Shropshire TF10 8NB, UK.

Introduction

Fusarium species are widely distributed both in temperate and tropical regions. They have been isolated from soil and plant material in Canada (Gordon, 1956), the USA (Cook, 1968, 1980), Europe (Maric, 1981; Cassini, 1981), Asia (Kelman and Cook, 1977), Africa (Tarr, 1962) and Australasia (Wearing and Burgess, 1977). Many of the species are saprophytes which survive on decaying aerial plant parts (Christensen and Kaufman, 1965), plant debris and other organic substrates (Booth, 1971). However, the greatest interest in the genus has been in their role as plant pathogens. Most genera of cultivated plants suffer diseases caused by *Fusarium* species. Several species of *Fusarium* are pathogenic to small grain cereals grown in temperate areas. Studies have been confined largely to *Fusarium* diseases of wheat and barley. On these crops *Fusarium* can infect either singly or as a complex of several species, causing disease at all stages of crop development. This chapter outlines *Fusarium* species affecting wheat and barley and discusses their interactions, both with each other and with their cereal host under the influence of environmental factors and fungicides.

The Pathogens

Smith (1884) made the earliest record of *Fusarium* disease of cereals in the UK. Later, Bennett (1935) isolated and identified 14 different species of

Fusarium from cereals throughout Britain. However, there are now recognized to be five *Fusarium* species which commonly cause disease in temperate cereal crops: *Fusarium culmorum*, *F. avenaceum* [*Gibberella avenacea*], *F. graminearum* [*G. zeae*], *F. poae* and *Microdochium nivale* [*Monographella nivalis*] formerly classified as *Fusarium nivale* (Mueller, 1977). Any of these species may be responsible for the three *Fusarium* diseases of cereals: seedling blight, fusarium foot rot and fusarium ear blight. An exception to this is *F. poae*, which although commonly isolated from ears, is rarely involved in either seedling blight or foot rot (Polley *et al.*, 1991).

In vitro studies of the growth of *Fusarium* species on osmotically adjusted agar under a range of temperatures (Cook and Christensen, 1975), showed that *F. graminearum*, *F. culmorum* and *F. avenaceum* grew optimally at low water potentials and at temperatures ranging between 20 and 30°C. English isolates of *F. culmorum* and *F. avenaceum* showed similar temperature optima for *in vitro* growth (Fig. 18.1). However, *M. nivale* preferred lower temperatures, being the fastest growing species at 5°C, with optimum growth between 15 and 20°C. These differences in temperature optima for growth are reflected in the relative distributions of *Fusarium* species. For example, *F. graminearum*, *F. culmorum* and *F. avenaceum* are species which have consistently been found to predominate in warmer, drier cereal growing

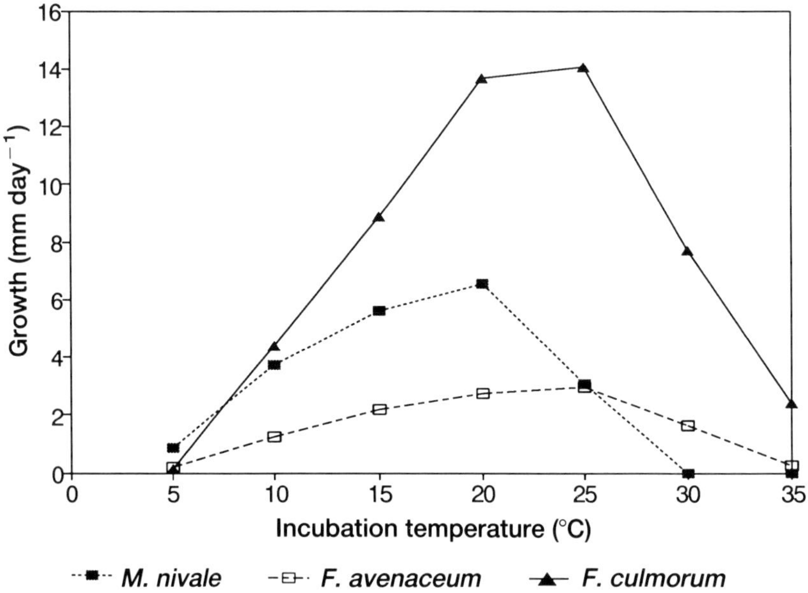

Fig. 18.1. Effect of temperature on the growth of *Fusarium avenaceum*, *F. culmorum* and *Microdochium nivale* on Potato Dextrose Agar.

areas such as exist in the USA (Cook, 1968) and Australia (Burgess *et al.*, 1975), whereas *M. nivale* is more commonly isolated from cereals grown in regions with a cooler, wetter growing season such as Scotland (Rennie *et al.*, 1983; Polley *et al.*, 1991).

Disease Cycle

In order to understand the succession of *Fusarium* diseases of cereals, it is helpful to consider the disease cycle as a whole (Fig. 18.2). Central to the disease cycle is the survival of *Fusarium* inoculum. The strategies for survival of the five cereal fusaria are given in Table 18.1. All of the species listed in Table 18.1 are capable of saprophytic colonization of crop debris. The ban on the practice of stubble burning introduced during 1992 in the UK may therefore lead to an increase in inoculum of these species on crop debris (J.F. Jenkyn, Rothamsted, 1992, personal communication and Chapter 17).

Although the presentation of a simplified disease cycle aids the understanding of the interaction between the *Fusarium* diseases in cereals, it does not take into account any possible antagonism or synergism that may occur

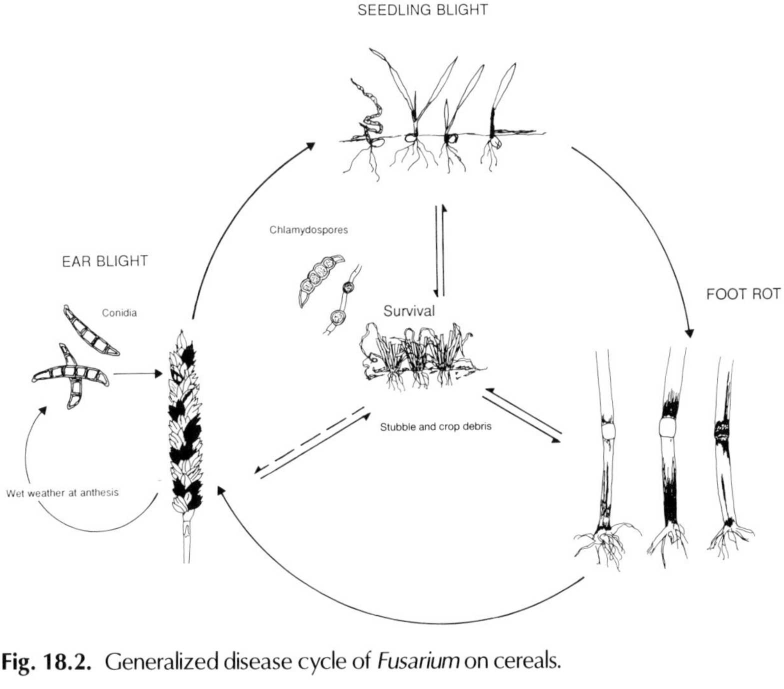

Fig. 18.2. Generalized disease cycle of *Fusarium* on cereals.

Table 18.1. Survival strategies employed by the four main *Fusarium* species and *Microdochium nivale* affecting temperate cereals.

Fusarium spp.		Survival strategy	Reference
F. avenaceum	(i)	Resting spores	Hargreaves and Fox (1977)
	(ii)	Saprophytic mycelium on crop debris	Snyder and Nash (1968)
	(iii)	Contaminated seed	Rennie *et al.* (1983) Hewett (1967) Polley *et al.* (1991)
F. culmorum	(i)	Resting spores (chlamydospores)	Snyder and Nash (1968) Cook (1968)
	(ii)	Saprophytic mycelium on crop debris	Cook (1980)
	(iii)	Contaminated seed	Rennie *et al.* (1983) Hewett (1967) Polley *et al.* (1991)
F. graminearum	(i)	Resting spores (chlamydospores) in soil	Sitton and Cook (1981)
	(ii)	Saprophytic mycelium on crop debris	Cook (1980)
	(iii)	Contaminated seed	Duthie and Hall (1987)
M. nivale	(i)	Saprophytic mycelium on crop debris	Booth and Taylor (1976) Bruehl *et al.* (1966)
	(ii)	Contaminated seed	Rennie *et al.* (1983) Hewett (1967) Polley *et al.* (1991)
F. poae	(i)	Contaminated seed	Polley *et al.* (1991) Hewett (1967)

between the species involved. There is also a break in our understanding of the cycle between the development of foot rot and the onset of fusarium ear blight epidemics. The source of inoculum for the latter disease is uncertain, although a strong case has been presented for the role of rain-splash dispersal of spores from sporodochia found on either crop debris or plant stem-bases (Jenkinson and Parry, 1994). The following sections describe some of the characteristics of temperate cereal *Fusarium* diseases and discuss recent epidemiological work pertinent to a fuller understanding of the disease cycle.

Seedling Blight

Fusarium seedling blight arises mainly as a result of seed-borne inoculum although soil-borne inoculum can also be responsible for the disease. Symptoms range from pre- and post-emergence death of seedlings to browning of the coleoptile and the development of superficial stem lesions on the emerged plants. In some cases, lens-shaped pale brown lesions can be seen on the first and second leaves. Often the first leaf takes on a shredded appearance or may be seen lying on the soil surface. *Microdochium nivale* has been recorded as the predominant species responsible for seedling blight of winter wheat and spring barley in Scotland (Rennie *et al.*, 1983). In the USA Warren and Kommedahl (1972) demonstrated that *F. graminearum* was the predominant species infecting wheat seedlings although *F. culmorum* and *F. avenaceum* were also isolated.

Environmental conditions have been shown to have a profound effect on the development of fusarium seedling blight. When caused by *M. nivale* this disease is favoured by cooler soil temperatures (Millar and Colhoun, 1969) whereas warmer, drier soils have been shown to be conducive to cereal seedling blights caused by *F. graminearum*, *F. culmorum* and *F. avenaceum* (Dickson, 1923; Colhoun and Park, 1964; Colhoun *et al.*, 1968). Similar results were obtained during recent work on winter wheat seedlings (cv. Mercia) (Table 18.2). After artificially inoculating seed with a conidial suspension of *F. culmorum*, seedling blight was most severe under warm (20°C) dry (–2.2 bar) soil conditions with up to 73% pre-emergence seedling death. Of those seedlings surviving, 63% were infected with *F. culmorum* of which nearly all exhibited visual symptoms (only 4% symptomless infection). In contrast, under cool (10°C) wet (–1.0 bar) soil conditions, there was only 24% pre-emergence seedling death. Of the surviving seedlings, 51% were infected but there were fewer seedlings with visual symptoms under the

Table 18.2. Infection in winter wheat seedlings cv. Mercia with *F. culmorum* in relation to temperature and soil water potential.

Water potential (bars)	Percentage pre-emergence death		Percentage infection in emerged seedlings		Percentage symptomless infection	
	10°C	20°C	10°C	20°C	10°C	20°C
−2.2 ('dry')	52	73	63	63	19	4
−1.6 ('medium')	39	52	48	56	25	23
−1.0 ('wet')	24	53	51	79	27	30

'cool wet' regime as indicated by the higher percentage of symptomless infections (27%). Such observations using artificially inoculated seed highlight the possibility of symptomless infections of cereal seedlings by *F. culmorum* in the field especially under cool wet soil conditions. Symptomless infections have often been detected in the stem-bases of cereal crops (Parry, 1990) and it appears possible that many of these infections are initiated at the seedling stage, perhaps only developing symptoms at a later stage of growth under the appropriate environmental conditions, for example drought stress (Papendick and Cook, 1974; Cook, 1980).

Foot Rot

Symptoms of foot rot on wheat and barley may first develop during spring. A brown decay of the lowest leaf sheaths occurs sometimes accompanied by a shredded outer leaf sheath which may become blackened and fall onto the soil surface. Often in these early attacks, the principal pathogen is *M. nivale* and following stem elongation, dark perithecia may appear in the brown areas of the lowest leaf sheaths. Later in the season, the incidence of *F. avenaceum* and *F. culmorum* increases. Stem-base infections, usually associated with the bottom three nodes of the plant, often become more apparent as the crop approaches anthesis. These are visible as a general greyish/brown discoloration which may girdle the lowest internode above the crown roots, or as vertical dark brown or black streaking or blotching (mottling) of the higher internodes. Browning of nodes is also common (Fig. 18.3). In severe cases, stem-bases may become rotten, often bearing the pink mycelium of the fungus. Such severely affected stems may break near the ground causing lodging of the crop or may lead to premature death of tillers and the development of sterile ears or 'whiteheads'.

Of the species responsible for foot rot of cereals in the UK, *M. nivale* has generally been shown to be the most common (Rennie *et al.*, 1983; Locke *et al.*, 1987; Parry, 1990; Polley *et al.*, 1991) although *F. culmorum* and *F. avenaceum* have also been isolated. In the warmer climates of Australia and the USA, *F. graminearum* and *F. culmorum* are the species most frequently responsible for fusarium foot rots of cereals (Burgess *et al.*, 1975; Cook, 1968, 1980).

In cereal stem-bases, *Fusarium* species are frequently found in multiple infections in association with each other and with other stem-base pathogens. The precise role of individual species in the expression of foot rot symptoms is not clearly understood, with symptoms produced by individual species being virtually indistinguishable. A succession of *Fusarium* species infecting cereal stem-bases through the growing season has been observed by several

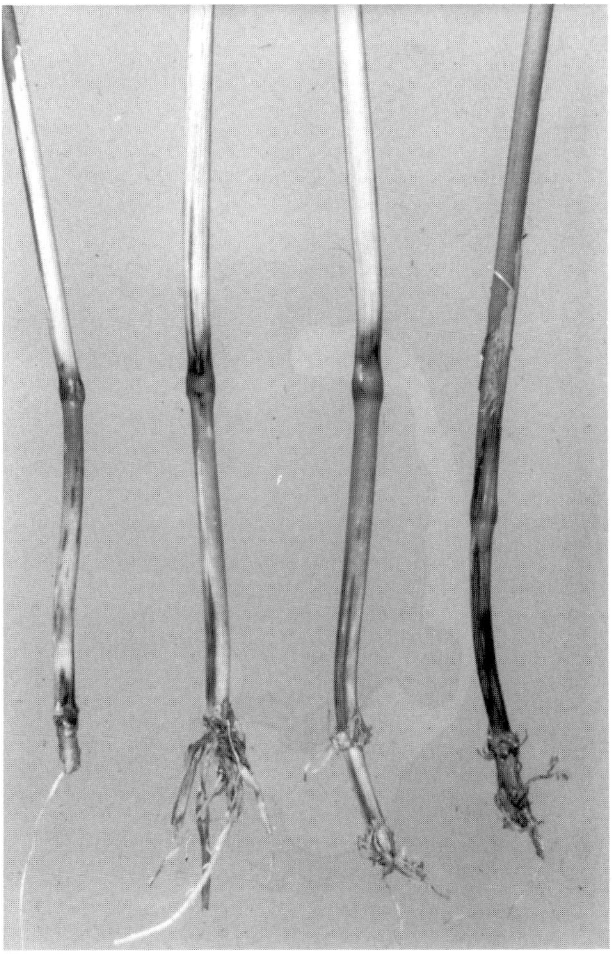

Fig. 18.3. Symptoms of fusarium foot rot of wheat.

workers (Duben and Fehrmann, 1979; Parry, 1990). Winter wheat crops in the West Midlands of England monitored over three growing seasons (1987–1989) showed marked fluctuations in the populations of individual *Fusarium* species infecting stem-bases (Parry, 1990). The incidence of *Fusarium* species varied not only from year to year but also within a single growing season. Figure 18.4 illustrates the results obtained for the incidence of *Fusarium* species isolated from stem-bases of two crops of the winter wheat cultivar Avalon. Crop (a) was harvested in 1987. The winter of 1986/87

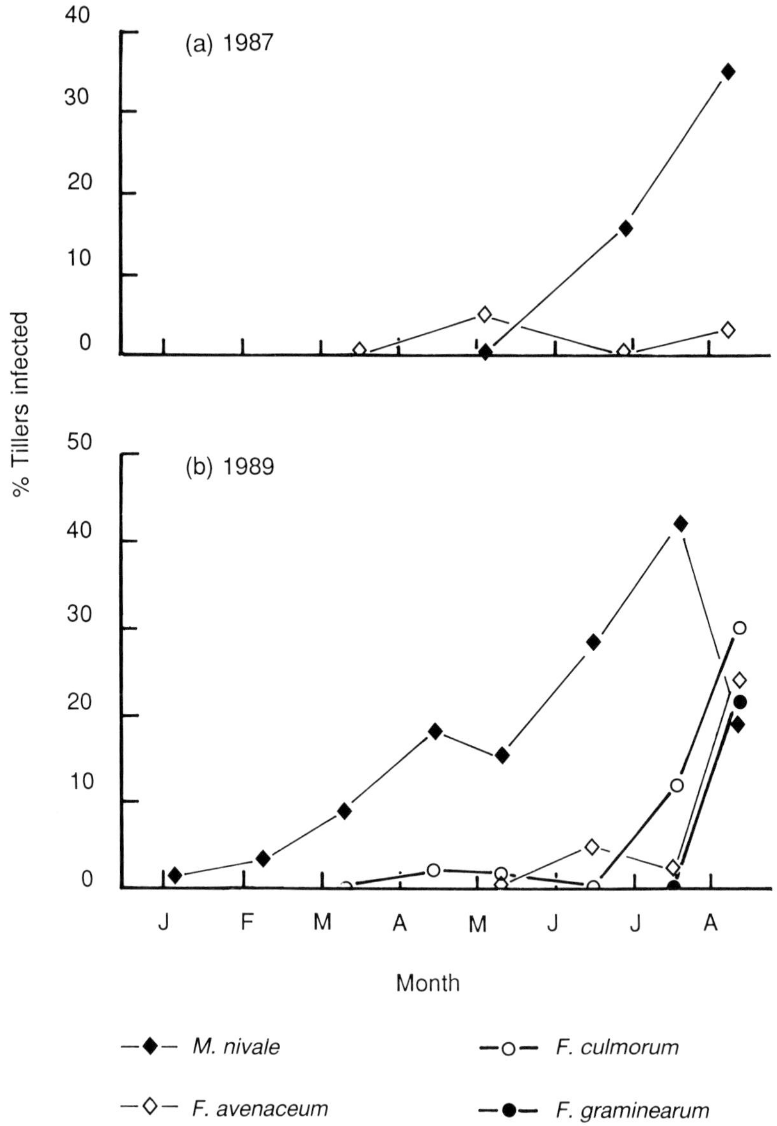

Fig. 18.4. Incidence of *Fusarium* species in stem-bases of winter wheat cv. Avalon. (After Parry, 1990.)

was particularly severe and was followed by a late spring. *Fusarium* species were not isolated from this crop to any great extent until early summer. In contrast, crop (b) was harvested in 1989. This growing season consisted of a mild winter followed by a warm summer and *M. nivale* was isolated throughout the winter months. During the warm summer of 1989, several

Table 18.3. Summer 1992 survey of *Fusarium* spp. in winter wheat stem bases using the isolation procedures of Pettitt *et al.* (1993).

Species isolated	Percentage of tillers infected
Fusarium culmorum	55.6
Microdochium nivale	34.0
F. avenaceum	9.9
F. poae	1.2
F. graminearum	0.2
Other *Fusarium* spp.	1.2
Mixtures of *F. culmorum* and *M. nivale*	12.2

Fusarium species were isolated which were not detected during 1987 including *F. culmorum* and *F. graminearum*. As the incidence of these 'warm-loving' species increased at the end of the growing season, a concurrent decline in the incidence of *M. nivale* was observed. Such a decline in *M. nivale* would suggest that this species may be out-competed by the warm-loving species under warmer, drier environmental conditions. However, recent work suggests that such an apparent decline in *M. nivale* may be a result of the isolation procedure used. The usual procedure for the isolation of cereal fusaria involves the use of either potato dextrose agar (PDA) or potato sucrose agar (PSA) as isolation media. However, on these media, *M. nivale* could be out-grown by other faster growing species. The relatively slow growth rate of *M. nivale* compared to *F. culmorum* was clearly demonstrated in Fig. 18.1. The incorporation of the benzimidazole fungicide benomyl (10 $\mu g\ ml^{-1}$) into these media inhibited the growth of *F. culmorum* and to a lesser extent *F. avenaceum* without inhibiting the growth of *M. nivale* (Pettitt *et al.*, 1993). This technique made use of the now widespread development of benzimidazole resistance in UK populations of *M. nivale* (Locke *et al.*, 1987). The usefulness of the benomyl-amended isolation media was evident in a recent survey carried out as an extension of the Agricultural Development and Advisory Service (ADAS) national cereal disease survey in collaboration with R.W. Polley at the UK Ministry of Agriculture, Fisheries and Food (MAFF), Central Science Laboratory. When recording the incidence of *Fusarium* in 3600 winter wheat tiller samples taken from 144 sites throughout England and Wales, a significant proportion of stem-bases that would only have shown *F. culmorum* infection using either PDA or PSA were shown to be infected by both *M. nivale* and *F. culmorum* when plated out onto the benomyl-amended medium (Table 18.3). The survey also revealed that *F. culmorum* was the predominant species isolated in many areas along with *M. nivale*. The incidence of *F. avenaceum* and *F. graminearum* was lower than would be expected in what was the third in a succession of comparatively warm years. When considered on a regional basis, the relative incidence of these *Fusarium* species appeared to be correlated with local temperatures and moisture conditions although this still remains to be fully assessed and will be published elsewhere.

Ear Blight

Early symptoms of fusarium ear blight consist of small brown water-soaked spots on the outer glumes. Under favourable conditions, florets may become infected or more usually the whole spikelet. The affected tissues lose their chlorophyll and take on a bleached straw-coloured appearance at a stage when unaffected heads are still quite green. In warm humid weather conditions, pinkish mycelium and conidia develop abundantly in infected spikelets and the infection spreads to adjacent spikelets or through the entire head. Purplish coloured perithecia may also develop on the bracts infected with *M. nivale*. With all species, the infected kernels become shrivelled and discoloured with white, pink or brown mycelial outgrowths.

The relative importance of each of the *Fusarium* species responsible for ear blight varies from country to country. In the USA, Wilcoxson *et al.* (1988) showed that *F. graminearum* was the predominant species accounting for up to 75% of *Fusarium* isolates collected from spring wheat throughout Minnesota. *Fusarium poae* accounted for 17% of isolations. In the UK, *F. poae* was considered by Polley *et al.* (1991) to be the predominant species responsible for ear blight. Previously, *F. culmorum* and *F. avenaceum* were considered as the predominant species on wheat (Parry *et al.*, 1985).

Fusarium ear blight and scab have been regarded as being of major importance throughout the world, especially during wet growing seasons (Martin and Johnston, 1982; Sutton, 1982). Contaminated grain resulting from fusarium ear blight can provide a primary source of inoculum for the development of seedling blight and foot rot (Hewett, 1983; Rennie *et al.*, 1983; Duthie and Hall, 1987). Also, if used as animal feed, contaminated cereal grain can, through the production of mycotoxins, produce adverse side effects in livestock (Hoerr *et al.*, 1982; Long *et al.*, 1982).

The source and dispersal of inoculum for the development of fusarium ear blight is not fully understood. The development of fusarium foot rot on the stem-bases of cereal plants, along with the presence of contaminated stubble, would appear to provide an obvious source of inoculum. However, how this inoculum is spread to the ears is unclear. Although transmission of inocula via arthropod vectors (Leach, 1940; Hardison, 1961) and systemic

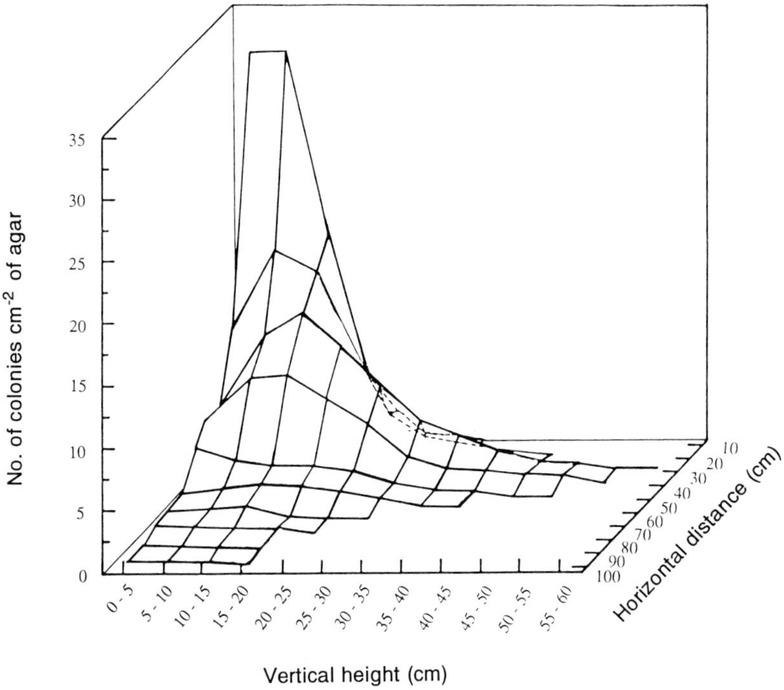

Fig. 18.5. Mean number of *Fusarium culmorum* colonies cm^{-2} of selective agar at various vertical heights (x – variable) and horizontal distances (z – variable) from the source following splash dispersal of conidia by simulated rain-drops, 5 mm in diameter, released from a height of 6 m. (From Jenkinson and Parry, 1994.)

growth of *Fusarium* from infected stem-bases to ears (Snijders, 1990; Hutcheon and Jordan, 1992) have been proposed, much of the evidence documented suggests that rain plays an important role in the dispersal of *Fusarium* inocula. Baker (1972) for example, whilst reviewing crop diseases of England and Wales between 1957 and 1968, noted that fusarium ear blights were a feature of wet growing seasons. Similarly, in Canada, Sutton (1982) found that epidemics of wheat ear blight caused by *F. graminearum* were associated with growing seasons which had a high annual rainfall. During a survey of fusarium ear blight in irrigated and non-irrigated wheat crops of the central Washington area of USA, Strausbaugh and Maloy (1986) found that those crops receiving overhead irrigation had a far higher incidence of *Fusarium*-infected ears (89% of ears infected) than crops which did not receive irrigation (no ears infected). Such observations indicate that rain plays an important role either in the dispersal of *Fusarium* inoculum to ears and/or in the initiation of ear blight disease. Jenkinson and Parry (1993) evaluated

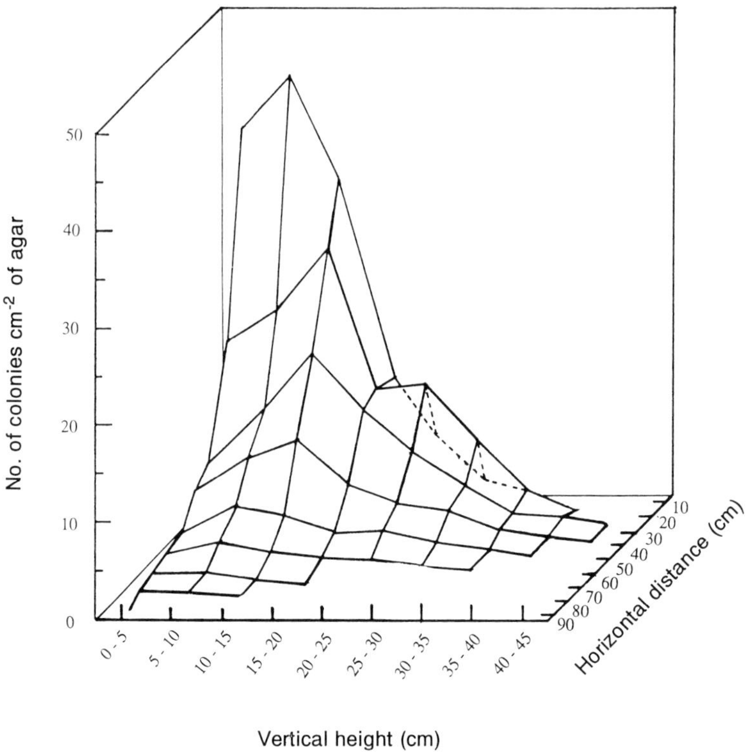

Fig. 18.6. Mean number of *Fusarium avenaceum* colonies cm^{-2} of selective agar at various vertical heights (x – variable) and horizontal distances (z – variable) from the source following splash dispersal of conidia by simulated raindrops, 5 mm in diameter, released from a height of 6 m. (From Jenkinson and Parry, 1994.)

the importance of splash droplets on the dispersal of conidia of *F. culmorum* and *F. avenaceum*. Splash dispersal patterns were studied in still air by allowing simulated raindrops (5 mm diam.) to fall on to stem-bases bearing sporodochia. The resulting splash-dispersed conidia were collected and grown on vertical strips of selective agar (100 cm × 3 cm) placed at distances ranging between 10 cm and 100 cm from the source. Microscopic examination of the selective agar strips revealed that, although the majority of *F. culmorum* and *F. avenaceum* conidia were collected within close proximity to the source (i.e. within a height of 20 cm and a horizontal distance of 30 cm), colonies of *F. culmorum* and *F. avenaceum* arising from dispersed conidia were found growing at maximum heights of 60 cm and 45 cm respectively and at maximum horizontal distances of >100 cm and 90 cm respectively (Figs 18.5 and 18.6). These results are consistent with splash

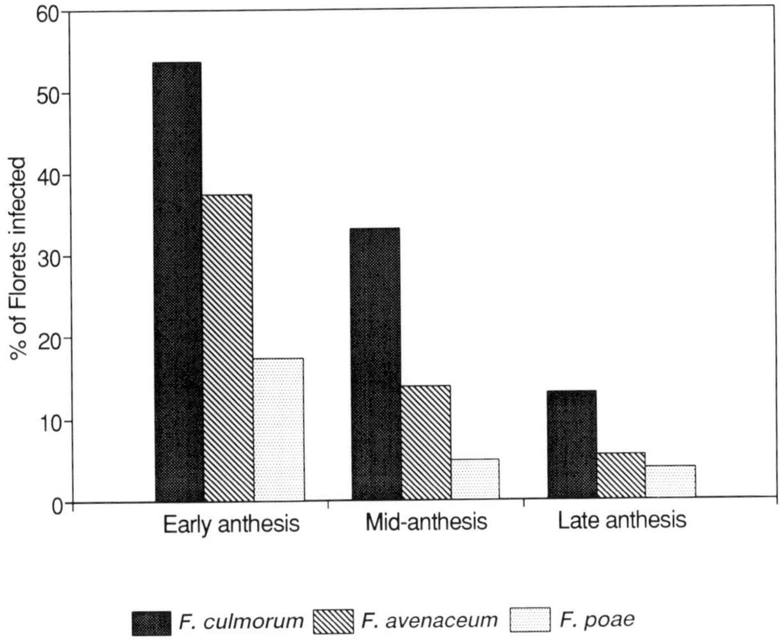

Fig. 18.7. Effect of growth stage of winter wheat cv. Longbow on the development of fusarium ear blight when artificially inoculated with *Fusarium avenaceum, F. culmorum* and *F. poae*.

dispersal patterns recorded for conidia of other pathogen species such as *Septoria nodorum* (Griffiths and Ao, 1975), *Rhynchosporium secalis* (Stedman, 1980; Fitt *et al*., 1988) and *Pseudocercosporella herpotrichoides* (Glynne, 1953; Fitt and Bainbridge, 1983). Our results clearly demonstrate the potential for *Fusarium* conidia to be dispersed in rain-splashed droplets. However, whether these conidia are dispersed directly to the ears of cereals or whether they are dispersed indirectly to the ears in a series of 'leaps' involving the previous infection of upper plant parts, such as flag leaves, remains uncertain.

The stage of host maturity has been shown to be a factor of considerable importance regarding the incidence and severity of fusarium ear blight. Both Pugh *et al*. (1933) and Anderson (1948) found that wheat ears became more susceptible to infection by *F. graminearum* as plants matured from anthesis. This was later shown to be due to the effect of anthers (McKay and Loughnane, 1945; Strange and Smith, 1971) and more specifically due to the presence of the fungal growth stimulants betaine and choline isolated from anther exudates (Strange and Smith, 1978). Our studies involving the

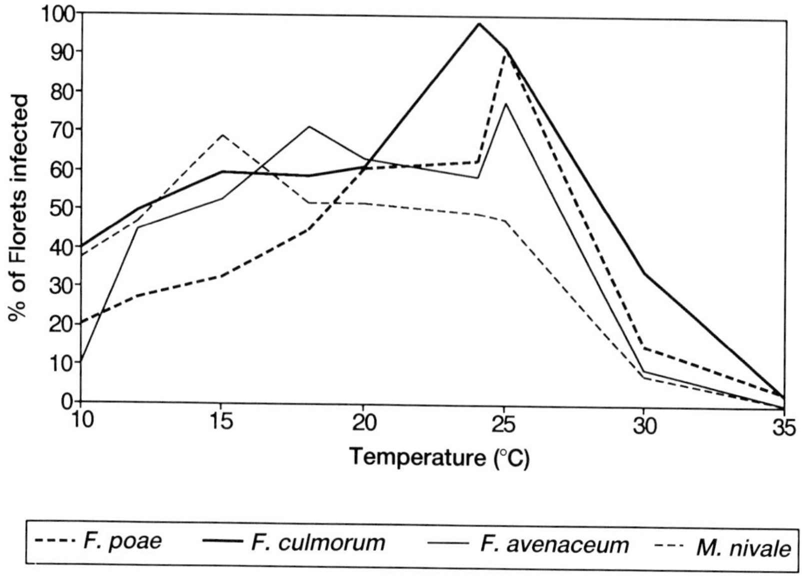

Fig. 18.8. Effect of temperature on the development of fusarium ear blight in winter wheat cv. Avalon when artificially inoculated with *Fusarium avenaceum, F. culmorum, F. poae* and *Microdochium nivale.*

inoculation of winter wheat ears (cv. Avalon) at a range of growth stages from early anthesis through to late anthesis showed that ears were most susceptible to infection by *F. culmorum*, *F. avenaceum* and *F. poae* during early anthesis (Fig. 18.7). As plants matured, ears became more resistant to infection. Further epidemiological studies into the effect of temperature on the development of ear blight showed that wheat ears were most susceptible to infection by either *F. culmorum*, *F. avenaceum* or *F. poae* at 25°C (Fig. 18.8). Infection of ears by *M. nivale* was optimum at the lower temperature of 15°C which may account for the higher incidence of this species isolated from seed grown in Scotland where ambient temperatures are lower (Polley *et al.*, 1991).

Fungicidal Control

There is no substantial evidence for effective and consistent control of stem base diseases of cereals caused by *Fusarium* using any fungicides currently available. Despite the lack of field performance, several workers including

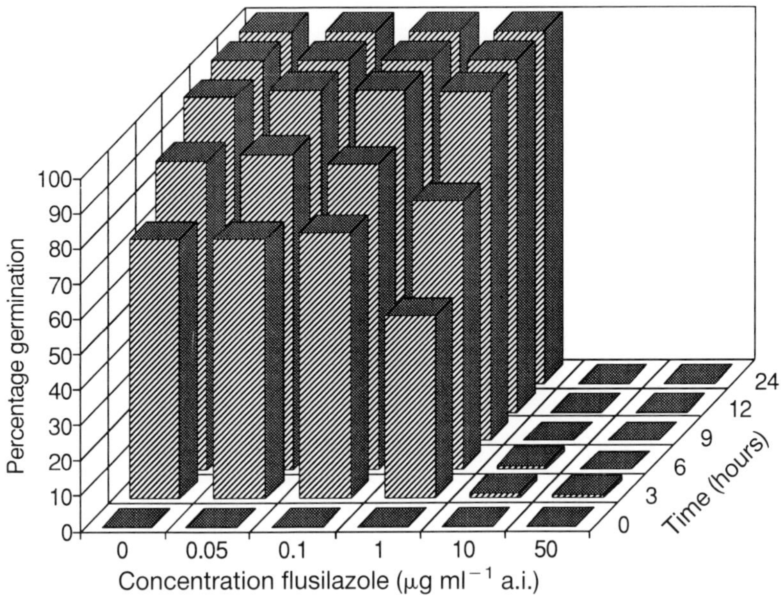

Fig. 18.9. Effect of varying concentration of flusilazole on the germination of macroconidia of *Microdochium nivale*.

Polley *et al.* (1991) have demonstrated *in vitro* inhibition of several cereal fusaria by ergosterol biosynthesis inhibiting fungicides including the azole flusilazole and the imidazole prochloraz. One of the possible explanations for the discrepancy between *in vitro* activity and field performance could be the differential response of each *Fusarium* species in the disease complex to any particular fungicide. Results of spore germination tests involving *M. nivale* and *F. culmorum* and an azole fungicide flusilazole clearly demonstrated the differential response of the *Fusarium* species to the fungicide *in vitro* (Figs 18.9 and 18.10). A concentration of 10 μg ml^{-1} flusilazole almost completely inhibited germination of macroconidia of *M. nivale* whereas the same concentration had little effect on the germination of macroconidia of *F. culmorum*.

A further documented example of the differential effects of fungicides on *Fusarium* species is that of benzimidazole resistance. In a recent survey of populations of fusaria isolated from stem bases of winter wheat, Pettitt *et al.* (1993) showed that between 88 and 100% of *M. nivale* isolates from different regions of England and Wales were resistant to benomyl at a concentration of 10 μg ml^{-1} in nutrient agar media. Resistance to benomyl was very rare in populations of *F. culmorum* and *F. avenaceum*.

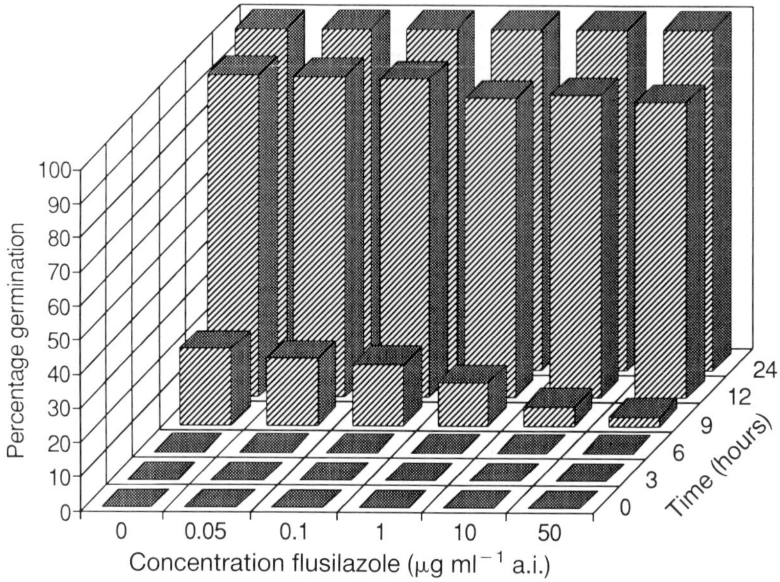

Fig. 18.10. Effect of varying concentration of flusilazole on the germination of macroconidia of *Fusarium culmorum*.

It is apparent that a combination of inherent differences in efficacy of fungicides against cereal fusaria, together with differences in fungicide resistance make predictions of field performance difficult. It is possible that the application of a particular fungicide to control fusarium foot rot results in a shift in the balance of the predominant species in the disease complex rather than a reduction in all of the *Fusarium* species involved and a concomitant reduction in disease severity.

Conclusions

It is clear from this brief review of past and current work on the cereal *Fusarium* complex that there are many areas of understanding which are lacking. These need to be addressed as it is possible that *Fusarium* diseases of cereals will assume a greater significance in years to come. If the current trend of increasing climatic warming (Hume and Cattle, 1990) continues, it is likely that the predominant species involved in the *Fusarium* disease complex in the UK will change to warm-loving and more pathogenic species such as *F. culmorum* and *F. graminearum*. An increase in trash-borne

Fusarium inoculum as a result of the recent ban on stubble burning together with uncertainties in fungicidal control (including the recent UK ban on organo-mercurial seed treatments) could result in an upsurge in all phases of *Fusarium* diseases of cereals.

The limitations of routine isolation techniques on artificial growth media together with recurrent problems experienced with classical *Fusarium* taxonomy have led to the employment of more precise molecular tools for the diagnosis and taxonomy of *Fusarium* species. Preliminary work has been started on the molecular taxonomy of cereal fusaria (Nicholson *et al.*, 1993). Molecular methods will undoubtedly play a part in the future by giving greater sensitivity to detection of *Fusarium* diseases in the field and species differentiation.

Acknowledgements

We acknowledge the assistance of Dr Tom Preece during this work and in the preparation of this manuscript. We also acknowledge the following bodies for funding: Ministry of Agriculture Fisheries and Food (T.R.P.); Science and Engineering Research Council and Schering Agrochemicals (P.J.); Home Grown Cereals Authority; and DuPont (A.K.L.).

References

Andersen, A.L. (1948) The development of *Gibberella zeae* head blight of wheat. *Phytopathology* 38, 595–611.

Baker, J.J. (1972) Report on diseases of cultivated plants in England and Wales for the years 1957–1968. In: *Technical Bulletin* 25, 9–19. Ministry of Agriculture, Fisheries and Food, London.

Bennett, F.T. (1935) *Fusarium* species on British cereals. *Annals of Applied Biology* 22, 479–501.

Booth, C. (1971) *The Genus Fusarium.* Commonwealth Agricultural Bureaux, London.

Booth, R.H. and Taylor, G.S. (1976) *Fusarium* diseases of cereals XI. Growth and saprophytic activity of *Fusarium nivale* in soil. *Transactions of the British Mycological Society* 66, 77–83.

Bruehl, G.W., Sprague, R., Fischer, W.R., Nagamitsu, M., Nelson, W.L. and Vogel, O.A. (1966) Snow mould of winter wheat in Washington. *Bulletin of the Washington Agricultural Experimental Station*, 677.

Burgess, L.W., Wearing, A.H. and Toussoun, T.A. (1975) Surveys of fusaria associated with crown rot of wheat in eastern Australia. *Australian Journal of Agricultural Research* 26, 791–799.

Cassini, R. (1981) *Fusarium* diseases of cereals in Western Europe. In: Nelson, P.E., Toussoun, T.A. and Cook, R.J. (eds), *Fusarium: Diseases, Biology and Taxonomy.* The Pennsylvania State University Press, University Park and London, pp. 58–63.

Christensen, C.M. and Kaufman, H.H. (1965) Deterioration of stored grain by fungi. *Annual Review of Phytopathology* 3, 69–84.

Colhoun, J. and Park, D. (1964) *Fusarium* diseases of cereals I. Infection of wheat plants, with particular reference to the effect of soil moisture and temperature on seedling infection. *Transactions of the British Mycological Society* 47, 559–572.

Colhoun, J., Taylor, G.S. and Tomlinson, R. (1968) *Fusarium* diseases of cereals II. Infection of seedlings of *F. culmorum* and *F. avenaceum* in relation to environmental factors. *Transactions of the British Mycological Society* 51, 397–404.

Cook, R.J. (1968) Fusarium root and foot rot of cereals in the Pacific Northwest. *Phytopathology* 58, 127–131.

Cook, R.J. (1980) Fusarium foot rot of wheat and its control in the Pacific Northwest. *Plant Disease* 64, 1061–1066.

Cook, R.J. and Christensen, A.A. (1975) Growth of cereal root rot fungi as affected by temperature-water potential interactions. *Phytopathology* 66, 193–197.

Dickson, J.G. (1923) Influence of soil temperature and moisture on the development of seedling blight of wheat and corn caused by *Gibberella saubinetii*. *Journal of Agricultural Research* 23, 837–869.

Duben, J. and Fehrmann, H. (1979) Vorkommen und Pathogenität von *Fusarium*-Arten am Winterweizen in der Bundesrepublik Deutschland. I Artenspektrum und jahreszeitliche Sukzession an der Halmbasis. *Zeitschrift für Pflanzenkrankheiten und Pflanzenschutz* 86, 638–652.

Duthie, J.A. and Hall, R. (1987) Transmission of *Fusarium graminearum* from seed to stems of winter wheat. *Plant Pathology* 36, 33–37.

Fitt, B.D.L. and Bainbridge, A. (1983) Dispersal of *Pseudocercosporella herpotrichoides* spores from infected wheat straw. *Phytopathologische Zeitschrift* 106, 214–225.

Fitt, B.D.L., McCartney, H.A., Creighton, N.F., Lacey, M.E. and Walklate, P.J. (1988) Dispersal of *Rhynchosporium secalis* conidia from infected barley leaves or straw by simulated rain. *Annals of Applied Biology* 112, 49–59.

Glynne, M.D. (1953) Production of spores by *Cercosporella herpotrichoides*. *Transactions of the British Mycological Society* 36, 46–51.

Gordon, W.L. (1956) The occurrence of *Fusarium* species in Canada V. Taxonomy and geographic distribution of *Fusarium* species in soil. *Canadian Journal of Botany* 34, 833–846.

Griffiths, E. and Ao, H.C. (1975) Dispersal of *Septoria nodorum* spores and spread of glume blotch of wheat in the field. *Transactions of the British Mycological Society* 67, 413–418.

Hardison, J.R. (1961) Evidence against *Fusarium poae* and *Sitroptes graminum* as causal agents of silver top of grasses. *Mycologia* 51, 712–718.

Hargreaves, A.J. and Fox, R.A. (1977) Survival of *Fusarium avenaceum* in soil. *Transactions of the British Mycological Society* 69, 425–428.

Hewett, P.D. (1967) A survey of seed-borne fungi of wheat II. The incidence of common species of *Fusarium*. *Transactions of the British Mycological Society* 50, 175–182.

Hewett, P.D. (1983) Seed-borne *Gerlachia nivalis* (*Fusarium nivale*) and reduced establishment of winter wheat. *Transactions of the British Mycological Society* 80, 185–186.

Hoerr, R.J., Carlton, W.W., Yagen, B. and Joffe, A.Z. (1982) Mycotoxicosis produced in broiler chickens by multiple doses of either T-2 toxin or diacetoxyscirpenol. *Avian Pathology* 11, 369–383.

Hume, C.J. and Cattle, H. (1990) The greenhouse effect – meteorological mechanisms and models. *Outlook on Agriculture* 19, 17–23.

Hutcheon, J.A. and Jordan, V.W.L. (1992) Fungicide timing and performance for *Fusarium* control in wheat. *Proceedings of the British Crop Protection Conference – Pests and Diseases*, Vol. 2, 633–638.

Jenkinson, P. and Parry, D.W. (1994) Splash dispersal of conidia of *Fusarium culmorum* and *Fusarium avenaceum*. *Mycological Research* (in press).

Kelman, A. and Cook, R.J. (1977) Plant pathology in The Peoples' Republic of China. *Annual Review of Phytopathology* 17, 409–429.

Leach, J.G. (1940) *Insect Transmission of Plant Diseases XVIII*. McGraw-Hill, London.

Locke, T., Moon, L.M. and Evans, J. (1987) Survey of benomyl resistance in *Fusarium* species in winter wheat in England and Wales in 1986. *Plant Pathology* 39, 619–622.

Long, G.G., Diekman, M., Tuite, J.F., Shannon, G.M. and Vesonder, R.F. (1982) Effects of *Fusarium roseum* corn culture containing zearalenone on early pregnancy in swine. *American Journal of Veterinary Research* 43, 1599–1603.

Maric, A. (1981) *Fusarium* diseases of wheat and corn in eastern Europe and the Soviet Union. In: Nelson, P.E., Toussoun, T.A. and Cook, R.J. (eds) *Fusarium: Diseases, Biology and Taxonomy*. The Pennsylvania State University Press, University Park and London, pp. 77–93.

Martin, R.A. and Johnston, H.W. (1982) Effects and control of cereal grains in the Atlantic Provinces. *Canadian Journal of Plant Pathology* 4, 210–216.

McKay, R. and Loughnane, J.B. (1945) Observations on *Gibberella saubinetii* (Mont.) Sacc. on cereals in Ireland in 1943 and 1944. *Scientific Procedures of the Royal Dublin Society* 24, 9–18.

Millar, C.S. and Colhoun, J. (1969) *Fusarium* diseases in cereals VI. Epidemiology of *Fusarium nivale* on wheat. *Transactions of the British Mycological Society* 52, 195–204.

Mueller, E. (1977) Die systematische stellung des "Schneeschmimmels". *Revue de Mycologie* 41, 129–134.

Nicholson, P., Jenkinson, P., Rezanoor, H.N. and Parry D.W. (1993) Restriction fragment length polymorphism analysis of variation in *Fusarium* species causing ear blight of cereals. *Plant Pathology* 42, 905–914.

Papendick, R.I. and Cook, R.J. (1974) Plant water stress and development of *Fusarium* foot rot in wheat subjected to different cultural practices. *Phytopathology* 64, 358–363.

Parry, D.W. (1990) The incidence of *Fusarium* spp. in stem bases of selected crops of winter wheat in the Midlands, UK. *Plant Pathology* 39, 619–622.

Parry, D.W., Bayles, R. and Priestley, R.H. (1985) Resistance of winter wheat varieties to ear blight caused by *Fusarium avenaceum* and *F. culmorum*. *Tests of Agrochemicals and Cultivars No. 6*. (*Annals of Applied Biology* 106, supplement), 465–468.

Pettitt, T.R., Parry, D.W. and Polley, R.W. (1993) Improved estimation of the incidence of *Microdochium nivale* in winter wheat stems in England and Wales, during 1992, by use of benomyl agar. *Mycological Research* 97, 1172–1174.

Polley, R.W., Turner, J.A., Cockerell, V., Robb, J., Scudamore, K.A., Sanders, M.F. and Magan, N. (1991) Survey of *Fusarium* species infecting winter wheat in England, Wales and Scotland, 1989–1990. *Home Grown Cereals Authority Project Report* 39.

Pugh, G.W., Johann, H. and Dickson, J.G. (1933) Factors affecting infection of wheat heads by *Gibberella saubinetii*. *Journal of Agricultural Research* 46, 771–797.

Rennie, W.J., Richardson, M.J. and Noble, M. (1983) Seed-borne pathogens and the production of quality seed in Scotland. *Seed Science and Technology* 11, 1115–1127.

Sitton, J.W. and Cook, R.J. (1981) Comparative morphology and survival of chlamydospores of *Fusarium roseum* 'Culmorum' and 'Avenaceum'. *Phytopathology* 71, 85–90.

Smith, W.G. (1884) *Diseases of Field and Garden Crops*. Macmillan, London.

Snijders, C.H.A. (1990) Systemic fungal growth of *Fusarium culmorum* in stems of winter wheat. *Journal of Phytopathology* 129, 133–140.

Snyder, W.C. and Nash, S.M. (1968) Relative incidence of *Fusarium* pathogens of cereals in rotation plots at Rothamsted. *Transactions of the British Mycological Society* 51, 417–425.

Stedman, O.J. (1980) Observations on the production and dispersal of spores and infection by *Rhynchosporium secalis*. *Annals of Applied Biology* 95, 163–175.

Strange, R.N. and Smith, H. (1971) A fungal growth stimulant in anthers which predisposes wheat to attack by *Fusarium avenaceum*. *Physiological Plant Pathology* 1, 141–150.

Strange, R.N. and Smith, H. (1978) Specificity of choline and betaine as stimulants of *Fusarium graminearum*. *Transactions of the British Mycological Society* 70, 187–192.

Strausbaugh, C.A. and Maloy, O.C. (1986) *Fusarium* scab of irrigated wheat in Central Washington. *Plant Disease* 70, 1104–1106.

Sutton, J.C. (1982) Epidemiology of wheat headblight and maize ear rot caused by *Fusarium graminearum*. *Canadian Journal of Plant Pathology* 4, 195–209.

Tarr, S.A.J. (1962) *Diseases of Sorghum, Sudan Grass and Broom Corn*, CMI, Kew, Surrey.

Warren, H.L. and Kommedahl, T. (1972) Fertilization and wheat refuse effects on *Fusarium* species associated with wheat roots in Minnesota. *Phytopathology* 63, 103–108.

Wearing, A.H. and Burgess, L.W. (1977) Distribution of *Fusarium roseum* "Graminearum" Group I and its mode of survival in eastern Australian wheat belt soils. *Transactions of the British Mycological Society* 69, 429–442.

Wilcoxson, R.D., Kommedahl, T., Ozman, E.A. and Windels, C.A. (1988) Occurrence of *Fusarium* species in scabby wheat from Minnesota and their pathogenicity to wheat. *Phytopathology* 78, 586–589.

19 Fungal Interactions on Living and Necrotic Leaves

J. Köhl and N.J. Fokkema

DLO Research Institute for Plant Protection (IPO-DLO) P.O. Box 9060, 6700 GW Wageningen, The Netherlands.

Introduction

This study of fungal interactions on leaves originates from our interest in the biological control of fungal foliar pathogens with the help of antagonistic fungi. Whereas research on biological control of soilborne pathogens made good progress during the last decade (Lumsden and Lewis, 1988), examples of successful biological control in the phyllosphere are rare. A main limitation of biological control of foliar diseases on green leaves is the limited time available for interactions between antagonistic microbial populations and the pathogen since penetration of the leaf tissue normally occurs within a few hours. Thereafter, the pathogen becomes independent of the environmental conditions in the phyllosphere and is protected from antagonists.

On the other hand, cheap and efficient fungicides are available and are easy to apply to the canopy to control fungal leaf pathogens. The expected restrictions on the use of fungicides by governments for environmental reasons will lead to the need for more research on the development of biological control agents (BCAs). Biocontrol in the phyllosphere is aimed at the prevention of infection by the pathogen or the reduction of pathogen inoculum by interfering with sporulation. To protect a leaf from infection, the antagonist and the pathogen must interact on the green leaf, either on intact or on wounded tissue. To reduce inoculum production by necrotrophic pathogens, interactions must take place in necrotic tissue prior to sporulation or even within the pathogen survival structure when antagonists attack sclerotia.

In this chapter, we will compare these two strategies for the biological control of foliar diseases caused by necrotrophic pathogens. Interactions on green leaves will be discussed only briefly to point out similarities and differences between the two ecological niches, the green and the necrotic leaf. The microbial ecology of the phyllosphere and biological control on green leaves are discussed in recent reviews (Blakeman and Fokkema, 1982; Gowdu and Balasubramanian, 1988; Andrews, 1992) and the proceedings of several symposia (Preece and Dickinson, 1971; Dickinson and Preece, 1976; Blakeman, 1981; Fokkema and Van den Heuvel, 1986; Andrews and Hirano, 1992). More emphasis will be laid on the interactions between antagonists and pathogens in necrotic tissue, which is a somewhat neglected field of research. Saprophytic antagonists may have advantages in necrotic leaf tissue compared to antagonists on green leaves. Sufficient nutrients are available to the antagonists for prolonged colonization of the necrotic substrate. The duration of the interaction period of the antagonists and the saprophytically growing mycelium or the survival structures of necrotrophs is relatively long and the pathogen cannot escape from the antagonistic interaction by infecting healthy tissue as on green leaves.

Interactions on Green Leaves

Abiotic factors

On living leaves, a limited amount of nutrients is available. Plant exudates, deposited pollen and aphid honeydew contain easily degradable sources of nutrients such as amino acids and sugars. Enhanced nutrient levels stimulate naturally-occurring saprophytes as well as necrotrophic pathogens. (Fokkema, 1971; Fokkema *et al.*, 1983; Dik, 1992). Conidial germination of some necrotrophs such as *Botrytis cinerea* or *Sclerotinia sclerotiorum* depends on the presence of exogenous nutrients.

Besides nutrients, fungal colonization of the phyllosphere is mainly restricted by rapid changes of the microclimate. High relative humidities (above 95%) in the boundary layer of a leaf (Burrage, 1971) or free water on the surface is a prerequisite for both spore germination of the pathogen and growth of the antagonist. Periods of high humidities or leaf wetness are frequently interrupted by dry periods. For successful infection of host tissue, pathogens such as *Botrytis squamosa* need no more than 6 h of leaf wetness (Alderman and Lacy, 1983). Thereafter, the pathogen, once inside the leaf, becomes independent of moisture on the surface. Consequently, interactions with antagonists are only possible during these short periods of leaf wetness. Temperature on the leaf surface and UV-radiation are other factors affecting pathogen/antagonist interaction (Burrage, 1971).

Fungal populations

Yeasts such as *Cryptococcus* spp., *Sporobolomyces* spp. and *Aureobasidium pullulans* and some filamentous fungi such as *Cladosporium* spp. are adapted to the phyllosphere (Blakeman, 1985). Population densities of naturally-occurring yeasts can reach 10^5 cells cm^{-2} on wheat leaves under field conditions (Fokkema *et al.*, 1979). Generally, densities depend on the microclimatic conditions and on the supply of nutrients (Dik *et al.*, 1992).

Fungal interactions

Conidia of pathogens which are deposited on the leaf surface will mostly remain ungerminated during dry periods. Most will lose vigour and the ability to penetrate the host, e.g. conidia of *Botrytis squamosa* survived for only 2–3 days on onion leaves at 60% RH (Alderman and Lacy, 1983).

Consumption of nutrients by yeasts restricts those pathogens whose spore germination or superficial mycelial growth prior to infection is stimulated by exogenous nutrients (Blakeman and Fokkema, 1982). Naturally-occurring yeast populations play an important role in controlling necrotrophs on green leaves, unless the system is disturbed by fungicides (Fokkema, 1988; Dik, 1992).

Wound protection

Generally, similar principles govern the protection of wounds on leaves from pathogen invasion as for the protection of the healthy leaf. When man-made wounds have to be protected, the use of BCAs is promising because the target and the optimum application time are clearly defined and protection is required for only a few days. Pathogen invasion of wounds of apples and other fruits can effectively be prevented by postharvest treatments with antagonists, mainly yeasts or bacteria (Janisiewicz and Roitman, 1988; Chalutz and Wilson, 1990). Examples of man-made wounds of leaves in the field result from the cutting of flowerbuds, e.g. of tulips, lilies or gladioli to increase bulb production or the cutting of green onion leaves before harvest. When *Trichoderma viride* was sprayed under field conditions on to these cut surfaces, entry of *Botrytis aclada* (*B. allii*), the cause of *Botrytis* neck rot of onions, was reduced from 35% to 24% (Köhl *et al.*, 1991).

Interactions in Necrotic Tissue

Abiotic factors

In some respects, the ecological niche presented by a necrotic leaf differs markedly from that of a living leaf. The most striking change is the amount and nature of the nutrient resource available to fungi. In necrotic tissue, mainly cell wall materials such as cellulose, hemicellulose or lignin are present, after amino acids and soluble sugars have been transported to tissues which are still green (Baddeley, 1971), leached from tissues or utilized by saprophytic microorganisms. Although the potential nutrient resource of necrotic tissue is high, compared to the surface of a green leaf, the degradation of cell wall material depends on the presence of saprophytes adapted to colonize and decompose this substrate.

As with green leaves, fungal growth is confined to periods when moisture is present. In contrast to the surface of a green leaf, dry necrotic tissue has a low affinity for moisture, and wetting of the tissue takes longer than for the surface of a leaf (James *et al.*, 1984). On the other hand, on return to dry conditions necrotic tissue remains wet for a somewhat longer time, because of the greater capacity of necrotic tissues to retain water. Although the duration of wetness periods essentially determines the development of fungi in necrotic tissues and their subsequent sporulation (e.g. *Botrytis squamosa*, Sutton *et al.*, 1983), limited information is available about water dynamics in necrotic tissue under field conditions. This might be due to a lack of suitable sensors for monitoring wetness within necrotic tissues (Sutton, 1988).

Still less information is available about temperatures in necrotic tissues, which may rise under dry conditions to above ambient air temperature. In a field experiment in summer 1992, a maximum temperature of 42°C was measured in dead onion leaf tips exposed to the sun, while the air temperature was 31°C. The effect of periods of high temperatures on populations of saprophytes and necrotrophic pathogens in dry necrotic tissues attached to living plants has not been previously investigated.

Fungal populations

Whereas the surface of green leaves is not intensively colonized by filamentous fungi compared to yeasts, such fungi become dominant in newly-senesced leaves (Dickinson, 1967). *Cladosporium* spp., *Alternaria* spp. and *Stemphylium* spp. were the predominant saprophytes on senescent *Pisum* leaves. Yeasts, which lack mycelial growth and suitable enzymes, may only be able to colonize necrotic tissue superficially. Hudson (1971) described *Cladosporium herbarum*, *Alternaria tenuis*, *Epicoccum nigrum*,

Aureobasidium pullulans and *Botrytis cinerea* as the common primary saprophytes colonizing senescent leaf tissues. Similar observations were reported by Hudson and Webster (1958) for decaying stems of *Agropyron repens*. On straw infested with *Pyrenophora tritici-repentis* Pfender and Wootke (1988) found *P. tritici-repentis*, *Septoria nodorum*, *Alternaria* spp. and *Cladosporium* spp. as primary colonizers, followed later in the succession by *Acremonium strictum*, *Diplodia* spp. and *Fusarium* spp. From dead onion leaf tips *Cladosporium herbarum*, *Stemphylium vesicarium*, *Alternaria alternata*, *A. tenuissima*, *Ulocladium chartarum* and *Epicoccum nigrum* were mainly isolated (Mathar *et al.*, unpublished).

Fungal interactions

Necrotrophic leaf pathogens colonize the necrotic tissue of lesions, dead leaves and crop debris. Colonizers of dead tissue must be able to produce cellulases and hemicellulases under varying environmental conditions. The production of chitinases enables antagonists to attack the hyphal cell walls of necrotrophs which are the primary invaders of the necrotic tissue (Pfender and Wootke, 1988).

Competition for limiting nutrients in the substrate is likely to be the main mechanism of antagonism. But also the physical characteristics of the substrate may be changed by the enzymatic activities of the saprophytes. During the decomposition of leaves by saprophytes, water availability increases in the substrate, which will promote the secondary colonization by fungi with higher water requirements (Dix, 1985). When *Athelia bombacina* was applied to apple leaves in an orchard in autumn, overwintered leaves reduced their strength by 60% compared to unattacked controls (Heye and Andrews, 1983; Miedtke and Kennel, 1990). Increased decomposition of the necrotic tissue by saprophytes shortens the time for pathogen survival or sporulation. Besides nutrient competition, antagonism by some saprophytes may result from direct hyperparasitism (Whipps and Gerlagh, 1992) or from the production of toxins, which may diffuse through the substrate.

Antagonists can interfere in three different stages of the life cycle of pathogens in necrotic tissue:

1. They may prevent colonization of healthy tissue from adjacent dead tissue. Pathogens such as *Botrytis cinerea* and *Sclerotinia sclerotiorum* often colonize dead plant parts before being able to penetrate healthy tissue.
2. Antagonists may reduce the production and viability of resting structures as primary inoculum in crop residues.
3. Antagonists may reduce the production and dissemination of spores as secondary inoculum during an epidemic of a pathogen.

Prevention of infections via necrotic tissue

Tomato fruits are invaded by *Botrytis cinerea* via adhering dead petals. Newhook (1957) sprayed saprophytes onto petals under greenhouse conditions with increased humidities and inoculated the tomato plants with *B. cinerea* 7 days later. In two trials, the percentage of rotted fruits was 31–42% in controls without antagonists, whereas only 0–3% of the tomatoes rotted after treatment with *Cladosporium herbarum* or *Penicillium* sp.

Ascospores from *Sclerotinia sclerotiorum* cannot infect healthy tissue without exogenous nutrients. Under field conditions, senescent petals are colonized first before mycelium can invade the intact leaf tissue. Boland and Hunter (1988) treated bean petals with *Alternaria alternata* or *Cladosporium cladosporioides*. The percentage of leaves of bean seedlings infected by *S. sclerotiorum* via such petals lying on the leaves was reduced from 62% in the control to 0% and 8%, respectively. Zhou and Reeleder (1991) found that *Epicoccum purpurascens* can colonize senescent and dead petals of snap bean (*Phaseolus vulgaris*) efficiently without having any deleterious effects on the plant. By this means, the colonization of petals by *S. sclerotiorum* was suppressed which resulted in less infection (Zhou and Reeleder, 1989). By spraying *E. purpurascens* in combination with 1% malt extract three to four times, the disease index was reduced compared to the control from 27% to 12% and from 28% to 13% in two field experiments.

Reduction of primary inoculum

The reduction of primary inoculum by antagonists may have a direct effect on the disease level of those diseases where no secondary inoculum is produced. Sclerotia of *Sclerotinia sclerotiorum*, surviving in soil for several years, produce apothecia containing ascospores. Besides ascospores, no further spores are formed to disseminate the disease. Thus, the application of antagonists to destroy sclerotia may reduce the disease in crops during the following years. A potential antagonist of sclerotia of *S. sclerotiorum* is *Coniothyrium minitans* which is able to infect and to destroy sclerotia over a wide range of temperatures (Trutman *et al.*, 1980). Soil applications of the antagonist resulted in a reduced recovery of sclerotia and fewer apothecia in celery and lettuce crops under glasshouse conditions (Whipps *et al.*, 1989). Gerlagh and Vos (1991) carried out foliar treatments with *Coniothyrium minitans* on a bean crop under field conditions, which reduced the viability of sclerotia of *S. sclerotiorum* by 85%. In an ongoing 5-year field experiment, the antagonist was sprayed on diseased bean crops which resulted in more than 90% of sclerotia becoming infested by *C. minitans* (Gerlagh *et al.*, 1993). The impact of this antagonism on the disease level will be investigated in the succeeding crops during the next few years.

The relation between the amount of primary inoculum and the disease progress of wheat tan spot (*Pyrenophora tritici-repentis*) was studied by Adee and Pfender (1989) in two field experiments under wet and dry weather conditions. The disease progress was mainly determined by the amount of locally-produced ascospores representing the source of primary inoculum and the amount of conidia as secondary inoculum initiated by the primary inoculum. Airborne secondary inoculum produced beyond the experimental field was less important under both climatic conditions. A reduction of primary inoculum by antagonists may therefore be effective in controlling a polycyclic disease. *Limonomyces roseipellis*, a secondary invader of wheat straw, suppressed the formation of ascocarps and production of ascospores of *P. tritici-repentis* on wheat straw under varying conditions (Pfender, 1988). When straw is colonized by airborne saprophytes such as *Penicillium*, *Aspergillus*, *Alternaria* or *Stemphylium* prior to incorporation into soil, the saprophytic colonization of the straw by the pathogen *Fusarium roseum* f.sp. *cerealis* 'Culmorum' after burial was prevented (Cook, 1970; see also Parry *et al.*, Chapter 18).

Venturia inaequalis produces pseudothecia on leaf litter as primary inoculum the year following foliar scab infection. The development of pseudothecia could be reduced either by spraying urea (5%) to stimulate naturally-occurring antagonistic saprophytes such as *Cladosporium*, *Alternaria* and *Fusarium* and to accelerate decomposition (Burchill and Cook, 1971), or by introducing an antagonist, *Athelia bombacina*, to competitively destroy leaf tissue (Heye and Andrews, 1983). It still has to be demonstrated whether such treatments may result in lower disease levels in the orchard when the conditions are favourable for rapid production and dissemination of conidia as the secondary inoculum.

In many cases, the reduction of primary inoculum alone will not be sufficient to control polycyclic diseases. Duthie and Campbell (1991) demonstrated that leaf spot diseases of alfalfa caused by *Phoma medicaginis*, *Stemphylium botryosum* and *Cercospora medicaginis* developed almost independently from the initial level of infected crop debris as a source of inoculum. The disease development was the same in plots where infested debris had been removed, had been left in the plot, or additional debris had been added. However, a significant effect of the initial inoculum density on the epidemic was found for *Leptosphaerulina trifolii*, but climatic factors had a modifying influence.

Reduction of secondary inoculum

Necrotrophs sporulate on necrotic lesions on the plant or on dead plant remains. In both cases, the amount of secondary inoculum disseminated on to the crop mainly determines the progress of an epidemic. The use of antagonists, which suppress the sporulation of a necrotroph, consequently may slow down an epidemic.

Peng and Sutton (1991) selected antagonists in a bioassay using strawberry leaf discs. They found *Alternaria alternata*, *Epicoccum purpurascens*, *Colletotrichum gloeosporioides*, *Gliocladium roseum*, *Myrothecium verrucaria*, *Penicillium* sp. and *Trichoderma viride* to be the most effective in the suppression of sporulation of *B. cinerea* in bioassays on detached strawberry leaves and under greenhouse conditions on whole strawberry plants. In field experiments, *G. roseum*, *T. viride* and *Penicillium* sp. suppressed sporulation on strawberry leaves as effectively as foliar applications of chlorothalonil (Peng and Sutton, 1990). Conidia produced on leaves are the source of inoculum for fruit infections (Braun and Sutton, 1987). Köhl *et al.* (unpublished) used segments of dead onion leaves, which were preinoculated with *Botrytis aclada*, to select for antagonistic saprophytes. Most isolates of *Alternaria alternata*, *Arthrinium* spp., *Chaetomium globosum*, *Gliocladium* spp., *Trichoderma viride* and *Ulocladium* spp. suppressed sporulation of the pathogen completely, whereas most isolates of *Cladosporium* spp., *Penicillium* spp., *Trichoderma harzianum* and yeasts were less suppressive.

Biles and Hill (1988) sprayed conidial suspensions of *Trichoderma harzianum* on wheat leaves showing necrotic lesions caused by *Cochliobolus sativus*. The sporulation capacity of *C. sativus* in lesions was significantly reduced by 47% by the antagonist after incubation in a wet chamber at 26°C, the optimum temperature for sporulation of the pathogen.

The suppression of sporulation as a means of biological control is only feasible if locally produced inoculum is responsible for most of the new infections during an epidemic. In these situations, airborne inoculum from outside the field should be of minor importance. Köhl *et al.* (1992a) conducted a field experiment with onions to estimate the reliability of biological control of *Botrytis* spp. by suppressing sporulation of the pathogen. The effect of antagonists was simulated by the artificial removal of necrotic tissue suitable for the sporulation of *Botrytis* spp. The removal of approximately 40% of the necrotic tissue from the plots led to a significantly slower increase in leaf spot epidemics in onions caused by *Botrytis squamosa* and *B. cinerea*, compared to the control. The spore load in the plots was also significantly reduced by the removal of necrotic tissue. These results support the hypothesis that the suppression of sporulation by antagonists can reduce disease development under field conditions.

Drought Tolerance as an Ecological Factor in the Performance of Antagonists in Necrotic Leaf Tissue

Under field conditions, one of the main factors limiting fungal growth in necrotic tissue is the rapid change in the water content of the substrate. Consequently, antagonists should not only be screened under optimum circumstances in moist chambers, but also under conditions of reduced water potentials.

The ability of saprophytes to colonize and to decompose plant residues differs under different water potentials (Magan and Lynch, 1986). *Penicillium* spp. and *Fusarium* sp. still grew on straw at –7 MPa, while *Trichoderma* spp., *Gliocladium* sp. and other saprophytes showed no growth. Pfender *et al.* (1991) tested the potential of saprophytes to reduce the colonization of wheat straw and the production of ascocarps by *Pyrenophora tritici-repentis* at several water potentials. Only a few saprophytes such as *Acremonium terricola*, *Epicoccum nigrum*, *Myrothecium roridum* and *Stachybotrys* sp. reduced colonization of straw by the pathogen even at –7 MPa in bioassays. Inhibition of ascocarp production by antagonists was found only at high water potentials.

Köhl *et al.* (1992b) compared the antagonism of *Gliocladium roseum* and *Trichoderma viride*, both highly efficient in moist chambers, against *Botrytis aclada* on onion leaf segments at low water potentials. *Gliocladium roseum* showed antagonistic activity against *Botrytis aclada* at –6.6 MPa whereas *Trichoderma viride* failed to suppress the sporulation of the pathogen at a similar water potential.

Other criteria, in addition to activity at low water potential, should be taken into account in selection of antagonists such as survival during dry periods and ability to grow rapidly thereafter. Saprophytes, which are strongly antagonistic under continuous wet conditions, were compared in bioassays on dead onion leaf segments after interrupted wetness periods (Fig. 19.1). When leaves were dried after 1 day following application of antagonists and kept dry for 6 h before rewetting, two out of four antagonists were still effective: *Alternaria alternata* and *Chaetomium globosum*. This was still the case after dry periods had been repeated up to three times. The isolate of *Gliocladium roseum* was very susceptible to dry periods, especially the first day after application. In a series of similar experiments, isolates of *Gliocladium* spp., *Trichoderma harzianum* and *Sesquicillium candelabrum* were highly sensitive to dry periods, whereas *Alternaria* spp., *Ulocladium* spp., *C. globosum* and *Aureobasidium pullulans*, even after repeated dry periods, were always as effective as on continuously wet leaves. In these experiments, antagonists were most sensitive to dryness during the early stage of colonization of the substrate, e.g. the first day after application. Germinating spores of antagonists are very sensitive to dryness (Diem, 1971) and will be further examined in future research.

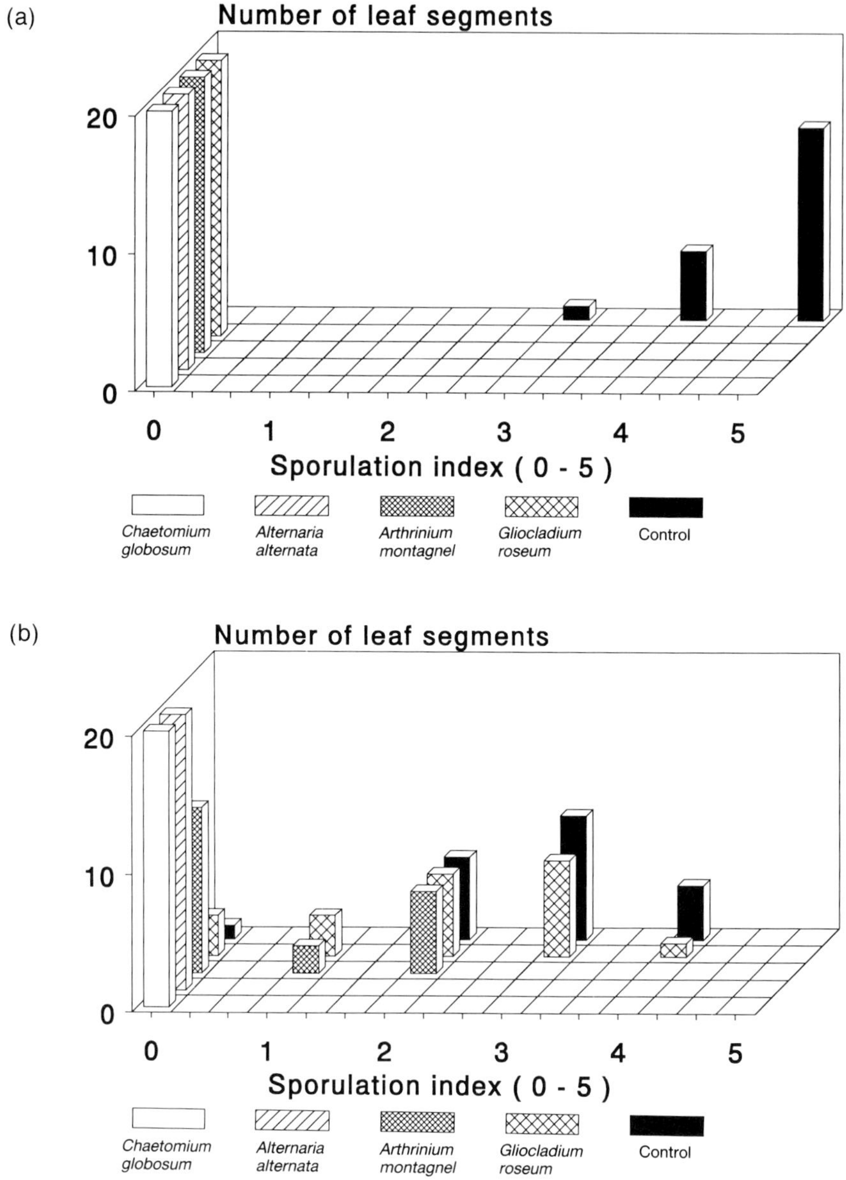

Fig. 19.1. Suppression of sporulation of *Botrytis aclada* in a bioassay on dead onion leaf segments by antagonists. Conidia of *B. aclada* were sprayed on to all leaves 24 h prior to antagonist treatment. (a) Without interrupted wetness periods. (b) With interrupted wetness periods at day 1, 2 and 3 after application of the antagonists. (Sporulation index: 0 = no sporulation of *B. aclada*, 1 = 1–5% leaf surface covered with conidiophores, 2 = >5–25%, 3 = >25–50%, 4 = >50–75%, 5 = >75–100%.)

References

Adee, E.A. and Pfender, W.F. (1989) The effect of primary inoculum level of *Pyrenophora tritici-repentis* on tan spot epidemic development in wheat. *Phytopathology* 79, 873–877.

Alderman, S.C. and Lacy, M.L. (1983) Influence of dew period and temperature on infection of onion leaves by dry conidia of *Botrytis squamosa*. *Phytopathology* 73, 1020–1023.

Andrews, J.H. (1992) Biological control in the phyllosphere. *Annual Review of Phytopathology* 30, 603–635.

Andrews, J.H. and Hirano, S.S. (1992) *Microbial Ecology of Leaves*. Springer, New York.

Baddeley, M.S. (1971) Biochemical aspects of senescence. In: Preece, T.F. and Dickinson, C.H. (eds), *Ecology of Leaf Surface Micro-organisms*. Academic Press, London, pp. 415–429.

Biles, C.L. and Hill, J.P. (1988) Effect of *Trichoderma harzianum* on sporulation of *Cochliobolus sativus* on excised wheat seedling leaves. *Phytopathology* 78, 656–659.

Blakeman, J.P. (1981) *Microbial Ecology of the Phylloplane*. Academic Press, London.

Blakeman, J.P. (1985) Ecological succession of leaf surface microorganisms in relation to biological control. In: Windels, C.E. and Lindow, S.E. (eds), *Biological Control on the Phylloplane*. American Phytopathological Society, St Paul, Minnesota, pp. 6–30.

Blakeman, J.P. and Fokkema, N.J. (1982) Potential for biological control of plant diseases on the phylloplane. *Annual Review of Phytopathology* 20, 167–192.

Boland, G.J. and Hunter, J.E. (1988) Influence of *Alternaria alternata* and *Cladosporium cladosporioides* on white mold of bean caused by *Sclerotinia sclerotiorum*. *Canadian Journal of Plant Pathology* 10, 172–177.

Braun, P.G. and Sutton, J.C. (1987) Inoculum sources of *Botrytis cinerea* in fruit rot of strawberries in Ontario. *Canadian Journal of Plant Pathology* 9, 1–5.

Burchill, R.T. and Cook, R.T.A. (1971) The interaction of urea and micro-organisms in suppressing the development of perithecia of *Venturia inaequalis* (Cke.) Wint. In: Preece, T.F. and Dickinson, C.H. (eds), *Ecology of Leaf Surface Micro-organisms*. Academic Press, London, pp. 471–483.

Burrage, S.W. (1971) The micro-climate at the leaf surface. In: Preece, T.F. and Dickinson, C.H. (eds), *Ecology of Leaf Surface Micro-organisms*. Academic Press, London, pp. 91–101.

Chalutz, E. and Wilson, C.L. (1990) Postharvest biocontrol of green and blue mold and sour rot of citrus fruit by *Debaryomyces hansenii*. *Plant Disease* 74, 134–137.

Cook, R.J. (1970) Factors affecting saprophytic colonization of wheat straw by *Fusarium roseum* f.sp. *cerealis* 'Culmorum'. *Phytopathology* 60, 1672–1676.

Dickinson, C.H. (1967) Fungal colonization of *Pisum* leaves. *Canadian Journal of Botany* 45, 915–927.

Dickinson, C.H. and Preece, T.F. (1976) *Microbiology of Aerial Plant Surfaces*. Academic Press, London.

Diem, H.G. (1971) Effect of low humidity on the survival of germinated spores commonly found in the phyllosphere. In: Preece, T.F. and Dickinson, C.H. (eds), *Ecology of Leaf Surface Micro-organisms*. Academic Press, London, pp. 211–219.

Dik, A.J. (1992) Interactions among fungicides, pathogens, yeasts, and nutrients in the phyllosphere. In: Andrews, J.H. and Hirano, S.S. (eds), *Microbial Ecology of Leaves*. Springer, New York, pp. 412–429.

Dik, A.J., Fokkema, N.J. and Van Pelt, J.A. (1992) Influence of climatic and nutritional factors on yeast population dynamics in the phyllosphere of wheat. *Microbial Ecology* 23, 41–52.

Dix, N.J. (1985) Changes in relationship between water content and water potential after decay and its significance for fungal successions. *Transactions of the British Mycological Society* 85, 649–653.

Duthie, J.A. and Campbell, C.L. (1991) Effects of plant debris on intensity of leaf spot diseases, incidence of pathogens, and growth of alfalfa. *Phytopathology* 81, 511–517.

Fokkema, N.J. (1971) The effect of pollen in the phyllosphere of rye on colonization by saprophytic fungi and on infection by *Helminthosporium sativum* and other leaf pathogens. *Netherlands Journal of Plant Pathology* 77, 1–60.

Fokkema, N.J. (1988) Agrochemicals and the beneficial role of phyllosphere yeasts in disease control. *Ecological Bulletins* 39, 91–93.

Fokkema, N.J. and van den Heuvel, J. (1986) *Microbiology of the Phyllosphere*. Cambridge University Press, Cambridge.

Fokkema, N.J., den Houter, J.G., Kosterman, Y.J.C. and Nelis, A.L. (1979) Manipulation of yeasts on field-grown wheat leaves and their antagonistic effect on *Cochliobolus sativus* and *Septoria nodorum*. *Transactions of the British Mycological Society* 72, 19–29.

Fokkema, N.J., Riphagen, I., Poot, R.J. and De Jong, C. (1983) Aphid honeydew, a potential stimulant of *Cochliobolus sativus* and *Septoria nodorum* and the competitive role of saprophytic microflora. *Transactions of the British Mycological Society* 81, 355–363.

Gerlagh, M. and Vos, I. (1991) Enrichment of soil with sclerotia to isolate antagonists of *Sclerotinia sclerotiorum*. In: Beemster, A.B.R., Bollen, G.J., Gerlagh, M., Ruissen, M.A., Schippers, B. and Tempel, A. (eds), *Biotic Interactions and Soil-borne Diseases*. Elsevier, Amsterdam, pp. 165–171.

Gerlagh, M., van de Geijn, H.M. and Verdam, B. (1993) Microbial suppression of viable sclerotia of *Sclerotinia sclerotiorum* and white mould in field crops. In: Fokkema, N.J., Köhl, J. and Elad, Y. (eds), *Biological Control of Foliar and Post-harvest Diseases. IOBC/WPRS Bulletin* 16, 64–68.

Gowdu, B.J. and Balasubramanian, R. (1988) Role of phylloplane micro-organisms in the biological control of foliar plant diseases. *Zeitschrift für Pflanzenkrankheiten und Pflanzenschutz* 95, 310–331.

Heye, C.C. and Andrews, J.H. (1983) Antagonism of *Athelia bombacina* and *Chaetomium globosum* to the apple scab pathogen, *Venturia inaequalis*. *Phytopathology* 73, 650–654.

Hudson, H.J. (1971) The development of the saprophytic fungal flora as leaves senesce and fall. In: Preece, T.F. and Dickinson, C.H. (eds), *Ecology of Leaf Surface Micro-organisms*. Academic Press, London, pp. 447–455.

Hudson, H.J. and Webster, J. (1958) Succession of fungi on decaying stems of *Agropyron repens*. *Transactions of the British Mycological Society* 41, 165–177.

James, T.D.W., Sutton, J.C. and Rowell, P.M. (1984) Monitoring wetness of dead onion leaves in relation to *Botrytis* leaf blight. *Proceedings of the British Crop Protection Conference*, pp. 627–632.

Janisiewicz, W.J. and Roitman, J. (1988) Biological control of blue mold and gray mold on apple and pear with *Pseudomonas cepacia*. *Phytopathology* 78, 1697–1700.

Köhl, J., Molhoek, W.M.L. and Fokkema, N.J. (1991) Biological control of onion neck rot. (*Botrytis aclada*): Protection of wounds made by leaf topping. *Biocontrol Science and Technology* 1, 261–269.

Köhl, J., Molhoek, W.M.L., van der Plas, C.H., Kessel, G.J.T. and Fokkema, N.J. (1992a) Biological control of *Botrytis* leaf blight in onions: significance of sporulation suppression. In: Verhoeff, K., Malathrakis, N.E. and Williamson, B. (eds), *Recent Advances in* Botrytis *Research*, Pudoc Scientific Publishers, Wageningen, pp. 192–196.

Köhl, J., Krijger, M.C. and Kessel, G.J.T. (1992b) Drought tolerance of *Botrytis squamosa, B. aclada* and potential antagonists. In: Verhoeff, K., Malathrakis, N.E. and Williamson, B. (eds), *Recent Advances in* Botrytis *Research*, Pudoc Scientific Publishers, Wageningen, pp. 206–210.

Lumsden, R.D. and Lewis, J.A. (1988) Selection, production, formulation and commercial use of plant disease biocontrol fungi: problems and progress. In: Whipps, J.M. and Lumsden, R.D. (eds), *Biotechnology of Fungi for Improving Plant Growth*. Cambridge University Press, Cambridge, pp. 171–190.

Magan, N. and Lynch, J.M. (1986) Water potential, growth and cellulolysis of fungi involved in decomposition of cereal residues. *Journal of General Microbiology* 132, 1181–1187.

Miedtke, U. and Kennel, W. (1990) *Athelia bombacina* and *Chaetomium globosum* as antagonists of the perfect stage of the apple scab pathogen (*Venturia inaequalis*) under field conditions. *Zeitschrift für Pflanzenkrankheiten und Pflanzenschutz* 97, 24–32.

Newhook, F.J. (1957) The relationship of saprophytic antagonism to control of *Botrytis cinerea* Pers. on tomatoes. *New Zealand Journal of Science and Technology* 38, 473–481.

Peng, G. and Sutton, J.C. (1990) Biological methods to control grey mould of strawberry. *Proceedings of the British Crop Protection Conference*, pp. 233–240.

Peng, G. and Sutton, J.C. (1991) Evaluation of microorganisms for biocontrol of *Botrytis cinerea* in strawberry. *Canadian Journal of Plant Pathology* 13, 247–257.

Pfender, W.F. (1988) Suppression of ascocarp formation in *Pyrenophora tritici-repentis* by *Limonomyces roseipellis*, a basidiomycete from reduced-tillage wheat straw. *Phytopathology* 78, 1254–1258.

Pfender, W.F. and Wootke, S.L. (1988) Microbial communities of *Pyrenophora*-infested wheat straw as examined by multivariate analysis. *Microbial Ecology* 15, 95–113.

Pfender, W.F., Sharma, U. and Zhang, W. (1991) Effect of water potential on microbial antagonism to *Pyrenophora tritici-repentis* in wheat residue. *Mycological Research* 95, 308–314.

Preece, T.F. and Dickinson, C.H. (1971) *Ecology of Leaf Surface Micro-organisms*. Academic Press, London.

Sutton, J.C. (1988) Predictive value of weather variables in the epidemiology and management of foliar diseases. *Fitopatologia Brasileira* 13, 305–312.

Sutton, J.C., James, T.D.W. and Rowell, P.M. (1983) Relation of weather and host factors to an epidemic of botrytis leaf blight in onions. *Canadian Journal of Plant Pathology* 5, 256–265.

Trutmann, P., Keane, P.J. and Merriman, P.R. (1980) Reduction of sclerotial inoculum of *Sclerotinia sclerotiorum* with *Coniothyrium minitans. Soil Biology and Biochemistry* 12, 461–465.

Whipps, J.M. and Gerlagh, M. (1992) Biology of *Coniothyrium minitans* and its potential for use in disease biocontrol. *Mycological Research* 96, 897–907.

Whipps, J.M., Budge, S.P. and Ebben, M.H. (1989) Effect of *Coniothyrium minitans* and *Trichoderma harzianum* on *Sclerotinia* disease of celery and lettuce in the glasshouse at a range of humidities. In: Cavalloro, R. and Pelerents, C. (eds), *Integrated Pest Management in Protected Vegetable Crops.* A.A. Balkema, Rotterdam, pp. 233–243.

Zhou, T. and Reeleder, R.D. (1989) Application of *Epicoccum purpurascens* spores to control white mold of snap bean. *Plant Disease* 73, 639–642.

Zhou, T. and Reeleder, R.D. (1991) Colonization of bean flowers by *Epicoccum purpurascens. Phytopathology,* 81, 774–778.

20 Biological Control of *Erwinia amylovora* with *Erwinia herbicola*

H.A.S. Epton[1], M. Wilson[2], S.L. Nicholson[1] and D.C. Sigee[1]

[1]*School of Biological Sciences, The University of Manchester, Oxford Road, Manchester M13 9PT, UK;* [2]*Department of Plant Pathology, University of California, Berkeley, California 94720, USA.*

Introduction

Host range and aetiology of fire blight

Erwinia amylovora has a host range of over 100 species of Rosaceae (van der Zwet and Keil, 1979) causing fire blight in most members of the subfamily Pomoideae, the most economically important genera being *Malus* (apple) and *Pyrus* (pear). The disease is now indigenous in most countries which produce apples or pears commercially, with the notable exceptions of Australia and South Africa. Among ornamental genera of the Pomoideae, *Pyracantha*, *Cotoneaster* and *Sorbus* are important hosts, while hawthorn (*Crataegus* spp.) is the most common wild reservoir for the pathogen in the UK.

Following overwintering of the pathogen in cankers, inoculum is carried to blossoms by insects or deposited by rain splash. Blossom infection results in reduced yield through the killing of fruit spurs, and may also lead to twig blight which can kill potential fruit-bearing wood. Losses also result from failure to set fruit by blighted blossoms, the blighting of fruit or the death of fruit due to the girdling of the branches bearing it. Young shoots may be similarly infected under conditions of high humidity. The disease may progress downwards from twigs and shoots resulting in the death of large limbs or the entire tree. Trunk, collar and root blight are less common but generally more destructive than blossom blight. Infection of nursery stock can cause severe losses since the adjacent, apparently healthy stock may be subject to quarantine restrictions. As well as direct disease losses, cultivation of pome fruit in some areas is no longer viable due to the threat of fire blight disease (Lopez and Fucikovsky, 1990). An extensive review of the aetiology, physiology and epidemiology of the disease is provided by van der Zwet and Keil (1979).

Control of fire blight

In the UK, control of fire blight is restricted to cultural practices, such as cutting-out of infected material and the removal of inoculum sources, such as hawthorn hedges. No chemical bactericides are approved for the control of fire blight in the UK (Anon., 1983) and in many European countries. In the United States control is mainly based on the use of streptomycin sprays, with the recommendation that oxytetracycline is used when streptomycin resistant strains develop (van der Zwet and Beer, 1992). However, many countries, such as the UK, France and Belgium, do not permit antibiotics of medical or veterinary significance to be used in plant protection (Deckers *et al.*, 1990), for fear of the development of resistance which may be transferred to human and animal pathogens. Streptomycin-resistant strains have developed in several areas of the USA (Moller *et al.*, 1981) resulting in poor fire blight control. This situation amply justifies the search for alternative, biological, means of control.

Biological Control

The earliest reports relating to biological control of fire blight are reviewed by van der Zwet and Keil (1979) beginning in the early 1930s when bacterial strains antagonistic to *E. amylovora* were isolated and shown to have a tendency to reduce the percentage of fire blight infection when applied to blossom. Subsequently many workers have investigated both Gram-positive (Farabee and Lockwood, 1958) and Gram-negative bacteria for their antagonistic properties towards *E. amylovora*, especially *Pseudomonas* spp. (McIntyre *et al.*, 1973; Wrather *et al.*, 1973; Thomson *et al.*, 1976; Lindow, 1985a) and *Erwinia herbicola* (Riggle and Klos, 1970, 1972; McIntyre *et al.*, 1973; Beer *et al.*, 1980).

The *E. herbicola* group contains saprophytic strains that are widespread in nature as epiphytic bacteria. Riggle and Klos (1970, 1972) obtained partial control of fire blight in the greenhouse and field plots by inoculation of pear blossom with *E. herbicola* 24 h before inoculation with a suspension of *E. amylovora*. Beer *et al.* (1980) suggested that a new control for fire blight might be developed based on the use of non-pathogenic strains of *E. herbicola*. They introduced suspensions of *E. herbicola* into apple blossoms 1 day before and 3 days after inoculation with *E. amylovora*. The degree of fire blight control was directly proportional to the bacterial concentration used in experiments on blossom and immature pear fruit (Beer, 1981a). Beer and Rundle (1983) described an immature pear fruit assay as a primary screen for antagonists. Sixteen strains of *E. herbicola* were shown to suppress *E. amylovora* development in immature pear fruits and this ability was significantly correlated with the suppression of blossom blight in the research orchard.

In experiments to control fire blight infections on ornamentals, Isenbeck and Schulz (1985) applied *E. herbicola* to *Cotoneaster bullatus*. When the antagonists were applied 24 h before the pathogen in the greenhouse, the control obtained was the same as that with 200 $\mu g\ cm^{-3}$ streptomycin. Under field conditions, following natural infections, the reduction in disease was about 20%.

Lindow (1985b) concluded that the level of control of fire blight achieved with applied antagonistic bacteria can approach that of frequent applications of bactericides. Fire blight of trees treated with antagonistic bacteria has been reduced from 20 to 85% in various studies and approaches the results obtained with treatments of streptomycin on a weekly basis.

Biocontrol of fire blight with non-pigmented strains of *E. herbicola*

Studies on *E. herbicola* in the United States have been concerned with yellow-pigmented strains of the organism which predominate as epiphytes over non-pigmented strains. One yellow-pigmented strain of US origin, Eh252, has been tested under field conditions both in North America and in Europe (Vanneste and Yu, 1990). When tested in orchards in the US, Eh252 was found to be as effective as streptomycin, whereas in France, while the strain reduced fire blight, it did not achieve the level of control obtained with streptomycin. This difference could have been explained by the use of an antagonist:pathogen ratio of 2.5:1, as opposed to a 10:1 ratio in the American experiments. In the UK non-pigmented strains appear to predominate (Billing and Baker, 1963) and so have been the focus of our studies on the biocontrol of fire blight.

Selection of strains and trials under protected conditions

Using a modification of the immature pear fruit assay developed by Eden-Green (1972) and Beer and Rundle (1983), Wilson *et al.* (1990a) obtained antagonistic non-pigmented isolates of *E. herbicola* from flowers and leaves of hawthorn. It was considered that as hawthorn is the main wild host, and therefore the plant most frequently infected with *E. amylovora* in the UK, the phylloplane of this plant may yield naturally-occurring antagonists more frequently than commercially important species, on which control measures for the pathogen are actively pursued. Under protected conditions some of the selected strains reduced artificially induced blossom blight and shoot blight and gave control of blossom blight equivalent to that provided by chemical agents (Table 20.1). *E. herbicola* strain WHL9, applied at a similar concentration to the pathogen, gave a level of control equivalent to that obtained with streptomycin sulphate at 100 µg cm^{-3} (over 80%), suggesting that it is probably similar in efficacy to the isolates described by Beer (1981a,b) and Beer *et al.* (1984, 1987). In a separate experiment this same strain, which was isolated from a hawthorn leaf, provided total control of shoot blight (Table 20.2). Three of the strains in this experiment, WHL9, WHL40 and WHL71, gave complete control of shoot blight when applied at a biological control agent/pathogen ratio of 10:1. Previously, such levels of

Table 20.1. Comparison of control of blossom-blight by *E. herbicola* and chemical control agents under protected conditions.

Strain/compound[1]	Inoculum level (cells cm^{-3}) or concentration of chemical control agent	Proportion blossoms blighted[2]	Percentage control[3]
WHF18	10^7	0.53 D	44.2
WHF18	10^8	0.39 CD	58.9
WHL9	10^7	0.42 CD	55.8
WHL9	10^8	0.18 BC	81.1
Agrimycin 17	100 mg a.i. l^{-1}	0.11 AB	88.4
Copac E	150 mg Cu l^{-1}	0.42 CD	55.8
S-0208	300 mg a.i. l^{-1}	0.02 A	97.9
JF4387	1 g l^{-1}	0.30 BCD	68.4
Ea519	10^8	0.95 E	0.0

[1]*E amylovora* Ea519 (10^8 cells cm^{-3}) applied 24 h after *E. herbicola* strain/compounds.
[2]Means of five replicate bouquets, assessed 5 days after inoculation with Ea519. Figures followed by the same letter are not significantly different ($P = 0.05$).
[3]Calculated from the mean proportions, compared to the pathogen-only control.
(Reproduced from: Wilson *et al.*, 1990a.)

Table 20.2. Comparison of the ability of *E. herbicola* strains to reduce the shoot blight disease index of hawthorn in a polythene tunnel.

Inoculum[1] strains (cells cm^{-3})	Shoot blight index[2]	Percentage control[3]
Ea519 (10^8)	21.8	0.0
Ea519 (10^8) + WHL7 (10^9)	8.7[4]	60.1
Ea519 (10^8) + WHL7 (10^8)	9.6[4]	56.0
Ea519 (10^8) + WHL9 (10^9)	0.0	100.0
Ea519 (10^8) + WHL9 (10^8)	12.8[4]	41.3
Ea519 (10^8) + WHL40 (10^9)	0.0	100.0
Ea519 (10^8) + WHL40 (10^8)	12.9[4]	40.8
Ea519 (10^8) + WHL71 (10^9)	0.0	100.0
Ea519 (10^8) + WHL71 (10^8)	12.2[4]	44.0

[1]*E. amylovora* strain Ea519 inoculated simultaneously with *E. herbicola* strains through a cut leaf.
[2]Nine replicate plants per treatment.
[3]Calculated from the mean disease index, compared to the pathogen-only control.

$$\text{Index} = \frac{\Sigma \, l/L \times 100}{a + b}$$

where a = no. of blighted shoots; *b* = no. of shoots without lesions; *l* = length of lesion; *L* = length of blighted shoot.
[4]Not significantly different from pathogen-only control ($P = 0.05$).
(Reproduced from: Wilson *et al.*, 1990a.)

control have been associated with a biological control agent/pathogen ratio of at least 100:1 (Goodman, 1965, 1967; McIntyre *et al.*, 1973; Isenbeck and Schulz, 1985), indicating that WHL9, WHL40 and WHL71 are highly effective antagonists.

Field trials with non-pigmented E. herbicola

The effectiveness of the *E. herbicola* strains selected by Wilson *et al.* (1990a) as biocontrol agents under orchard conditions was examined on perry pears, which are among the most susceptible hosts of *E. amylovora* in the UK. As the previously mentioned strains had been isolated from hawthorn and tested against *E. amylovora* infection on hawthorn, additional isolates of *E. herbicola* were selected from perry pear (Nicholson, 1992), using the techniques described by Wilson *et al.* (1990a). In addition to biocontrol efficacy, these strains were also assessed for epiphytic fitness in the orchard, by determining their longevity following application.

It has been suggested that an antagonist isolated from the target host is more likely to be able to become established and be effective than an antagonist isolated from another host (Crosse, 1971; Isenbeck and Schulz, 1985). The epiphytic fitness of a biological control agent is vitally important to its effectiveness. An antagonist unable to sustain a sufficient population density after application would require repeated applications which could make it uneconomic. On the other hand, a biological control agent which could persist for long periods and colonize fruit surfaces in large numbers might be unacceptable to the consumer or to regulatory authorities.

Biological control experiments on perry pears

Nicholson (1992) conducted field trials near Taunton, Somerset, UK on the late flowering cultivar 'Red Pear'. Blossoms were spray-inoculated with suspensions containing 10^8 cells cm^{-3} of *E. herbicola* strains WHL9, WHF18, from hawthorn leaves and flowers respectively (Wilson *et al.*, 1990a), and strains NL18 and NL19 from perry pear leaves. *E. amylovora* Ea519 was

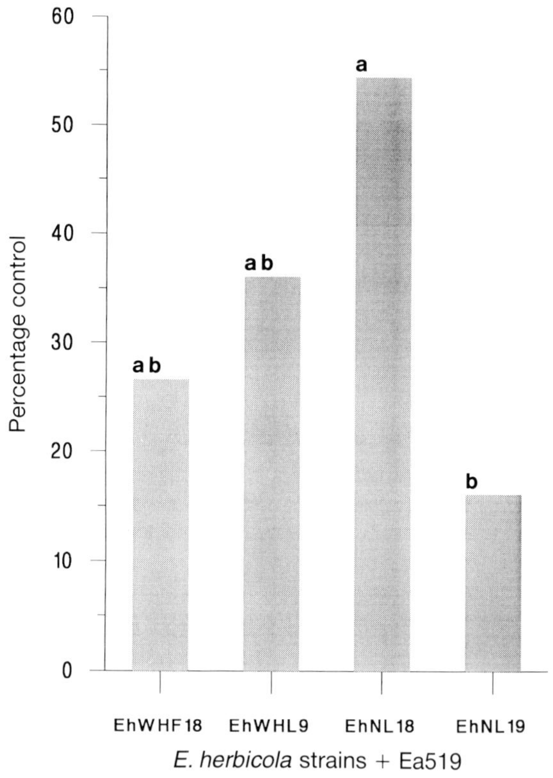

Fig. 20.1. Percentage control of fire blight symptoms on blossom of perry pear c.v. Red Pear, 20 days after application of *Erwinia herbicola* strains (Eh) and 19 days after inoculation with *Erwinia amylovora* (Ea519). Columns marked with the same letter are not significantly different.

applied by spray 24 h later, at the same concentration. Disease was assessed by scoring individual blossoms of experimental branches for fire blight symptoms. The percentage control of disease 19 days after infection is shown in Fig. 20.1. A significant level of disease control was obtained compared to the pathogen-only treatment and NL18 gave significantly better control than the NL19 treatment; however, there were no significant differences between any other treatments.

The proportion of blossom which had set fruit at 19 days after inoculation was assessed as another indication of disease severity. The application of all the *E. herbicola* strains significantly increased fruit set compared to the trees inoculated with the pathogen alone.

Assessment of the percentage of fruit/blossom spurs showing fire blight symptoms 135 days after infection showed a fairly high mean percentage of blighted fruit/blossom spurs on all treatments where the pathogen was applied, whether alone or after application of an *E. herbicola* strain. However, the mean percentage of blighted fruit/blossom spurs was reduced by application of the *E. herbicola* strains. Strain NL18 was most effective in the prevention of blight of fruit/blossom spurs.

An assessment 134 days after infection of the yield of healthy fruit expressed as a proportion of the number of blossoms inoculated, showed no

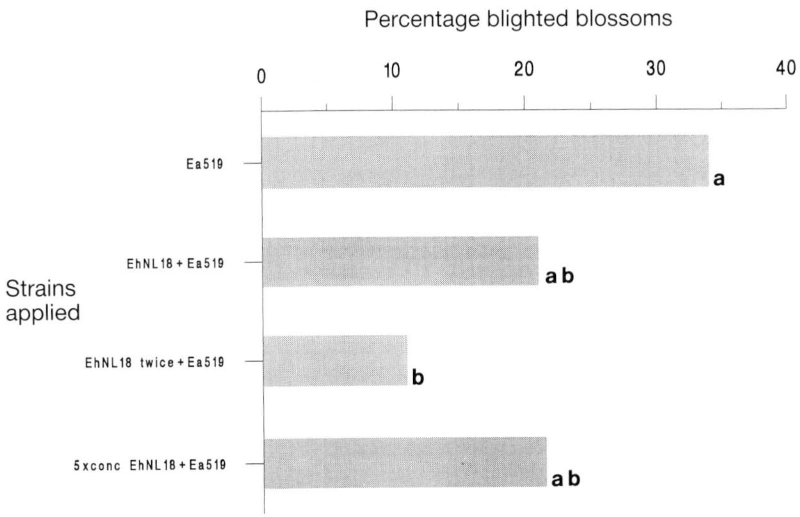

Fig. 20.2. Percentage of blighted blossoms on perry pear cv. Winnals Longden 24 days after application of *Erwinia herbicola* (Eh) strain NL18 and 23 days after inoculation with *Erwinia amylovora* (Ea519). Columns marked with the same letter are not significantly different.

significant difference between the *E. herbicola* + pathogen treatments and the pathogen-alone treatment.

While application of antagonistic *E. herbicola* strains reduced the severity of blossom blight, this reduction in disease levels did not result in an improved fruit yield. The antagonists which appeared to have a significant effect when assessed 19 days after infection may have merely delayed the onset of fire blight symptoms. None of the antagonists completely prevented disease development.

A further trial was performed using the most effective strain, NL18. Nicholson (1992) compared the effect of increased application rate with increased frequency of application, to determine whether control efficacy could be improved. *E. herbicola* strain NL18 was first applied to trees (cv. Winnals Longden) either at 10^8 or 5×10^8 cells cm^{-3}, and one treatment received a second application of NL18 at 10^8 cells cm^{-3} 5 days after the application of the pathogen. The effect on blossom blight, assessed 23 days after inoculation, is shown in Fig. 20.2, which indicates that while there was no significant difference between the treatment with the pathogen alone and a single application at both concentrations of *E. herbicola*, two applications of the biocontrol agent significantly reduced blossom blight.

Mechanisms of Biological Control of Fire Blight

The mechanisms by which biocontrol agents act against plant pathogens are discussed by Blakeman and Brodie (1976), Blakeman and Fokkema (1982) and Napoli and Staskawicz (1985). Biological control agents may act by competition for nutrients or space, by the production of inhibitory substances such as antibiotics or acids, by the stimulation of host defences, by parasitism or by predation.

Competition for space and/or nutrients

Riggle and Klos (1972) demonstrated that *E. herbicola* could consume all organic nitrogen when grown on media containing sugar concentrations comparable to pear nectar, and they suggested that this could partly account for the biological control of *E. amylovora*. The motility and chemotaxis of *E. herbicola* and its effect on *E. amylovora* was studied by Klopmeyer and Ries (1987) who suggested that the non-specific attraction of *E. herbicola* to all components of nectar extract (from apple cv. Jonathan) might give *E. herbicola* an advantage over other microorganisms, including *E. amylovora*, in locating suitable niches during blossom colonization. The similarity in

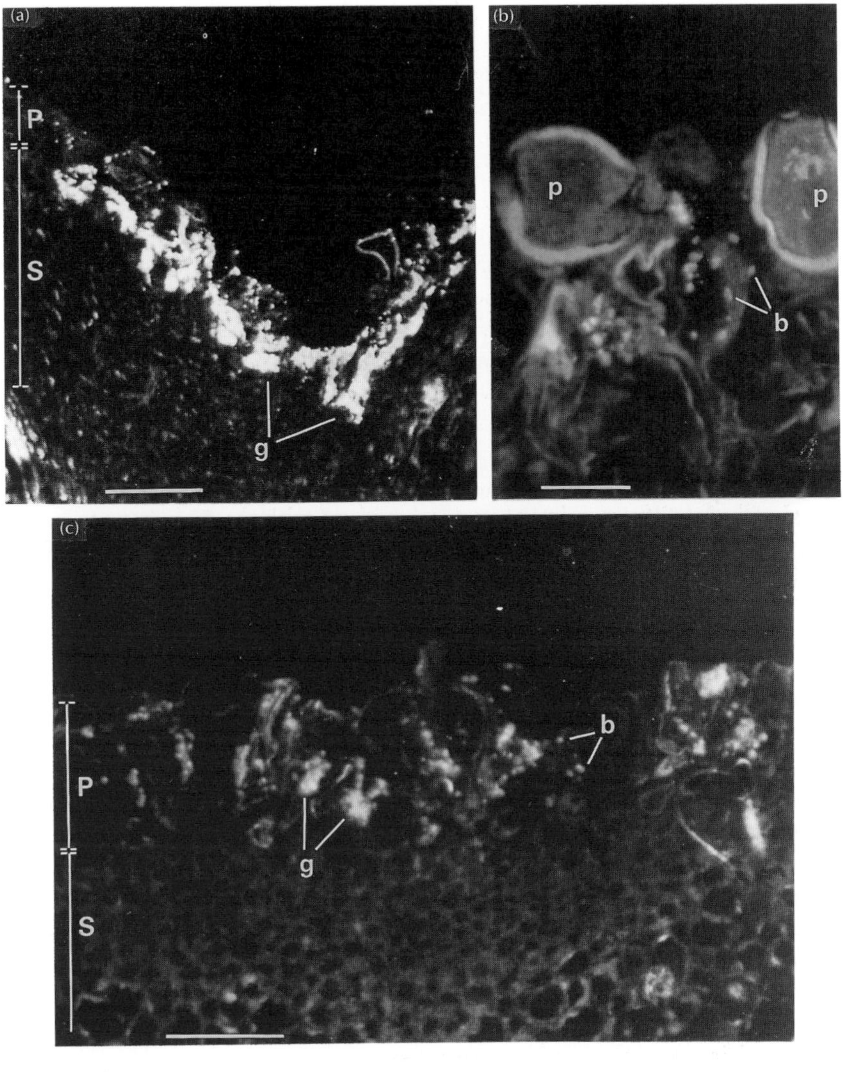

Fig. 20.3a–c. Immunofluorescent micrographs of sections of aldehyde-fixed, resin-embedded stigmas of hawthorn, labelled with rabbit primary antiserum to *Erwinia amylovora* (a) and *Erwinia herbicola* (b and c), followed by FITC-conjugated anti-rabbit IgG. (a) 72 h after inoculation with *Erwinia amylovora*. The pathogen has colonized the zone of collapsed papillae (P) and the upper levels of secretory tissue (S), where fluorescent groups of bacteria (g) are clearly visible. Bar marker = 50 μm. (b) 24 h after inoculation with *Erwinia herbicola*. Cells of the biological control agent (b) are sparsely distributed over and between the papillae, some of which (p) show autofluorescence. Bar marker = 20 μm. (c) 48 h after inoculation with *Erwinia herbicola*. The biological control agent has colonized the papillary zone (P), where it can be seen labelled in groups (g) and as single cells (b). Colonization has not yet proceeded to the secretory zone (S). Bar marker = 50 μm.

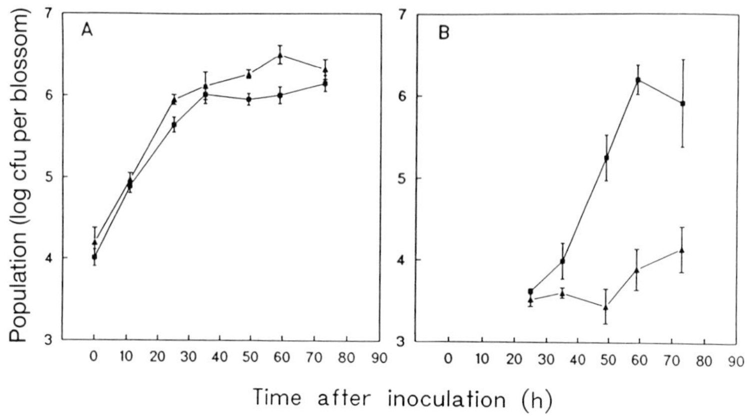

Fig. 20.4. Inoculation of *Erwinia herbicola* EhWHL9N13 on the stigma of the pistil of hawthorn blossom 24 h before *E. amylovora* Ea519Rif.
A) Population of *E. herbicola* EhWHL9N13 when inoculated alone (■) and when inoculated 24 h before *E. amylovora* Ea519Rif (▲). B) Population of *E. amylovora* Ea519Rif when inoculated alone (■) and when inoculated 24 h after *E. herbicola* EhWHL9N13 (▲). Bars represent one standard error of the mean. (Reproduced from: Wilson *et al.*, 1992.)

taxis of *E. herbicola* and *E. amylovora* towards organic acids, specifically malate, may be a clue to the mechanism by which the saprophyte inhibits infection by the pathogen. In addition, *E. herbicola* may catabolize the chemoattractants of the pathogen, thereby affecting pathogen invasion and subsequent infection.

Following SEM studies of apple blossoms inoculated by spraying with suspensions of *E. herbicola* and *E. amylovora*, Hattingh *et al.* (1986) suggested that the effective antagonist restricts the pathogen on apple flowers by competing for the same site on the stigmatic surface. Rundle and Beer (1987) studied the population dynamics of *E. amylovora* and *E. herbicola* on apple blossom parts inoculated by spraying. They concluded that when apple blossoms are inoculated by spraying bacterial suspensions, infection by *E. amylovora* takes place primarily through the stigma. The importance of the stigma in fire blight infections was further demonstrated by Thomson (1986) who found that, under field conditions, epiphytic populations of *E. amylovora* develop on the stigmas of *Pyrus communis*, *Malus sylvestris*, *Pyracantha* spp., *Crataegus* spp. and *Cotoneaster* spp. These epiphytic populations led to blossom infection when inoculum was transferred to the nectarial surface by the action of rain or dew.

Hattingh *et al.* (1986) suggested that preemptive exclusion of *E. amylovora* by *E. herbicola* Eh252 on the stigma of apple occurred because the two strains occupied a similar ecological niche and that prior colonization of

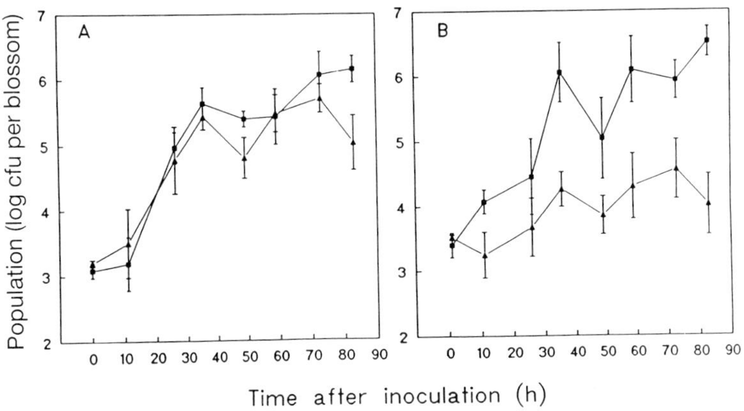

Fig. 20.5. Coinoculation of *Erwinia herbicola* EhWHL9N13 and *E. amylovora* Ea519Rif on the stigma of the pistil of hawthorn blossom.
A) Population of *E. herbicola* EhWHL9N13 when inoculated alone (■) and when coinoculated with *E. amylovora* Ea519Rif (▲). B) Population of *E. amylovora* Ea519Rif when inoculated alone (■) and when coinoculated with *E. herbicola* EhWHL9N13 (▲). Bars represent one standard error of the mean. (Reproduced from: Wilson *et al.*, 1992.)

stigmatic sites by Eh252 prevented subsequent occupation of those sites by *E. amylovora*. Rundle and Beer (1987) concluded that the stigmas were the apparent site of interaction between the pathogen and effective biological control agents, since *E. herbicola* appeared to multiply only on the stigma, where populations of *E. amylovora* were suppressed.

Studies by Wilson (1989), using scanning electron microscopy and phase contrast/immunofluorescent light microscopy, confirmed that *E. amylovora* and *E. herbicola* compete for the same sites on the stigma of hawthorn blossom. Colonization of the stigmatic surface by both the pathogen (Fig. 20.3a) and the biological control agent, strain WHL9Nal1, a nalidixic acid-resistant mutant of WHL9 (Fig. 20.3b,c), involved initial invasion of the outer papillary layer. In a parallel study (Wilson *et al.*, 1992), the population development of another nalidixic acid-resistant mutant WHL9N13 and a rifampicin-resistant mutant of *E. amylovora* were observed following inoculation of the stigmatic surface of hawthorn flowers. When *E. herbicola* was applied to the stigma 24 h before *E. amylovora* (Fig. 20.4), the population of *E. herbicola* was not significantly affected by subsequent inoculation with *E. amylovora* (Fig. 20.4A). Growth of the pathogen was significantly reduced when stigmas were inoculated with *E. amylovora* 24 h after inoculation with *E. herbicola* (Fig. 20.4B). When inoculations were simultaneous, using a mixed suspension of *E. amylovora* and *E. herbicola* (Fig. 20.5), the population of *E. herbicola* was not significantly affected by the presence of *E. amylovora* (Fig. 20.5A), but the population of *E. amylovora* was significantly reduced by the presence of *E. herbicola* (Fig. 20.5B), at all time points (except 48 h) from 36 h after inoculation onwards. These results suggest that *E. herbicola* and *E. amylovora* can occupy a similar ecological niche on the hawthorn stigma, colonize the same physical sites and compete for the same growth-limiting resources (Wilson *et al.*, 1992). These findings, however, do not preclude the involvement of antibiotics in the interactions between *E. herbicola* and *E. amylovora*.

Production of antimicrobial compounds

Antibiotics

E. herbicola has been shown to produce antibiotics which may be responsible for antagonism against a wide range of microorganisms (Winkelmann *et al.*, 1980; Isenbeck and Schulz, 1986; Ishimaru *et al.*, 1988; Kempf and Wolf, 1989; Vanneste *et al.*, 1990, 1992). Experiments on immature pear fruits using non-antibiotic-producing mutants of *E. herbicola* Eh252 (Vanneste *et al.*, 1990) supported the view that antibiotic production is involved in the control of fire blight. It appears that *E. herbicola* strains produce at least two types of antibiotics. Wodzinski *et al.* (1987b) showed that Eh252 produced an antibiotic in the exponential/stationary phases, which was inhibited by histidine and destroyed at pH 3.0 or 100°C, while Eh318 produced an antibiotic in the exponential phase, which was inhibited by arginine and histidine, but stable at pH 3.0 and 100°C. Similar results have been obtained by Ishimaru and Klos (1984) and Ishimaru *et al.* (1988) with *E. herbicola* strain C9-1, which produces two types of antibiotic, herbicolin O and I. Herbicolin O is a broad spectrum antibiotic, inhibited by L-histidine and destroyed by extremes of pH, whilst herbicolin I is a narrow spectrum antibiotic, not inhibited by L-histidine and destroyed only at high pH.

The herbicolacin, 112Y, produced by *E. herbicola* strain Eh112Y, which was shown by Wodzinski *et al.* (1987a) to be inhibited by complex nitrogen, was considered by Ishimaru *et al.* (1988) to be similar to herbicolin I. Ishimaru *et al.* (1988) believed that the confusion over the identity of the toxin H112Y, may have arisen because of its narrow spectrum of activity.

An isogenic, stable, antibiotic-deficient mutant of Eh252 was produced by Vanneste *et al.* (1987), using Tn5 mutagenesis. The mutant multiplied on blossom at a similar rate to the wild-type and, although the biological control ability of the mutant was reduced compared to Eh252, it was not entirely lost, indicating that antibiotics are only partly responsible for biological control activity. Ishimaru *et al.* (1988) also reached the conclusion that herbicolins could not account for the full protective effect of *E. herbicola* strain C9-1.

Wilson *et al.* (1990b) showed that *E. herbicola* strains, which showed some success in biological control of fire blight on hawthorn, produced a broad spectrum antibiotic *in vitro*, but it is not known whether this same compound was produced *in planta*. They suggested site and/or nutrient competition as the mechanism of biological control, and that the competitive advantage might be due to the production of an inhibitor *in situ*. Results from experiments with *E. herbicola* WHL9 suggested production of an antibacterial toxin *in situ* (Wilson *et al.*, 1990a). Although WHL9 produced a broad spectrum antibiotic on PDA, no correlations were observed between antibiotic production *in vitro* and control of blossom blight in several *E. herbicola* strains. This work provided further confirmation that *in vitro* antibiosis cannot be used as the sole basis for selection of antagonists (Weller *et al.*, 1985). Antibiotics produced *in vitro* may not be produced *in vivo*, or breakdown may occur by other epiphytes or inactivation by adsorption to plant surfaces.

An antibiotic from *E. herbicola* Eh318 has been identified which delays disease development in immature pear fruit by *E. amylovora* Ea273 (Wodzinski *et al.*, 1990a,b). The antibiotic may work by inhibiting the preparation of an enzyme which is involved in arginine biosynthesis since the purified antibiotic was not inhibitory to Ea273 in the presence of arginine and histidine.

Bacteriocins

Bacteriocins may be defined as 'antibiotic compounds with bactericidal specificity restricted to bacterial strains closely related to the producer. They are a subclass of antibiotics' (Fravel, 1988). Bacteriocins are typically high molecular weight polypeptides. The chief advantage of using bacteriocin-producing strains in biological control is the presumed occupation of the same ecological niche as the pathogen (Vidaver, 1983). An *Erwinia* sp. which reduced colonization and infection of pear flowers by *E. amylovora* was found to produce a bacteriocin *in vitro* which was lethal to *E. amylovora* (Thomson *et al.*, 1976). Beer and Rundle (1980) showed that several strains of *E. herbicola* and some unidentified strains isolated from pome-fruit trees produced bacteriocin-like substances which inhibited *E. amylovora*. Of the *E. herbicola* strains selected by Beer and Vidaver (1978), strain Eh112Y was found to produce a soluble, bacteriocin-like substance which inhibited Ea273 *in vitro* (Hodges *et al.*, 1980). The bacteriocin was partially purified and designated herbicolacin 112Y by Stein and Beer (1980). Although strain Eh112Y was moderately effective in suppressing fire blight in the immature pear fruit assay and in the orchard, a bacteriocin-deficient mutant of it was equally effective (Beer *et al.*, 1980; Beer, 1981a). This finding called into question the role of bacteriocins in biological control of fire blight and further tests with a variety of *E. herbicola* strains found no correlation between bacteriocin production and suppression of fire blight (Beer *et al.*, 1984).

Habitat modification

Farabee and Lockwood (1958) compared the inhibitory activity of a *Bacterium* sp. in buffered and unbuffered media and showed that it inhibited *E. amylovora* by increasing the acidity of the culture medium to a degree unfavourable for growth of the pathogen. Studies of unidentified bacteria isolated from ornamentals showed that a great reduction in pH occurred in the media of isolates antagonistic to *E. amylovora* (Isenbeck and Schulz, 1986). No inhibitor formation could be measured when the bacteria were grown in buffered media. Therefore it was suggested that the saprophytes either did not produce inhibitory substances at neutral pH (which is optimal for *E. amylovora*) or that the substances were inactivated under these conditions by adsorption or decomposition.

It has been shown that *E. herbicola* can reduce the pH of culture media to levels inhibitory to *E. amylovora* (Riggle and Klos, 1970, 1972; Erskine and Lopatecki, 1975). The ability of *E. herbicola* to cause a reduction in pH has also been suggested as the cause of its antagonism towards *Xanthomonas campestris* pv. *oryzae* in the biological control of bacterial leaf blight of rice (Santhi *et al.*, 1987).

Phenolic compounds

Several workers have considered that *E. herbicola* might induce a physiological response in the host plant, causing release of phenolic compounds which are toxic to *E. amylovora*. Hildebrand and Schroth (1963) demonstrated that *E. herbicola* exhibits β-glucosidase activity, capable of cleaving arbutin from pear tissue to release hydroquinone, which is toxic to *E. amylovora in vitro*. Chatterjee *et al.* (1969) showed that hydroquinone inhibits the oxidative metabolism of *E. amylovora* and concluded that physiological events associated with the plant's response to invasion induced by *E. herbicola* contributed to resistance by inhibiting metabolic activities of *E. amylovora*. Phloridzin, present in apple tissue, is cleaved by *E. herbicola* to phloretin and subsequently to phloroglucinol and phloretic acid (Chatterjee and Gibbins, 1969). Although this has not been demonstrated *in planta*, a similar system might operate in apple to that proposed in pear by Chatterjee *et al.* (1969).

Prospects for Biological Control with *E. herbicola*

For regulatory authorities to consider the approval of biocontrol agents such as *E. herbicola*, many gaps need to be filled with regard to our knowledge of the interactions involved. Further understanding of the mode of action is a clear fundamental requirement, not only for registration purposes but also to determine optimal conditions for the effective use of such agents. Where antibiotics may be produced by biocontrol agents, techniques must be developed to enable the assessment of the production of these compounds on plant surfaces or within invaded tissues, whether or not the antibiotics are involved in the mode of action. Consideration must be given to the interaction of the applied agents with the existing microflora, with regard to both the longevity of the control agent and the possibility of harmful effects on indigenous microorganisms. The essential properties of biocontrol agents which will permit their integration into existing programmes of pest and disease control must also be addressed. While some of these topics are currently being pursued, much work still remains to be done.

References

Anon. (1983) *Top Fruit Growers Guide to the Use of Chemical Sprays 1983*. Ministry of Agriculture, Fisheries and Food, Her Majesty's Stationery Office, London UK.

Beer, S.V. (1981a) Towards biological control of fire blight. *Phytopathology* 71, 859 (Abstract).

Beer, S.V. (1981b) Biological control of fire blight. *Acta Horticulturae* 117, 123 (Abstract).

Beer, S.V. and Rundle, J.R. (1980) Inhibition of *Erwinia amylovora* by bacteriocin-like substances. *Phytopathology* 70, 459.

Beer, S.V. and Rundle, J.R. (1983) Suppression of *Erwinia amylovora* by *Erwinia herbicola* in immature pear fruits. *Phytopathology* 73, 1346.

Beer, S.V. and Vidaver, A.K. (1978) Bacteriocins produced by *Erwinia herbicola* inhibit *Erwinia amylovora. Abstracts of Papers, Third International Congress of Plant Pathology*, p. 75, Paul Parey, Berlin.

Beer, S.V., Norelli, J.R., Rundle, J.R., Hodges, S.S., Palmer, J.R., Stein, J.I. and Aldwinckle, H.S. (1980) Control of fire blight with non-pathogenic bacteria. *Phytopathology* 70, 459.

Beer, S.V., Rundle, J.R. and Norelli, J.L. (1984) Recent progress in the development of biological control of fire blight – a review. *Acta Horticulturae* 151, 195–201.

Beer, S.V., Rundle, J.R. and Norelli, J.L. (1987) Orchard evaluation of five strains of *Erwinia herbicola* for control of blossom infection. *Acta Horticulturae* 217, 219.

Billing, E. and Baker, L.A.E. (1963) Characteristics of *Erwinia*-like organisms found in plant material. *Journal of Applied Bacteriology* 26, 58–65.

Blakeman, J.P. and Brodie, I.D.S. (1976) Inhibition of pathogens by epiphytic bacteria on aerial plant surfaces. In: Dickinson, C.H. and Preece, T.F. (eds), *Microbiology of Aerial Plant Surfaces.* Academic Press, London, pp. 529–558.

Blakeman, J.P. and Fokkema, N.J. (1982) Potential for biological control of plant diseases on the phylloplane. *Annual Review of Phytopathology* 20, 167–192.

Chatterjee, A.K. and Gibbins, L.N. (1969) Metabolism of phloridzin by *Erwinia herbicola*: nature of the degradation products and the purification and properties of phloretin hydrolase. *Journal of Bacteriology* 100, 594–600.

Chatterjee, A.K., Gibbins, L.N. and Carpenter, J.A. (1969) Some observations on the physiology of *Erwinia herbicola* and its possible implications as a factor antagonistic to *Erwinia amylovora* in the 'fire blight' syndrome. *Canadian Journal of Microbiology* 15, 640–642.

Crosse, J.E. (1971) Interactions between saprophytic and pathogenic bacteria in plant disease. In: Preece, T.F. and Dickinson, C.H. (eds), *Ecology of Leaf Surface Micro-organisms.* Academic Press, London, pp. 283–290.

Deckers, T., Porreye, W. and Maertens, P. (1990) Three years of experience in chemical control of fire blight in pear orchards in Belgium. *Acta Horticulturae* 273, 367–376.

Eden-Green, S.J. (1972) Studies in fire blight disease of apple, pear and hawthorn [*Erwinia amylovora* (Burrill) Winslow *et al.*]. Unpublished PhD Thesis, University of London.

Erskine, J.M. and Lopatecki, L.E. (1975) *In vitro* and *in vivo* interactions between *Erwinia amylovora* and related saprophytic bacteria. *Canadian Journal of Microbiology* 21, 35–41.

Farabee, G.J. and Lockwood, J.L. (1958) Inhibition of *Erwinia amylovora* by *Bacterium* sp. isolated from fire blight cankers. *Phytopathology* 48, 209–211.

Fravel, D.R. (1988) Role of antibiosis in the biocontrol of plant diseases. *Annual Review of Phytopathology* 26, 75–91.

Goodman, R.N. (1965) *In vitro* and *in vivo* interactions between components of mixed bacterial cultures isolated from apple buds. *Phytopathology* 55, 217–221.

Goodman, R.N. (1967) Protection of apple stem tissue against *Erwinia amylovora* infection by avirulent strains and three other bacterial species. *Phytopathology* 57, 22–24.

Hattingh, M.J., Beer, S.V. and Lawson, E.W. (1986) Scanning electron microscopy of apple blossoms colonised by *Erwinia amylovora* and *Erwinia herbicola. Phytopathology* 76, 900–904.

Hildebrand, D.C. and Schroth, M.N. (1963) Relation of arbutin-hydroquinone in pear blossoms to invasion by *Erwinia amylovora. Nature* 197, 153.

Hodges, S.S., Beer, S.V. and Rundle, J.R. (1980) Effects of a bacteriocin produced by *Erwinia herbicola* on *Erwinia amylovora. Phytopathology* 70, 463.

Isenbeck, M. and Schulz, F.A. (1985) Biological control of fire blight *Erwinia amylovora* on ornamentals. I. Control of the pathogen by antagonistic bacteria. *Phytopathologische Zeitschrift* 113, 324–333.

Isenbeck, M. and Schulz, F.A. (1986) Biological control of fire blight *Erwinia amylovora* on ornamentals. II. Investigation about the mode of action of the antagonistic bacteria. *Journal of Phytopathology* 116, 308–314.

Ishimaru, C. and Klos, E.J. (1984) New medium for detecting *Erwinia amylovora* and its use in epidemiological studies. *Phytopathology* 74, 1342–1345.

Ishimaru, C.A., Klos, E.J. and Brubaker, R.R. (1988) Multiple antibiotic production by *Erwinia herbicola*. *Phytopathology* 78, 746–750.

Kempf, H.J. and Wolf, G. (1989) *Erwinia herbicola* as a biocontrol agent of *Fusarium culmorum* and *Puccinia recondita* f.sp. *tritici* of wheat. *Phytopathology* 79, 990–994.

Klopmeyer, M.J. and Ries, S.M. (1987) Motility and chemotaxis of *Erwinia herbicola* and its effects on *Erwinia amylovora*. *Phytopathology* 77, 909–914.

Lindow, S.E. (1985a) Integrated control and role of antibiosis in biological control of fire blight and frost injury. In: Windels, C.E. and Lindow, S.E. (eds), *Biological Control on the Phylloplane*. American Phytopathological Society, St. Paul, Minnesota, pp. 45–62.

Lindow, S.E. (1985b) Foliar antagonists: status and prospects. In: Hoy, M.A. and Herzog, D.C. (eds), *Biological Control in Agricultural IPM Systems*. Academic Press, London, pp. 395–413.

Lopez, C. and Fucikovsky, L. (1990) Distribution of fire blight in Mexico and the identification of the bacteria on pear and pyracantha. *Acta Horticulturae* 273, 33–36.

McIntyre, J.L., Kuć, J. and Williams, E.B. (1973) Protection of pear against fire blight by bacteria and bacterial sonicates. *Phytopathology* 63, 872–877.

Moller, W.G., Schroth, M.N. and Thomson, S.V. (1981) The scenario of fire blight and streptomycin resistance. *Plant Disease* 65, 563–568.

Napoli, C. and Staskawicz, B. (1985) Molecular genetics of biological control agents of plant pathogens: status and prospects. In: Hoy, M.A. and Herzog, D.C. (eds), *Biological Control in Agricultural IPM Systems*. Academic Press, London, pp. 455–463.

Nicholson, S.L. (1992) Biological control of fire blight of perry pear. Unpublished PhD Thesis, University of Manchester.

Riggle, J.H. and Klos, E.J. (1970) Inhibition of *Erwinia amylovora* by *Erwinia herbicola*. *Phytopathology* 60, 1310.

Riggle, J.H. and Klos, E.J. (1972) Relationship of *Erwinia herbicola* to *Erwinia amylovora*. *Canadian Journal of Botany* 50, 1077–1083.

Rundle, J.R. and Beer, S.V. (1987) Population dynamics of *Erwinia amylovora* and a biological control agent, *Erwinia herbicola*, on apple blossom parts. *Acta Horticulturae* 217, 221–222.

Santhi, D.P., Unnamalai, N. and Gnanamanickam, S.S. (1987) Epiphytic association of *Erwinia herbicola* with rice leaves infected by *Xanthomonas campestris* pv. *oryzae* and its interaction with the pathogen. *Indian Phytopathology* 40, 327–332.

Stein, J.I. and Beer, S.V. (1980) Partial purification of a bacteriocin from *Erwinia herbicola*. *Phytopathology* 70, 459 (Abstract).

Thomson, S.V. (1986) The role of the stigma in fire blight infections. *Phytopathology* 76, 476–482.

Thomson, S.V., Schroth, M.N., Moller, W.J. and Reil, W.O. (1976) Efficacy of bactericides and saprophytic bacteria in reducing colonization and infection of pear flowers by *Erwinia amylovora*. *Phytopathology* 66, 1457–1459.

van der Zwet, T. and Beer, S.V. (1992) *Fire Blight – Its nature, Prevention, and Control: a Practical Guide to Integrated Disease Management*. US Department of Agriculture, Agriculture Information Bulletin No. 631, 83 pp.

van der Zwet, T. and Keil, H.L. (1979) *Fire Blight: A Bacterial Disease of Rosaceous Plants.* Agriculture Handbook 510. USDA Science and Education Administration. 200 pp.

Vanneste, J.L. and Yu, J. (1990) Control of fire blight by *Erwinia herbicola* Eh252 in an experimental orchard in Dax (Southwest of France). *Acta Horticulturae* 273, 409–410.

Vanneste, J.L., Smart, L.B. and Beer, S.V. (1987) Role of antibiotic production by *Erwinia herbicola* strain Eh252 in the control of fire blight. In: *Abstracts of the Fallen Leaf Lake Conference on* Erwinia, South Lake Tahoe, USA.

Vanneste, J.L., Smart, L.B., Zumoff, C.H., Yu, J. and Beer, S.V. (1990) Control of fire blight by *Erwinia herbicola* Eh252: role of antibiotic production. *Acta Horticulturae* 273, 393–394.

Vanneste, J.L., Yu, J. and Beer, S.V. (1992) Role of antibiotic production by *Erwinia herbicola* Eh252 in biological control of *Erwinia amylovora. Journal of Bacteriology* 174, 2785–2796.

Vidaver, A.K. (1983) Bacteriocins: The lure and the reality. *Plant Disease* 67, 471–475.

Weller, D.M., Zhang, B. and Cook, R.J. (1985) Application of a rapid screening test for selection of bacteria suppressive to take-all of wheat. *Plant Disease* 69, 710–713.

Wilson, M. (1989) Epidemiology and biological control of fire blight of hawthorn. Unpublished PhD Thesis, University of Manchester, UK.

Wilson, M., Epton, H.A.S. and Sigee, D.C. (1990a) Biological control of fire blight of hawthorn (*Crataegus monogyna*) with *Erwinia herbicola* under protected conditions. *Plant Pathology* 39, 301–308.

Wilson, M., Epton, H.A.S. and Sigee, D.C. (1990b) Biological control of fire blight of hawthorn. *Acta Horticulturae* 273, 363–365.

Wilson, M., Epton, H.A.S. and Sigee, D.C. (1992) Interactions between *Erwinia herbicola* and *E. amylovora* on the stigma of hawthorn blossoms. *Phytopathology* 82, 914–918.

Winkelmann, G., Lupp, R. and Jung, G. (1980) Herbicolins: new peptide antibiotics from *Erwinia herbicola. The Journal of Antibiotics* 33, 353–358.

Wodzinski, R.S., Sobiczewski, P. and Beer, S.V. (1987a) Survival of an introduced strain and natural populations of *Erwinia herbicola* on apple. *Acta Horticulturae* 217, 245–251.

Wodzinski, R.S., Umholtz, T.E., Mudgett, M.B. and Beer, S.V. (1987b) Involvement of antibiotics in the mechanism by which *Erwinia herbicola* inhibits *Erwinia amylovora*. In: *Abstracts of the Fallen Leaf Lake Conference on* Erwinia, South Lake Tahoe, USA.

Wodzinski, R.S., Clardy, J.C., Coval, S.J., Beer, S.V. and Zumoff, C.H. (1990a) Antibiotics produced by strains of *Erwinia herbicola* that are highly effective in suppressing fire blight. *Acta Horticulturae* 273, 411–412.

Wodzinski, R.S., Mudgett, M.B. and Beer, S.V. (1990b) Mechanism by which the antibiotic of *Erwinia herbicola* Eh318 inhibits *Erwinia amylovora* Ea273. *Acta Horticulturae* 273, 390.

Wrather, J.A., Kuć, J. and Williams, E.B. (1973) Protection of apple and pear fruit against fire blight with non-pathogenic bacteria. *Phytopathology* 63, 1075–1076.

Index

absolute humidity 85
acetophenones 53, 57
ACMV 179–80, 269–70
appearance 270–1
complexity of ecology 273–4
incidence and symptom expression 271
source and transmission of infection 272–3
spread 270–1
factors influencing 275–8
Acremonium strictum on necrotic leaves 325
aerobactin 48
African Cassava Mosaic Virus/Disease *see* ACMV
Agaricus bisporus production affected by *Trichoderma* spp. 135–41
Agrobacterium rhizogenes, pathogenic interaction with host plant 55–8
Agrobacterium tumefaciens, pathogenic interaction with host plant 55–8
agrobactin 48
air flow, effects on humidity 85–8
Alternaria spp., spore dispersal 178
Alternaria alternata, on necrotic leaves 325
Alternaria dauci, effects of humidity on 80
Alternaria porri, effects of humidity on 80
Alternaria tenuis, on necrotic leaves 324
Alternaria tenuissima, on necrotic leaves 325
Ambrosia artemisiifolia, pollen dispersal 175
Ammiad project 70–1
anthracnose *see Colletotrichum* spp.
antibiosis 104
antibiotics
bacterial 47–8, 104, 106
production by *Erwinia herbicola* 346–7
Aphanomycetes spp., chemotaxis induced by root exudates *41*
appressoria 191, 198
arbutin 348
Armillaria bulbosa, phalanx growth form 9
Armillaria mellea, methods of detection 119–29
Ascochyta phaseolorum, effect of intercropping on 65
Athelia bombacina 325, 327
atmospheric humidity *see* humidity
Aureobasidium pullulans
on green leaves 323
on necrotic leaves 325
autecology
Crinipellis perniciosa 245–63
definition 209–10

autecology *contd*
- *Gaeumannomyces graminis* var. *tritici* 210–24
- *Pseudomonas syringae* group 227–39

auto-inhibition 190
avidin 127
Azorhizobium spp., chemical signal exchange in legume symbiosis 49–55

Bacillus spp., as biological control agents 103–4, 106–14
Bacillus brevis, as biological control agent 103, 106–12
Bacillus cereus, as biological control agent 103
Bacillus licheniformis, as biological control agent 103
Bacillus megaterium, as biological control agent 103
Bacillus pumilis, as biological control agent 103
Bacillus subtilis
- as biological control agent 103
- rainsplash dispersal 175

bacteria
- as modular organisms 5
- somatic mutations 7

bacteriocins, production by *Erwinia herbicola* 347–8
basidiomycetes, outcrossing 262
bean common mosaic potyvirus (BCMV)
- alternative hosts 67
- effect of intercropping on 65

beet necrotic yellow vein virus, quantification in soil 161–6
benomyl 315
benzimidazole 315
benzoic acid 199
Beta vulgaris, pollen dispersal 175
Betula pubescens, pollen dispersal 175
biological control 101
- bacterial species as agents 103–4
- future prospects 113–14
- interactions involved in systems 102
- introduction of foreign organisms 105–12
- in the phyllosphere 321
- studies 104–5

biotin 127
biotrophs, interaction with necrotrophs 27–8
blackcurrant, 'running-off' 195
BNYVV, quantification in soil 161–6
Botrytis aclada, on necrotic leaves 329–30
Botrytis cinerea
- biocontrol 108–12, 328
- effects of humidity on 80
- on green leaves 322
- interaction with *Puccinia lagenophorae* 27–8
- latent infection 188, 192–3
  - of cut flowers 193
  - of fruit plant flowers 194–6
  - polygalacturonases in 197
  - proanthocyanidins in 200
- on necrotic leaves 325, 326

Botrytis fabae, effects of humidity on 80, 82–3
Botrytis squamosa 322, 323, 324
- biological control 328

Botrytis tulipae 196
Bovista plumbea, spore dispersal 175
Bradyrhizobium spp., chemical signal exchange in legume symbiosis 49–55
Bradyrhizobium japonicum, chemotaxis induced by root exudates *41*, 42
Brassica napus, pollen dispersal 175, 176

capsicannol 200
capsidiol 200
cassava 270
- in epidemiology of ACMV 273
- production and maintenance of ACMV-free 272
- reversion 274
- *see also* ACMV

Cercospora medicaginis 327

cereal mixtures, effect on pathogens 66–8
chalcones 57
chocolate spot of bean, effects of humidity on 80, 82–3
cinnamic acids 57
Cladosporium spp.
- on green leaves 323
- on necrotic leaves 324–5, 326
- spore dispersal 176

climate, and spread of ACMV 275–8
clonal growth 5
- potential 6

Cochliobolus miyabeanus, effects of humidity on 80
Cochliobolus sativus, biological control 328
cocoa, witches' broom disease *see* *Crinipellis perniciosa*
Colletotrichum spp., quiescent infection on fruit 197–200
Colletotrichum gloeosporioides
- quiescent infection 199, 200
- resistance to 71–2

Colletotrichum lindemuthianum
- on bean mixtures 67
- effect of intercropping on 65
- evaluation in perennial mixed pastures 68, 69

Colletotrichum musae 198, 199–200
competition between pathogens 26–31
complex ecosystems 63–4
- characteristics 64
- evaluation of plant pathogens in 65–73

conifer leaves, latent and quiescent infections 190–2
Coniothyrium minitans 326
Corylus avellana, pollen dispersal 175
coumarins 53
Crinipellis eggersii 250
Crinipellis perniciosa
- biotrophic establishment 250
- biotypes 246–50, 260–3
- breeding biology 254–6
- competition strategy 252–4
- disease history 245–6
- ecophysiology 250–4
- genetic variation 259–60, 261
- host specificity 246–9
- population structure 256–9, 260
- pseudosclerotial pads 251–2
- related species 250
- saprotrophic growth 251–2
- somatic compatibility groupings 257–8, 259
- subspecies taxonomy 249

Crinipellis siparunae 250
Crinipellis stipitaria 250
Crinipellis trinitatis 250
crops
- effect of varietal mixtures on pathogens 66–8
- wild relatives as sources of resistance 70–3

Cryptococcus spp. 323
cucurbitacins 201
cut flowers, latent infection on 193
cyanidin 201

delphinidin 201
Deschampsia caespitosa, phalanx growth form 8
dew point meters 93
diffusion theory models of spore dispersal 178
dilution method 162–6
Diplodia spp. on necrotic leaves 325
disease resistance, wild relatives as sources of 70–3
disease severity, prediction of 18–20
Drechslera teres, effect of straw disposal method on 288–9
drought tolerance and antagonism on necrotic leaf tissue 329–30

ecologically obligate parasite 210
ELISA, in identification of *Armillaria* 125–6
Endoconidiophora fagacearum, effects of humidity on 80
endogenous dormancy 190
endophytes 192

enzyme-linked immunosorbent assay, in identification of *Armillaria* 125–6
Epicoccum nigrum, on necrotic leaves 324, 325
Epicoccum purpurascens, on necrotic leaves 326
Erwinia amylovora
  aetiology 335–6
  biological control 336–49
  chemotaxis induced by root exudates 42
  control 336
  host range 335
Erwinia carotovora
  spore dispersal 175
  spore removal 173
Erwinia herbicola
  antibiotic production 346–7
  bacteriocin production 347–8
  in biological control of *Erwinia amylovora* 336–49
  phenolic compound production 348
Erysiphe graminis
  effects of humidity on 80, 83
  interaction with *Septoria nodorum* 28
  resistance to 71
  spore dispersal 172–3

fire blight *see Erwinia amylovora*
flavanols 53, 58
flavanones 53
flavones 53, 58
flusilazole 315, 316
Fragaria vesca, guerrilla growth form 8
frequency-dependent disease dynamics 24
fruits, fungal quiescence in 196–201
fungi
  dimorphism and nutrient acquisition 8
  limits to size 10
  as modular organisms 5
  somatic mutations 6–7
  spore dispersal 171–81
fungicides
  effects on *Fusarium* spp. 314–16
  problems with 101
Fusarium spp. 301, 316–17
  as cereal pathogens 301–3
  disease cycle on cereals 303–4
  ear blight 310–14
  effect of straw disposal method on 291
  foot rot 306–10
  fungicidal control 314–16
  germination stimulation by root exudates 43, *44*, 45
  on necrotic leaves 325, 329
  seedling blight 305–6
Fusarium avenaceum 302
  benomyl resistance 315
  and ear blight 310–14
  effect of temperature on growth 302
  and foot rot 306–10
  and seedling blight 305
  survival strategies 304
Fusarium culmorum 302, 316
  benomyl resistance 315
  and ear blight 310–14
  effect of temperature on growth 302
  and foot rot 306–10
  fungicidal control 316
  and seedling blight 305
  survival strategies 304
Fusarium graminearum 302, 316
  and ear blight 310–14
  and foot rot 306–10
  and seedling blight 305–6
  survival strategies 304
Fusarium poae 302
  and ear blight 310–14
  survival strategies 304
Fusarium solani
  secretion of cutinase 58
  spore dispersal 173
fusion, benefits and hazards 10–11

Gaeumannomyces graminis var. *tritici*
  autecology 210–24

Gaeumannomyces graminis var. *tritici contd*
effect of straw disposal method on 295–7
see also Gaeumannomyces–Phialophora complex
Gaeumannomyces-Phialophora complex 145–6
PCR identification of isolates from 146–8
RAPD analysis 152–4
RFLP analysis 149–52
Gaussian Plume models of spore dispersal 178
gel electrophoresis in identification of *Armillaria* 124–5
genet 3, 21
Gliocladium, on necrotic leaves 329
gramicidin S 106–8, 112
grazing in perennial mixed pastures 69–70
guerrilla growth forms 8–9

halo-blight *see Pseudomonas syringae* pv. *phaseolicola*
Helminthosporium maydis, spore dispersal 172
Hendersonia acicola, pine needle infection by 191
herbicolacin 112Y 346, 347
herbicolin I 346
herbicolin O 346
Heterobasidion annosum, growth inhibition by root exudates 45
honey fungus, methods of detection 119–29
host plant
interactions with pathogen 20–6
as mediator 30
host resistance 34
humidistat 90
humidity
control 88–91
disease favoured by low 83
effect on *Botrytis fabae* infection 80, 82
effect on *Phytophthora infestans* infection 80, 82, 87–8
effect of transpiration and air flow on 85–8
importance of 79–82
leaf-related factors affecting 88
measurement 91–3
physics of 83–5
hydrogen cyanide 47
hydrogen sulphide 47
hydroquinone 348
hyperparasites, interactions with pathogens 31–4

IAA 55
immunological techniques in identification of *Armillaria* 125–6
increased *nod* gene inducing activity 55
incubation period, definition 189
indeterminate growth 5
infra-red gas analysers 93
INI 55
intercropping, effect on pathogens 65–6
iron, role of rhizosphere microorganisms in uptake 48
isoflavones 53

laccase 200–1
latent infection
on cut flowers 193
definition 189–90
on flowers of fruit plants 194–6
on leaves 190–3
latent period
definition 188–9
long 190–3
leaves
drought tolerance and antagonism on necrotic tissue 329–30
fungal interactions on 321–3
necrotic tissue 324–8
latent infection of 190–3
legumes
chemical signal exchange in symbiosis with rhizobia 49–55
weedy species as alternative hosts to bean pathogens 67–8

Leptosphaeria maculans, effects of surface water on 81
Leptosphaeria nodorum, effects of surface water on 81
Leptosphaerulina trifolii 327
Limonomyces roseipellis 327
Lophodermella conjuncta, pine needle infection by 191
Lophodermella sulcigena, pine needle infection by 190–1
Lotka-Volterra equations 28

Macrophominia phaseolina, germination stimulation by root exudates 43, *44*
Melampsora lini, effects of humidity on 80
microbial metabolites, effects on plant growth 47–8
Microdochium nivale 302–3
- benomyl resistance 315
- and ear blight 310–14
- effect of temperature on growth 302
- and foot rot 306–10
- fungicidal control 315
- and seedling blight 305
- survival strategies 304

modular organisms 3–5
- analogies based on growth of 7–11
  - predictions arising from *12*
  - pros and cons 12–13
- division 9–10
- foraging strategies 8–9
- fusion 10–11
- implications of lifestyle 6–13
- limits to size 9–10
- major attributes *4*
- resource depletion zones 7–8, 10
- somatic mutations 6–7

module 4
monolignols 57
most probable number (MPN) method 162–6
mushroom compost, identification of *Trichoderma* spp. colonizing forms isolated from 135–41
mutations in modular organisms 6–7
Mycosphaerella graminicola, effects of humidity on 80

6n-pentyl-*2H*-pyran-2-one 47–8
necrotrophs, interaction with biotrophs 27–8
Nectria gelligena, quiescent infection by 199
nitrogen cycle, role of rhizosphere microorganisms 48
nitrogen fixation, chemical signal exchange in 49–55
Nod factors 53–5
nod genes 42, 49–55
NodD proteins 49–53
nucleic acid techniques in identification of *Armillaria* 126–9

Oncobasidium theobromae, effects of surface water on 81
outcrossing in basidiomycetes 262
oxytetracycline 336

pathogens
- interactions with higher trophic levels 31–4
- interactions with host plant 20–6
- interactions with other pathogens 26–31

PCR *see* polymerase chain reaction
pectins 196
pelargonidin 200
Penicillium spp., on necrotic leaves 326, 329
Penicillium simplicissimum, effects on root exudation 46, 47
perennial mixed pastures, evaluation of pathogens 68–70
Peronospora tabacina
- effects of humidity on 80
- spore release 172

PG 196–7
PGIP 197

Phaeoisariopsis griseola, effect of intercropping on 65
phalanx growth forms 8–9
Phaseolus vulgaris, effect of landrace mixtures on pathogens 66–7
phenolic compounds
 and induction of *nod* genes 50–1, 52
 production by *Erwinia herbicola* 348
Phialophora spp. *see* *Gaeumannomyces–Phialophora* complex
phloretic acid 348
phloretin 348
phloridzin 348
phloroglucinol 348
Phoma medicaginis 327
Phragmidium rubi-idaei
 latent period 188
 overwintering 190
phytoalexins, in quiescent infection of fruit 199–200
Phytophthora spp.
 chemotaxis induced by root exudates *41*, 42
 germination stimulation by root exudates 43, *44*, 45
Phytophthora cactorum
 effects of surface water on 81
 spore dispersal 173
Phytophthora infestans, effects of humidity on 80, 82–3, 87–8
phytosiderophores 48
Picea abies, pollen dispersal 175
pine needles, latent and quiescent infections 190–2
pisatin 47
Podaxis pistillaris, spore dispersal 175
Podosphaera clandestina, effects of surface water on 81
pollen, wind dispersal 174–8
polygalacturonase-inhibiting proteins 197
polygalacturonases 196–7
polymerase chain reaction (PCR) 143, 154–6
 applications 145–54
 principles and methods 144–5
Polymyxa betae, quantification in soil 161–6
population dynamics
 and the biological environment 17–34
 Pseudomonas syringae 234–7
potato blight, effects of humidity on 80, 82–3, 87–8
predators, interaction with pathogens 31–4
proanthocyanidins, in quiescent infection 200–1
prochloraz 315
pseudobactin 48
Pseudocercosporella herpotrichoides
 effects of straw disposal method on 289–95
 interaction with other pathogens 26, 28
 spore dispersal 175
Pseudomonas spp., as biological control agents 103
Pseudomonas lachrymans, chemotaxis induced by root exudates 42
Pseudomonas syringae
 epiphytic populations 229–34
 heterogeneity 228
 lesion formation and ecological success 237–9
 regulation of population dynamics by physical environment 234–7
Pseudomonas syringae pv. *garcae*, epiphytic populations 230
Pseudomonas syringae pv. *glycinea*, epiphytic populations 229, 230
Pseudomonas syringae pv. *morsprunorum* 228
 epiphytic populations 230
Pseudomonas syringae pv. *papulans*
 effect of rain on growth of 236
 epiphytic populations 230
Pseudomonas syringae pv. *phaseolicola* 228
 alternative host 67
 on bean mixtures 67
 effect of intercropping on 65
 epiphytic populations 229

Pseudomonas syringae pv. *savastanoi* 228
epiphytic populations 229, 230
Pseudomonas syringae pv. *syringae* 227–8
effect of rain on growth of 234–7
epiphytic population and disease 229–34
lesion formation and ecological success 237–9
as model system 239
Pseudomonas syringae pv. *tabaci* 228
Pseudomonas syringae pv. *tomato*
effect of rain on growth of 236
epiphytic populations 229, 230
psychrometers 93
pterostilbene 200
Puccinia arachidis, effects of surface water on 81
Puccinia coronata, resistance to 72
Puccinia coronata f.sp. *lolii*, in perennial mixed pastures 69
Puccinia graminis f.sp. *tritici*, resistance to 71
Puccinia lagenophorae, interaction with *Botrytis cinerea* 27–8
Puccinia punctiformis, germination stimulation by root exudates *44*
Puccinia recondita f.sp. *tritici*, resistance to 71
Puccinia striiformis, effects of surface water on 81
Pyrenopeziza brassicae
spore dispersal 175, 176–7
spore release 172
Pyrenophora teres, effects of humidity on 80
Pyrenophora tritici-repentis
interaction with *Septoria nodorum* 27
on necrotic leaves 325, 327
Pythium spp., germination stimulation by root exudates 43, *44*, 45
Pythium aphanidermatum, chemotaxis induced by root exudates *41*

quiescent infection
definition 189
of fruit plant flowers 195
in fruits 196–201
of leaves 191–2, 193

rain
effect on population dynamics and pathogenicity of *Pseudomonas syringae* 234–7
role in spore dispersal 175, 178–9
role in spore removal 173–4
ramet 4, 5, 20–1
random walk models of spore dispersal 178
randomly amplified polymorphic DNA analysis *see* RAPD analysis
Ranunculus repens, guerrilla growth form 8
RAPD analysis
Gaeumannomyces– Phialophora complex 152–4
Trichoderma spp. 138–9
rDNA *see* ribosomal DNA
relative humidity 84–5
resource depletion zones 7–8, 10
rest periods 190
restriction fragment length polymorphism analysis *see* RFLP analysis
resveratrol 200
RFLP analysis
of PCR-amplified rDNA 149–51
Trichoderma spp. 136–8
Rhabdocline parkeri, pine needle infection by 191–2
Rhabdocline pseudotsugae, pine needle infection by 191–2
Rhizobium spp.
chemotaxis induced by root exudates *41*, 42
symbiosis with legumes, chemical signal exchange in 49–55
Rhizoctonia spp., evaluation in perennial mixed pastures 69

Rhizoctonia cerealis
effects of straw disposal method on 290–5
interaction with *Pseudocercosporella herpotrichoides* 26
Rhizopus stolonifer, effects of humidity on 80
rhizosphere microorganisms
chemotaxis induced by root exudates 40–2
effects on plant growth 47–8
effects of root exudates on propagule germination 42–5
effects on root exudation 46–8
interactions with roots and root exudates 39–40
preferential assimilation of root exudate components 45–6
Rhynchosporium secalis
effect of straw disposal method on 289
spore dispersal 173
ribosomal DNA (rDNA) 148
PCR-amplified
as probe 151–2
RFLP analysis 149–51
sequencing 148-9
root exudates 39, 58–9
diffusion away from roots 40
effects on propagule germination 42–5
effects of rhizosphere microorganisms on production 46–8
and induction of chemotaxis 40–2
interactions with rhizosphere microorganisms 39–40
and nitrogen fixation 49–55
preferential assimilation by rhizosphere microorganisms 45–6
rusts, unrecognized regulation by hyperparasites 31–2

Saccharomyces cerevisiae, dimorphism 8
sandwich assay 127
saturation water vapour pressure 84–5
Sclerotinia fructicola, effects of humidity on 80
Sclerotinia sclerotiorum
on green leaves 322
guerrilla growth form 9
on necrotic leaves 325, 326
spore ejection 172
Sclerotium cepivorum, germination stimulation by root exudates 43, *44*
Scrophularia nodosa, phalanx growth form 8
SDS PAGE in identification of *Armillaria* 124–5
semi-obligate parasite 210
Septoria nodorum
effect of straw disposal method on 288
interactions with other pathogens 26–7, 28
on necrotic leaves 325
spore dispersal 175, 179
Septoria tritici
effect of straw disposal method on 288
interaction with *Septoria nodorum* 26–7
spore dispersal 179
sequence analysis, *Trichoderma* spp. 139
siderophores 48
Solidago canadensis, phalanx growth form 8
Sphaerotheca pannosa, effects of humidity on 80, 83
spores
deposition 180–1
dispersal 174–9
removal 171–4
Sporobolomyces
on green leaves 323
spore dispersal 176
Stemphylium on necrotic leaves 324, 325, 327
Stokes' number 180
Straw disposal 285–7, 297–9
effects on leaf diseases 288–9

Straw disposal *contd*
effects on stem base diseases 289–95
effects on take-all 295–7
Streptomyces spp. as biological control agents 103
streptomycin 47, 336
Striga spp., interaction with host plant 58
Stylosanthes spp. resistance to anthracnose 71–2
subsistence farming systems, evaluation of pathogens in 65–8
sugarbeet rhizomania, quantification of pathogenic organisms in soil 161–6
surface water, importance of 79, 81–2
Synchytrium desmodii, evaluation in perennial mixed pastures 69

take-all fungus *see Gaeumannomyces graminis* var. *tritici*
Taphrina deformans, effects of humidity on 80
temperature and humidity control 83–5
Thielaviopsis basicola, germination stimulation by root exudates 43, *44*
Tilletia caries, spore dispersal 175
Tilletia tritici, spore dispersal 175
transpiration, effects on humidity 85–8
Trichoderma spp.
identification of colonizing forms isolated from mushroom compost 135–41
on necrotic leaves 329
Trichoderma harzianum, effects on root exudation 47
Trifolium repens, guerrilla growth form 8
Triticum turgidum var. *dicoccoides*, disease resistance 70–1
trophic levels 17
tropical perennial mixed pastures, evaluation of pathogens 68–70
tuliposides 195–6

Ulocladium chartarum on necrotic leaves 325
unitary organisms 3–5
major attributes *4*
Uromyces appendiculatus, effect of intercropping on 65
Uromyces fabae, effects of humidity on 83
Uromyces phaseoli, effects of humidity on 80
Ustilago maydis, germination stimulation by root exudates *44*

Venturia inaequalis
effects of surface water on 81
on necrotic leaves 327
Verticillium spp., germination stimulation by root exudates 43, *44*
vir genes, induction 55–8
virus pathogens, effects on root exudation 47

Wilsonomyces carpophilus, effects of surface water on 81
wind
role in spore dispersal 174–8
role in spore removal 172–3
witches' broom disease of cocoa *see Crinipellis perniciosa*

Zea mays, pollen dispersal 175
zeatin 55
Zygophiala jamaicensis, effects of humidity on 80